Applied Statistics in the Pharmaceutical Industry

T0188822

Springer
New York
Berlin
Heidelberg
Barcelona
Hong Kong
London
Milan
Paris
Singapore
Tokyo

Steven P. Millard   Andreas Krause

Editors

Applied Statistics in the Pharmaceutical Industry

With Case Studies Using S-Plus

With 131 Illustrations

 Springer

Steven P. Millard
Probability, Statistics & Information
7723 44th Avenue NE
Seattle, WA 98115-5117
USA

Andreas Krause
Novartis Pharma AG
Biostatistics
P.O. Box
4002 Basel
Switzerland

Library of Congress Cataloging-in-Publication Data
Applied statistics in the pharmaceutical industry : with case studies using S-PLUS /
editors, Steven P. Millard, Andreas Krause.
    p. cm.
    Includes bibliographical references and index.
    ISBN 978-1-4419-3166-5
    1. Drugs—Research—Statistical methods. I. Millard, Steven P. II. Krause, Andreas.
RS122 .A665 2001
615′.19′0727—dc21   00-053767

Printed on acid-free paper.

S-PLUS is a registered trademark of Insightful Corporation.

© 2010 Springer-Verlag New York, Inc.
All rights reserved. This work may not be translated or copied in whole or in part without the
written permission of the publisher (Springer-Verlag New York, Inc., 175 Fifth Avenue, New York,
NY 10010, USA), except for brief excerpts in connection with reviews or scholarly analysis. Use
in connection with any form of information storage and retrieval, electronic adaptation, computer
software, or by similar or dissimilar methodology now known or hereafter developed is forbidden.
The use of general descriptive names, trade names, trademarks, etc., in this publication, even if the
former are not especially identified, is not to be taken as a sign that such names, as understood by
the Trade Marks and Merchandise Marks Act, may accordingly be used freely by anyone.

Production managed by Timothy Taylor; manufacturing supervised by Jerome Basma.
Camera-ready copy prepared from the authors' LaTeX2e and Microsoft Word files.

Printed in the United States of America.

9 8 7 6 5 4 3 2 1

Springer-Verlag New York Berlin Heidelberg
A member of BertelsmannSpringer Science+Business Media GmbH

Preface

Each year, hundreds of new drugs are approved for the marketplace. The approval of a single new drug is the result of years of screening tens of thousands of compounds, performing preclinical research on their effects, and designing, implementing, and analyzing the results of clinical trials. Statisticians are involved in every phase of this process. Between the years 1960 and 2000, the number of statisticians working in the pharmaceutical industry grew from less than 100 to over 2500.

This book provides a general (but not exhaustive) guide to statistical methods used in the pharmaceutical industry, and illustrates how to use S-PLUS to implement these methods. Specifically, each chapter in this book:

- Illustrates statistical applications in the pharmaceutical industry.
- Illustrates how the statistical applications can be carried out using S-PLUS (each chapter except the first contains an appendix with S-PLUS code).
- Illustrates why S-PLUS is a useful software package for carrying out these applications.
- Discusses the results and implications of a particular application.

The target audience for this book is very broad, including:

- Graduate students in biostatistics.
- Statisticians who are involved in the industry as research scientists, regulators, academics, and/or consultants who want to know more about how to use S-PLUS and learn about other subfields within the industry that they may not be familiar with.
- Statisticians in other fields who want to know more about statistical applications in the pharmaceutical industry.

The data and code from each chapter are available on the following web site: http://www2.active.ch/~krause.a/doc/statistics-in-pharma/. (Due to confidentiality, the raw data are not available for some chapters.)

Part 1 of this book includes Chapter 1, which is an introductory chapter explaining the history and current state of statistics in the drug development process. Following Chapter 1, this book is divided into six more sections that follow the sequence of the drug development process (see Figure 1.2). Part 2 encompasses basic research and preclinical studies and Chapter 2 within this section discusses one-factor comparative studies. Part 3 covers preclinical safety assessment. Chapter 3 discusses analysis of animal carcinogenicity data, and Chapter 4 toxicokinetic and pharmacokinetic data. Part 4 involves Phase I studies. Chapter 5 discusses the analysis of pharmacokinetic data in humans, Chapter 6 illustrates graphical presentations of single patient results, Chapter 7 explains the design and analysis of Phase I oncology trials, Chapter 8 discusses the analysis of analgesic trials, Chapter 9 points out problems with patient compliance and how this affects pharmacokinetic analysis, and Chapter 10 discusses the classic 2,2,2 crossover design. Part 5 includes chapters on Phase II and III clinical trials. Chapter 11 discusses power and sample size calculations, Chapter 12 compares how S-PLUS and SAS handle analysis of variance, Chapter 13 explains a technique for sample size reestimation, Chapter 14 discusses permutation tests for Phase III clinical trials, Chapter 15 discusses comparing two treatments in a Phase III clinical trial, and Chapter 16 covers meta-analysis of clinical trials. Part 6 covers Phase IV studies, and includes Chapter 17 on the analysis of health economic data. Finally, Part 7 encompasses manufacturing and production. Chapter 18 discusses the decimal reduction time of a sterilization process, and Chapter 19 covers acceptance sampling plans by attributes.

A Note about S-PLUS Versions

The chapters in this book were written when the current versions of S-PLUS were Version 3.4 and then 5.1 for UNIX, and Version 4.5 and then 2000 for Windows. The Windows version of S-PLUS includes a graphical user interface (GUI) with pull-down menus and buttons. By the time this book is published, both S-PLUS 6.0 for UNIX and S-PLUS 6.0 for Windows will be available, and both of these versions of S-PLUS will have GUIs. All of the chapters in this book, however, illustrate how to use S-PLUS by writing S-PLUS commands and functions, rather than using pull-down menus.

Typographic Conventions

Throughout this book, S-PLUS commands and functions are displayed in a fixed-width font, for example: `summary(aov.psize)`. Within chapters, S-PLUS commands are preceded with > (the "greater than" sign), which is the default S-PLUS prompt. Rather than use the S-PLUS continuation prompt (+ by default) for continuation lines, we follow Venables and Ripley (1999) and use indentation instead. Within appendices, S-PLUS prompts are omitted.

Companion Web Site for This Book

For more information about the authors, datasets, software, and related links, please refer to http://www2.active.ch/~krause.a/doc/statistics-in-pharma/.

Information on S-PLUS

For information on S-PLUS, please contact Insightful Corporation:

Insightful Corporation
1700 Westlake Ave N, Suite 500
Seattle, WA 98109-3044 USA
800-569-0123
sales@insightful.com
www.insightful.com

Insightful Corporation
Knightway House
Park Street
Bagshot, Surrey
GU19 5AQ
United Kingdom
+44 1276 452 299
sales@uk.insightful.com
www.uk.insightful.com

Acknowledgments

In the Spring of 1997, one of us (Steve) taught a course on using S-PLUS at two different branches of Merck Research Laboratories. The course was organized by Charles Liss and Thomas Bradstreet of Merck, and Charles (Chuck) Taylor of MathSoft, Inc. (now Insightful Corporation), and revolved around using S-PLUS to analyze three different data sets from the pharmaceutical industry. That course was the seed for this book.

Thomas Bradstreet put Steve in touch with several statisticians and provided early guidance on the scope and direction of this book. Andreas, who works in the pharmaceutical industry, had already published a book with Melvin Olson on using S-PLUS (*The Basics of S and S-PLUS*), and therefore knew several people who became contributors to this book. We are grateful to all of the authors who generously donated their time and efforts to writing chapters for this book. We would also like to thank John Kimmel for his help and guidance throughout the process of editing this book. Steve Millard would like to thank Aleksandr Aravkin, Sam Coskey, and Jeannine Silkey for their help in formatting many of the chapters. Finally, we would like to thank all of the statisticians in the pharmaceutical industry; you have devoted yourselves to improving the health of the world.

Contents

List of Contributors

The following is a list of contact information for chapter authors that was current at the time this book was published. Note that author affiliation may differ between what is listed here and what is listed at the beginning of a chapter. Affiliation listed at the beginning of a chapter reflects author affiliation at the time the chapter was written.

Hongshik Ahn
*Department of Applied Mathematics
  and Statistics
State University of New York
Stony Brook, NY 11794-3600, USA
hahn@ams.sunysb.edu*

Darrel L. Allen
*Toxicology and Drug Disposition
  Research Laboratories
Lilly Research Laboratories
2001 West Main Street
Indianapolis, IN 46140, USA
allen_darrel_l@lilly.com*

Dhammika Amaratunga
*The R.W. Johnson Pharmaceutical
  Research Institute
OMPC Building, Room 2196,
  Route 202
P.O. Box 300
Raritan, NJ 08869-0602, USA
damaratu@prius.jnj.com*

Axel Benner
*Biostatistics Unit
German Cancer Research Center
Heidelberg, Germany
benner@dkfz-heidelberg.de*

Vance Berger
*Biometry Research Group, DCP
National Cancer Institute
Executive Plaza North, Suite 344
Bethesda, MD 20892-7354, USA
vance.berger@nih.gov*

Jürgen Bock
*F. Hoffmann-La Roche AG
PDBS 74/OG-W, 4070
Basel, Switzerland
juergen.bock@roche.com*

Tom Bradstreet
*Department of Clinical
  Pharmacology Statistics
Merck Research Laboratories
10 Sentry Parkway, BL 3-2
Blue Bell, PA 19422, USA
thomas_bradstreet@merck.com*

George W. Carides
*Department of Health Economic
  Statistics
Merck Research Laboratories
10 Sentry Parkway, BL 2-3
Blue Bell, PA 19422, USA
george_carides@merck.com*

David Carlin
*MedImmune, Inc.
35 West Watkins Mill Road
Gaithersburg, MD 20878, USA
carlind@medimmune.com*

James Z. Chou
*TAP Pharmaceutical Products Inc.
675 North Field Drive
Lake Forest, IL 60045, USA
james.chou@tap.com*

John R. Cook
Department of Health Economic
  Statistics
Merck Research Laboratories
10 Sentry Parkway, BL 2-3
Blue Bell, PA 19422, USA
john_cook@merck.com

Erik J. Dasbach
Department of Health Economic
  Statistics
Merck Research Laboratories
10 Sentry Parkway, BL 2-3
Blue Bell, PA 19422, USA
erik_dasbach@merck.com

Lutz Edler
Biostatistics Unit
German Cancer Research Center
Heidelberg, Germany
edler@dkfz.de

Jeffrey Eisele
Department of Biostatistics
Novartis Pharma AG
CH-4002 Basel, Switzerland
jeffrey.eisele@pharma.novartis.com

Ene I. Ette
Clinical Pharmacology
Vertex Pharmaceuticals, Inc.
Cambridge, MA 02139, USA
ette@vpharm.com

Mauro Gasparini
Dipartimento di Matematica
Politecnico di Torino
I-10129 Torino, Italy
gasparin@calvino.polito.it

Gernot Hartung
Oncological Center
Mannheim Medical School
Heidelberg, Germany
gernot.hartung@med3.ma.uni-heidelberg.de

Michael A. Heathman
Department of Pharmacokinetics and
  Pharmacodynamics
Lilly Research Laboratories
2001 West Main Street
Indianapolis, IN 46140, USA
heathman_michael_a@lilly.com

Wherly P. Hoffman
Division of Statistics and
  Information Sciences
Lilly Research Laboratories
2001 West Main Street
Indianapolis, IN 46140, USA
hoffman_wherly_p@lilly.com

Anastasia Ivanova
Department of Biostatistics
School of Public Health, CB# 7400
University of North Carolina
Chapel Hill, NC 27599-7400, USA
aivanova@bios.unc.edu

Michaela Jahn
F. Hoffmann-La Roche AG
PDBS, Building 74, 4-W, 4070
Basel, Switzerland
michaela.jahn@roche.com

Ralph L. Kodell
Division of Biometry and
  Risk Assessment
National Center for Toxicological
  Research
Food and Drug Administration
Jefferson, AR 72079, USA
rkodell@nctr.fda.gov

Andreas Krause
Novartis Pharma AG
P.O. Box
4002 Basel
Switzerland
andreas.krause@pharma.novartis.com

Peter Lockwood
Department of Pharmacokinetics,
  Dynamics and Metabolism
Pfizer Global Research and
  Development
2800 Plymouth Road
Ann Arbor, MI 48105, USA
peter.lockwood@pfizer.com

Jaap Mandema
Pharsight Corporation
800 West El Camino Real, Suite 200
Mountain View, CA 94040, USA
jaapm@pharsight.com

Steven P. Millard
Probability, Statistics & Information
7723 44th Avenue NE
Seattle, WA 98115-5117, USA
smillard@probstatinfo.com

Raymond Miller
Department of Pharmacokinetics,
  Dynamics and Metabolism
Pfizer Global Research and
  Development
2800 Plymouth Road
Ann Arbor, MI 48105, USA
raymond.miller@pfizer.com

Ha Nguyen
Global Medical Affairs
Aventis Pharmaceuticals, Inc.
P.O. Box 6800
Bridgewater, NJ 08807, USA
ha.h.nguyen@aventis.com

Keith O'Rourke
Clinical Epidemiology Unit and
  Department of Surgery
Ottawa Hospital
University of Ottawa
Ottawa, Ontario K1Y-4E9, Canada
korourke@lri.ca

Melvin Olson
Allergan, Inc.
2525 Dupont Drive
Irvine, CA 92623-9534, USA
olson_melvin@allergan.com

Bill Pikounis
Biometrics Research
Merck Research Laboratories
126 East Lincoln Avenue
Mail Stop RY70-38
Rahway, NJ 07065-0900, USA
v_bill_pikounis@merck.com

Amy Racine
Department of Biostatistics
Novartis Pharma AG
CH-4002 Basel, Switzerland
amy.racine@pharma.novartis.com

Bruce Rodda
Health Services & Research
Southwest Texas State University
601 University Drive
San Marcos, TX 78666-4616, USA
bruce_rodda@msn.com

Beverley Shea
Clinical Epidemiology Unit and
  Department of Medicine
Ottawa Hospital
University of Ottawa
Ottawa, Ontario K1Y-4E9, Canada
bshea@lri.ca

Wenping Wang
Pharsight Corporation
5520 Dillard Drive, Suite 210
Cary, NC 27511, USA
wwang@pharsight.com

George A. Wells
Department of Epidemiology and
  Community Medicine
University of Ottawa
Ottawa, Ontario K1Y-4E9, Canada
gwells@uottawa.ca

Harry Yang
MedImmune, Inc.
35 West Watkins Mill Road
Gaithersburg, MD 20878, USA
yangh@medimmune.com

Part 1:

Introduction

1

Statistics and the Drug Development Process

Bruce Rodda
Southwest Texas State University, San Marcos, TX, USA

Steven P. Millard
Probability, Statistics & Information, Seattle, WA, USA

Andreas Krause
Novartis Pharma AG, Basel, Switzerland

1.1 Introduction

This introductory chapter begins with a brief history of the drug development process and the evolution of statistics and statisticians in that process. This is followed by a more detailed summary of the various elements that comprise the discovery, development, approval, and manufacture of new therapeutic entities.

1.2 History

The evolution of the statistics and information management professions in the pharmaceutical industry has had two primary drivers. The first was the change from an industry that was fragmented and focused on the manufacture of nostrums and cure-alls compared with the cutting edge, highly scientific, research-oriented industry that exists today. The second was the growth of governmental control over the pharmaceutical industry in the latter half of the twentieth century. Figure 1.1 depicts significant drug regulatory milestones in the United States.

Prior to the twentieth century, there was little control over the content and claim structure of medicinal products. To address this issue, the U.S. Congress passed the Pure Food and Drug Act in 1906. This legislation was designed to protect the population against unscrupulous individuals and companies who would intentionally misbrand or adulterate the therapies of that time. Although well intended, neither this legislation, nor subsequent legislation during the ensuing 25 years had any clearly positive effect on improving the quality, safety, or effectiveness of medicinal products available to the population.

SIGNIFICANT DRUG REGULATORY MILESTONES IN THE U.S.

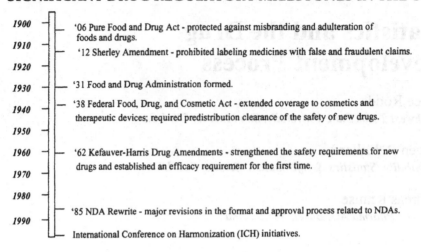

Figure 1.1. Significant drug regulatory milestones in the United States. Modified from Johnson (1987, p. 8), by courtesy of Marcel Dekker, Inc.

The year 1931 marked the creation of the Food and Drug Administration (FDA) in the United States and concentrated the responsibility and authority for regulation of pharmaceuticals into a specific federal agency. However, the most important drug legislation of this period was to come 7 years later, in 1938. In this year, Congress passed the Food, Drug, and Cosmetic Act which, for the first time, required premarketing approval of the *safety* of new drugs. This was the first legislation that provided the FDA with the authority to prohibit the marketing of new medicinal products (including devices and cosmetics), unless the manufacturer demonstrated that the product had an acceptable safety profile for its intended use. Prior to this legislation, a manufacturer could sell its products without performing any studies on health effects, and detection of any safety problems was only possible after they had manifested themselves in the patient population.

It was also about this time that the first statisticians were employed in the pharmaceutical industry. During the ensuing quarter of a century, the number of statisticians and their influence grew slowly. Initially, statisticians had little authority and were primarily employed as analysts in nonclinical research and manufacturing settings.

In the 1950s a substance called thalidomide was developed in Germany and promoted as a sleeping medication and a treatment for morning sickness in pregnant women. Clinical testing of the drug began in Western Europe in 1955, and later in the United States. The drug was marketed under various names in different countries, including Contergan (Germany), Distavan (England), and Kevadon (Canada). In the early 1960s it became apparent that thalidomide was teratogenic and linked to thousands of birth defects in Western Europe. The FDA denied approval of the drug, and public interest in drug regulation was aroused. In response, in October of 1962, the United States Congress passed the Kefauver–Harris Drug Amendments, which strengthened drug safety requirements by requiring extensive animal pharmacological and toxicological testing before a drug could be tested in humans (see Drews, 1999, pp. 142–145, and www.fda.gov/fdac/special/newdrug/benlaw.html).

The Kefauver–Harris Drug Amendments clearly established the future of statisticians and statistics in the pharmaceutical industry. Although preapproval documentation of safety had already been required for approval of new therapies, there were no requirements that new drugs needed to be efficacious. This legislation not only strengthened drug safety requirements, but also, for the first time, required premarketing approval of the *efficacy* of new medicines. The consequence of this requirement was the creation of large organizations within the pharmaceutical industry (and in the FDA) that were dedicated to the design, implementation, and evaluation of clinical trials. Since these trials were now required for the approval of new drugs, the number of statisticians needed to participate in this responsibility grew at an extraordinary pace. In 1962, the pharmaceutical industry employed less than 100 statisticians; by the year 2000, this figure had increased to more than 3000. For this reason, the Kefauver–Harris legislation has been affectionately called "The Statisticians' Full Employment Act" by pharmaceutical statisticians.

This new opportunity for statisticians completely changed the nature of their roles and responsibilities. Where their roles had previously been in nonclinical environments, clinical statisticians became increasingly more common. Today, approximately 90% of all statisticians in the pharmaceutical industry are involved in clinical trials.

During the succeeding 20 years after the passage of the Kefauver–Harris Drug Amendments, both the industry and the FDA suffered inefficiencies due to a need for structure and guidance with respect to the submission and approval process for new drugs. This was addressed in the mid-1980s by a document referred to as the "FDA Rewrite." This manuscript clarified the format and content of submissions in support of new drugs. Approximately two-thirds of this document discusses issues of importance to statisticians.

About this time, the need for some codification of "Good Clinical Practices" (GCP) became apparent. Although the U.S. regulations defined basic scientific requirements for acceptable evidence in support of new drugs, many countries relied on testimonials from respected physicians as critical elements of their review. These inconsistencies were addressed during the 1990s by an organized cooperation among regulatory, industrial, and academic representatives from the European Union, Japan, and the United States. The initiatives developed by the International Conference on Harmonization (ICH) address many critical areas and have benefited greatly from substantial expert statistical input. In fact, one complete guidance (*E-9, Statistical Principles for Clinical Trials*) provides an excellent treatise on this topic.

The past 40 years have seen an enormous change in the roles, responsibilities, and authority of statisticians in the pharmaceutical industry. The profession has evolved from a small group of analysts to a fraternity of several thousand scientists involved in all aspects of discovery, development, manufacturing, and education.

1.3   The Drug Development Process

The discovery and development of new therapeutic entities is a highly technical, multidisciplinary process that presents a plethora of challenges that demand solutions from many scientific and technical disciplines. Although each of several specialties plays a critical role at any specific stage, statistics and information management are essential throughout the process.

The evolution of a new drug from concept to patient consists of both sequential and parallel processes (Figure 1.2). The process begins with efforts to discover a new therapeutic molecule. Although the ultimate target is man, the potential agent must be tested in selected animal models to determine whether it may have potential therapeutic activity. If the results of this screening process are positive, the compound is then subjected to a variety of safety evaluations in various animal species to determine whether its projected safety profile is consistent with the risks of administering it to humans.

The clinical component of the research process is further refined by three relatively well-defined phases (Phases I, II, and III) that characterize the stage of the product in its clinical development prior to registration. These first three phases and Phase IV (conducted after approval of the product) are more or less sequential, providing an opportunity to evaluate the current information before proceeding to the next phase of research. Since the statistician is usually more familiar with the data, its analysis, and its interpretation than any other individual on the project team, his or her contributions to each phase of these processes are critical.

Concurrent with the clinical development stage of the process are those components associated with the expected manufacture of the product after approval by regulatory agencies. The basic active ingredient is never given to patients alone. The active moiety must be formulated with other chemicals and compounds to optimize its ability to reach its therapeutic destination. Thus development of the final formulation is a critical step. This must be done as early in the process as possible, since the majority of Phase II and III studies must be done with the final formulation to obviate the need for additional studies and extending the development timeline.

Once the final formulation is defined and characterized, processes for the economical and efficient manufacturing of the product must be developed. Concurrent with this activity, various stability studies must be conducted to determine how long a product will retain its therapeutic potency after manufacture.

DRUG DEVELOPMENT PROCESS

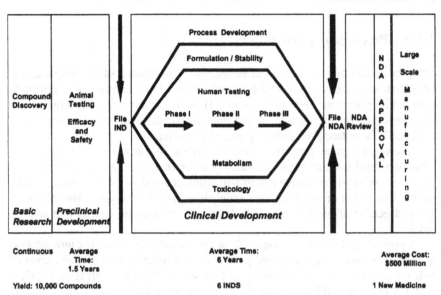

Figure 1.2. Schema for the drug development process.

1.3.1 Discovery

As recently as the 1970s, drug discovery was a very inefficient process. At that time, much of the new drug discovery process was based on experience. Molecule modification, rather than novel molecule discovery, was the rule. The consequence was that truly novel therapeutic entities were relatively uncommon.

Today, genomics, high throughput screening, informatics, robotics, and other highly technical advances are used by scientists to identify the tens of thousands of new molecules required to produce one new lead. However, the capability of identifying large numbers of potential new leads requires state-of-the-art screening procedures to identify the small number that will ultimately be used in clinical trials. In addition to effective selection procedures, sophisticated modeling processes, such as computer-assisted drug design, are used to modify the molecule to optimize its potential therapeutic benefit.

1.3.2 Preclinical Pharmacology

Introducing a new chemical into a human being carries a significant risk. If the potential agent does not demonstrate the expected benefit, no risk is acceptable. To determine whether the new drug should proceed to the next step in development, carefully designed studies, in well-defined animal models, are conducted.

These studies must be definitive regarding the therapeutic benefit expected in man, must yield a targeted initial range of doses for man, but must minimize the utilization of animals.

1.3.3 Preclinical Toxicology

Drug therapies must not only be effective; they must have a safety profile that is consistent with their benefit and their indication. For example, an ointment to treat a mild rash must be totally benign. However, many cancer therapies have been approved for use despite having poor safety profiles, because their potential benefit far outweighed their unpleasant side effects.

Safety assessment programs differ according to the nature of the therapy. Drugs designed to be administered acutely and infrequently will require a different battery of safety assessment than will a product designed for a patient to take for the rest of his or her life. The anticipated patient exposure, class of agent, age and gender of patient, etc., will determine the specific studies that will be required.

A portion of these evaluations will be conducted before the first introduction of the drug in man. As mentioned previously, the product must give evidence of efficacy and have acceptable safety to proceed. If the entire preclinical safety profile were required before human testing began, there could be an unacceptable delay in the development process and ultimate availability to the practicing physician. For this reason, safety assessment in animals continues throughout the clinical programs in humans.

1.3.4 Clinical Research

Clinical development of new therapeutic agents is traditionally partitioned into four sequential, but overlapping, phases. The clinical research portion of new drug development varies in duration with the nature of the disease being treated, the duration of any positive or negative effects of the drug, and the intended period of administration. In general, 2 years would be an absolute minimum for the clinical phase of a new agent, while products such as hormonal contraceptives would require a much longer period of clinical development.

Phase 1: Clinical Pharmacology

The initial introduction of a new agent into humans is a scientifically exciting period in its development. The goals of this stage of development include determining initial tolerability, developing a range of doses to pursue in patients, and characterizing the drug's pharmacokinetics. Phase I studies are commonly conducted in normal, male, volunteers who do not possess the disease that will be studied in later phases. The reason for using male normals at this stage is threefold. First, there is not enough evidence of therapeutic benefit to substitute the new agent for a patient's therapy. Second, normal subjects provide a more homogeneous population, unaffected by concurrent therapies and diseases that might confound the interpretation of results in patients. Third, women of child-bearing potential are generally excluded because any risk associated with a potential pregnancy is unacceptable at this stage of development.

To obtain initial safety and tolerance information, a group of subjects is given the lowest dose suggested for humans by the animal toxicology studies. If the tolerability is acceptable in these subjects, they (or a different group) are given a higher dose. The process continues until unacceptable toxicity is observed. For products in which the safety profile is expected to be poor (such as cytotoxic agents), these studies will be conducted in patients. When the studies in this phase have been completed, a range of doses will have been defined that possesses an acceptable safety profile to permit introduction to patients.

Although one generally thinks about the way a drug treats the patient, pharmacokinetics characterizes the manner in which the patient treats the drug. The broad components of pharmacokinetics include absorption, distribution, metabolism, and excretion (ADME) of the product and its metabolites. These processes are determined by sophisticated mathematical modeling and complex computer programs.

Studies to garner this information are usually designed to administer the drug to 10–20 subjects and collect blood and urine samples at selected time points over several hours. The resultant profile is then carefully modeled to estimate its component parameters. These kinds of results are essential for determining dosing regimens and characterizing how different populations may vary with respect to their ultimate therapeutic response.

A variation of this type of study, the bioavailability or bioequivalence study, is conducted in various forms throughout the development process to determine whether changes in the formulation, dosage form, manufacturing process, etc., will affect the way the body handles the drug. These types of studies are also sensitive for determining the potential effects of food, concomitant therapies, and concurrent illnesses, such as compromised liver function.

Phase II: Early Studies in Patients

When a tolerable dose range has been identified and the pharmacokinetics have been determined, there is sufficient evidence to proceed to patients. These early studies in patients are usually decisive regarding whether the product will continue to the large scale, expensive, studies of Phase III.

These initial studies in patients with the disease of interest have three related objectives. Similar to the Phase I tolerance studies in normal volunteers, the lowest dose which provides clinical benefit must be determined. This is determined in a manner similar to the tolerance studies of Phase I, but doses are increased until a therapeutic effect is observed, rather than toxicity.

Once the bottom of the dose–response curve is estimated, studies are designed to determine the effect of administering higher doses. Different studies address two similar, but different, questions. First, are higher doses more effective, per se, i.e., if a higher does is given, does it evoke a better effect? Second, for patients in whom a specific dose fails, will increasing the dose provide benefit? The designs of these studies are different, and in general, a study designed to answer one of the questions will not answer the other.

At this stage in the development process, pharmacokinetic studies will be conducted in the target patient population to identify any differences in pharmacokinetics that may attend the disease, or its common concomitant medications, or whether there are changes in functional physiology that may be associated with the condition.

Phase III: Confirmation Studies

Although a substantial amount of time, effort, patient exposure, and resources have been expended in the development process to this point, Phase III and the attendant process elements from this point forward will dwarf what has occurred previously. All available information regarding efficacy, safety, pharmacokinetics, animal toxicology, etc., will be intellectually integrated to determine whether proceeding to the implementation of Phase III is warranted.

Phases I and II research is designed to provide a basis upon which to determine whether to proceed to Phase III. For this reason, early phase research is highly focused and exposes a minimal number of subjects and patients to the experimental therapy. Phases I and II could total as few as 100 subjects/patients; each followed for a short period. This consequently limits the number of patients who have access to the drug, the variety of populations that have received it, the duration of exposure, and long-term observation for continuing efficacy or tachyphylaxis. In addition, little information is available regarding adverse experiences and the long-term safety of the drug.

Phase III is typically comprised of a set of several studies. Since Phase III will provide the primary clinical basis upon which regulatory agencies will make a decision to approve or reject the compound, the goals of these studies are to provide conclusive evidence of efficacy, to fully characterize the adverse experience profile, to gain experience in a truly clinical environment with many of its uncontrolled elements, and to obtain safety and efficacy information in special groups or patients. These large, and (often) long, studies regularly compare the new agent against the current standard of care, are usually multicenter, often multinational, and routinely enroll hundreds of patients per study. The entire registration dossier often comprises information from well over 1000 patients.

At the conclusion of the Phase III studies, the information from them and all other sources is compiled into a dossier for submission to regulatory agencies. This will be described in more detail in Section 1.4.

Phase IV: Post-Approval Studies

The clinical research process does not end with the submission and approval of the New Drug Application (NDA) or dossier. Research continues for three very important reasons. First, despite the enormous quantity of information that is submitted in support of registration, patient exposure is tiny compared to the product's ultimate utilization. It is for this reason that pharmaceutical companies conduct additional studies in thousands of patients in an effort to learn as much as possible about the adverse experiences of their drug and to make this information available to the medical community. These studies are usually epidemiological in nature and often continue for several years after the product is first introduced.

Critical diseases often demand taking more risk by using aggressive therapeutic approaches. This concept demands that regulatory agencies provide conditional approval for new drugs for cancer, AIDS, and other life-threatening illnesses in a shorter period of time and without the more complete knowledge of a product's safety and efficacy than would be required routinely. In these circumstances, approval may be granted conditional upon the sponsor's agreement to conduct additional studies after approval to completely characterize the product's safety and efficacy.

The third objective for conducting post-approval studies is to provide a sound foundation for marketing the new product. These studies are often very large or very small. The former are designed to provide large numbers of potential prescribers with the new drug; the latter are specific studies to characterize the new product in special populations or in unique clinical situations. In addition, these studies are often conducted in individual countries to focus on certain country-specific issues, such as competing products, different medical practices, pricing, and insurance related issues.

1.3.5   Concurrent Processes

The process described to this point has ignored a critical point in new drug development, i.e., pharmaceutical products do not leap from the head of the bench chemist to immediate introduction into the clinical research process. The sample product submitted with a New Drug Application (NDA) required its own evolution and development. Some of this development was required before the first unit was administered to humans. The remainder included continuing developments to optimize the product in various ways, characterize its properties, and define its manufacture.

Chemicals are not given to patients. The active component must be administered with a variety of other agents which give it desirable properties. Some of these properties are associated with its therapeutic effects. For example, the formulation should optimize absorption. Certain excipients may improve absorption; some may have a negative effect. Other considerations include the manufacture of the dosage form. Tablets are uniquely challenging, because the powder that forms the basis of the tablet must flow freely in the equipment, yet have the proper compressibility to form a tablet without crumbling.

Stability is a critical issue for all products. The approved useful life of all drugs is indicated on the label. Obviously, long stability without unusual storage conditions is important, both for ease of storage and administration, and to minimize the rate at which material becomes obsolete and unusable.

A pharmaceutical dosage form is a complex mixture of a number of components, each of which serves a different purpose in the product. The goal is to optimize the proper amount of each element to produce the best (by some definition) product. This optimization evolution continues throughout the development process. The results of this process are as essential to the approval of a new product as the results of the clinical program.

1.4   Submission and Review

The submissions may be called NDAs (New Drug Applications), NDSs (New Drug Submissions), registration dossiers, PMAs (premarketing applications), or many other terms. The specifics of these submissions vary from country to country, but the general document is an integrated summary of all safety and efficacy information, accompanied by the actual data or database.

The New Drug Application or registration dossier must provide information that is sufficient for a regulatory agency to determine whether the new product will be safe and effective if used as indicated, whether the sponsor has provided labeling that is consistent with the recommended use, and whether the sponsor can manufacture the new agent according to Good Manufacturing Practice guidelines which ensure quality, purity, and composition.

These documents are extraordinarily large and can easily exceed 100,000 pages if case reports or their tabulations are included. For this reason, most submissions are now provided in an electronic version (as well as paper, in many cases). Compact disks and hyperlinked submissions have made reviewing these documents more efficient. Links among clinical summaries, databases, statistical analyses, and case report forms have provided reviewers the capability to browse for important associations more readily than in the past. This has reduced the review time from years to months and has been responsible for more complete and valuable reviews.

1.5   Manufacturing

Pharmaceuticals are products. They are highly pure, highly regulated, and exquisitely manufactured products. They have clearly specified expirations, have specific instructions regarding their safety, efficacy, and instructions for use. And, in the end, they are a product like no other—a product that demands the precise requirements under which it is manufactured, because the public health is its measure.

Although the safety and efficacy of pharmaceuticals is required, their price is an important consideration to their affordability, and to some people, their availability. For this reason, optimization is of the highest priority in the manufacture of pharmaceuticals. Pharmaceutical scientists work hand in hand with engineers and statisticians to develop manufacturing processes which not only bring important products to the health of mankind, but which continually improve these products to raise their value and reduce their cost.

1.6 Summary

The discovery, development, regulatory approval, manufacture, and distribution of pharmaceutical products is a complex, time-consuming, and scientifically rigorous process. Unfortunately, because of its essential contribution to the public health, it can also have political elements that detract from its success. Historically, the process was evolutionary and followed a well-defined, yet pedantic, track toward approval. In the past, clinical research was equated with the practice of medicine, and although medical practice will always be an important component of clinical research, today's research depends to an enormous degree on information capture, management, and interpretation. The paradigm has changed, and the future of drug development will be more revolutionary than evolutionary. The use of statistics, computers, software, the internet, and other information management tools and concepts will not just be supportive of the process, they will be the process.

1.7 References

1.7.1 Print References

Drews, Jürgen. (1999). *In Quest of Tomorrow's Medicines*. Springer-Verlag, New York.

Guarino, R.A. (1987). *New Drug Approval Process: Clinical and Regulatory Management*. Marcel Dekker, New York.

Johnson, J. (1987). Past and present regulatory aspects of drug development. In: Peace, K., ed., *Biopharmaceutical Statistics for Drug Development*. Marcel Dekker, New York, Chap. 1.

Mathieu, M. (1990). *New Drug Development: A Regulatory Overview*. PAREXEL International Corp., Waltham, MA.

Mathieu, M. (1997). *Biologics Development: A Regulatory Overview*. PAREXEL International Corp., Waltham, MA.

Peace, K., ed. (1987). *Biopharmaceutical Statistics for Drug Development*. Marcel Dekker, New York.

1.7.2 Electronic References

http://www.fda.gov/
  U.S. Food and Drug Administration

http://www.emea.eu.int/
  European Agency for the Evaluation of Medicinal Products

http://www.fda.gov/fdac/special/newdrug/benlaw.html
Evolution of U.S. drug law, including the Kefauver–Harris amendment

http://www.mcclurenet.com/
Provides the ICH guidelines

http://www.ifpma.org/ich1.html
Official website for the ICH

http://clinicaltrials.gov/ct/gui/c/w2b/
NIH website on clinical trials

Part 2:
Basic Research and
Preclinical Studies

2

One-Factor Comparative Studies

Bill Pikounis

Merck Research Laboratories, Rahway, NJ, USA

2.1 Introduction

Virtually every kind of educational text on fundamental statistical methods covers the Student t-test for comparing two groups with respect to some response. It is arguably the most frequently applied statistical inference technique to data, helped along by its inclusion into practically any computer software that provides statistical capabilities. A considerable subset of these texts and softwares also covers data layouts with more than two groups.

The one-factor design of two or more groups produces a (if not the) most common data structure resulting from comparative experiments in the preclinical[1] stages of drug research and development. Unlike the clinical phases, there do not exist any governmental regulatory agency mandates that require statisticians to be involved in studies conducted in the preclinical phases. There are at best a handful of statisticians to help a population of hundreds or even thousands with statistical analysis needs in basic research. Therefore most data analyses are performed without any statistician involvement at all. This imposes a reliance on formal education for a research scientist, such as a statistics course or two taken back in college. More increasingly prominent in this author's observation, however, is a reliance on available software for learning about and practicing statistics. Unfortunately these conditions lead to a statistical analysis practice[2] that often is only concerned with the calculation of p-values, which at best is an inefficient extraction of information and value from the study data, and at worst a misleading account of interesting or real effects or lack thereof.

This chapter is concerned with an analysis approach for one-factor studies that is intended to be thorough so that information and value can be maximized from the data. The approach relies on exploration and exposition through good data graphs, analysis that focuses on magnitudes and expressions of effects, the assessment of validity and sensitivity of statistical methodology, and guidance on planning subsequent studies. All these aspects are ill-treated or ignored when an analysis only focuses on p-values, and interpretation suffers as a consequence. S-PLUS provides an excellent computing environment to readily ad-

---

[1] Preclinical refers to all stages of drug discovery and development that take place prior to human testing.

[2] Practitioners include statisticians as well as nonstatisticians!

dress these issues with built-in functions that can also be extended as needed for such purposes.

Section 2.2 introduces an example used throughout and sets the stage for the evaluations and interpretations in Section 2.3. Guidance on use of the results to help plan another study is covered in Section 2.4. We finish up with summary remarks in Section 2.5, followed by references and an Appendix of the programming code.

2.2   Example Study Data

2.2.1   Background

A canine Benign Prostatic Hyperplasia (BPH) study (Rhodes et al., 2000) was performed where prostate size was evaluated under five different treatments. BPH is the formal name for the condition of male prostate enlargement, and it is known that particular hormone levels have a direct impact on prostate size. The primary purpose of the experiment was to evaluate the effect of a physiological dose of estradiol on prostate growth in dogs. An endpoint of prostate volume (cm^3) was measured by transrectal ultrasonography. In subsequent discussions, we also refer to this endpoint with the label "prostate size."

Twenty-five animals were randomized to the five treatment groups in a balanced fashion, so that there were five subjects per group. The study protocol actually produced multiple measurements on each subject over the course of the 12-week study. The handling of such repeated measures data are outside the intended scope of this chapter, including the usual complexities that are introduced in choosing a tractable analysis that is scientifically meaningful.

We start with a simplified version of the data that was created by taking the average of the six measurements within each subject.[3] This produced a summary measure of one observation per subject that we will use for the basis of the one-factor evaluation.

2.2.2   Importing the Data

The data were saved in a comma-separated-value text formatted file psdata.csv:

```
AE,9.132
AE,10.07
```

...(middle 21 records)

---

[3] Given the context of a dataset in general, other types of appropriate summary measures could be used, such as Area Under the Curve (AUC), Maximum, etc.

```
NC,14.336
NC,25.102
```

and read into S-PLUS with

```
> psize.df <- read.table("psdata.csv", sep = ",",
    col.names = c("grp", "size"))
```

The following data frame listing occurs when the name is typed at the prompt:

```
> psize.df
    grp    size
 1   AE   9.132
 2   AE  10.070
 3   AE  20.077
 4   AE  14.691
 5   AE  23.698
 6   E1  10.356
 7   E1   6.313
 8   E1  21.708
 9   E1  12.651
10   E1  15.464
11   E2  37.200
12   E2  12.639
13   E2  16.791
14   E2  36.996
15   E2  22.808
16   CC   1.975
17   CC   3.125
18   CC   4.433
19   CC   6.154
20   CC   4.175
21   NC   9.301
22   NC  13.531
23   NC  12.840
24   NC  14.336
25   NC  25.102
```

The group labels are defined and described in Table 2.1. The ordering of levels in the grp factor are, by default, done alphabetically. For S-PLUS to use the Table 2.1 order of factor levels for later operations instead, we redefine the grp factor with

```
> psize.df$grp <- factor(psize.df$grp,
    levels = c("CC", "NC", "AE", "E1", "E2"))
```

Table 2.1. Factor levels.

Treatment group	Description
CC	Castrated controls
NC	Noncastrated controls
AE	Castrated and treated with androstandiol and estradiol
E1	Castrated and low dose of estradiol
E2	Castrated and high dose of estradiol

2.3 Evaluation

The approach considers three stages of evaluation[4] on the dataset:

- **Exploration**. Descriptive and graphical.
- **Analysis**. Formal statistics.
- **Diagnostics**. Validity checking of statistical assumptions.

2.3.1 Exploration

The term "exploration" refers to a branch of statistical analysis techniques that complements formal analysis procedures, in the spirit of Tukey's (1977) exploratory data analysis (EDA). Unlike formal methods based on goals of confirmation, exploration methods are not directly concerned with p-values, statistical significance, confidence percentages, and the like. Rather, they are based on looking at the data as a whole, typically with graphs or tables, so the sophisticated capabilities of the eye–brain system can be effectively used to summarize information, look at detailed features, and thereby gain insight into patterns and relationships in the data.

For easier access to the variables we attach the `psize.df` data frame created in Section 2.2.2. We also set the graphical parameter `pty` to create graphs with a square region rather than the system default rectangular layout, and set `mfrow` to create two graphs in a single row:

```
> attach(psize.df)
> par(pty = "s", mfrow=c(1, 2))
> plot(jitter(as.numeric(grp)), size,
    ylab = "Prostate Volume (cm3)", xlab = "",
    xlim = c(0.5, 5.5), axes = F, main = "Jittered")
> axis(1, at = 1:5, label = levels(grp))
> axis(2)
> box()
```

---

[4] "Evaluation" is felt by this author to better represent the broad intent of the chapter, rather than the word "analysis," which is often too narrowly regarded as synonymous with formal techniques such as hypothesis testing.

```
> plot(as.numeric(grp), size,
    ylab = "Prostate Volume (cm3)",xlab = "",
    xlim = c(0.5, 5.5), axes = F, main = "Not Jittered")
> axis(1, at = 1:5, label = levels(grp))
> axis(2)
> box()
> # Restore to one graph per row layout for Fig 2.2
> par(mfrow=c(1, 1))
```

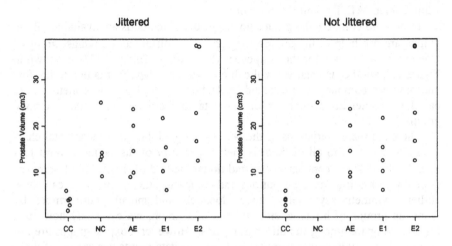

Figure 2.1. Point graphs with and without jittering.

In both panels of Figure 2.1 there is a display of a point graph,[5] which is just a special case of the scatterplot, the fundamental graphical tool for analyzing the relationship between two variables. It is simply a Cartesian plot with the x-axis variable of factor levels versus the y-axis variable of response values.

In the left panel of Figure 2.1 we applied the technique of "jittering" to the x-values that represent the groups. Jittering adds random noise to one or both variables before plotting their values in a scatterplot graph. The amount of noise is chosen so as not to substantively distort the data (Chambers et al., 1983, pp. 106–107). Jittering is a useful method to alleviate the overlap of plotted data points that occurs when observed values are very similar or equal. Otherwise, a single dot on a scatterplot may represent one, two, or more points. Note, in the right panel, a couple of points in the E2 and CC groups that are hard to distinguish, even though open circle plot symbols are used.[6] While these are clearly distinguished in the left panel version, it remains clear which data points come

---

[5] Other descriptions are used for this type of plot, most notably strip plot (Cleveland, 1993). There is an S-PLUS built-in Trellis function called `stripplot` that alternatively could be used.

[6] Solid circle symbols, which are often used, make discrimination of overlapping points dramatically harder or impossible.

from which group. Indeed for the factor-response type of x versus y data in a one-factor layout, only the x-variable needs to be jittered. The y-values (observed response) remain unmodified.

The point graph allows us to see all of the data, get a feel for the individual group distributions, and gauge expected or unexpected data behavior. For instance, we can see the two largest values in the E2 group that are far away from the rest of the data. We also get a sense of which groups tend toward higher, lower, or similar responses. The castrated controls (CC) group clearly has lower prostate sizes due to the absence of naturally occurring (NC group) or treatment administered (AE, E1, and E2) hormones.

In order to compare the groups however, better methods are available. Point graphs are not helpful in gauging magnitudes of differences amongst groups.[7] One graphical method for this purpose is the boxplot (Tukey, 1977) as shown in Figure 2.2; another is what we will call an error bar graph that is tied to the formal pairwise comparisons presented in Section 2.3.2. Both these methods are based on numerical summaries of the data to facilitate comparisons among groups.

There are many variations of the boxplot method, but the fundamental characteristic common to all of them is their superiority of displaying several features of the data, in contrast to the traditional method of the mean ±1 standard error for portraying data distributions and comparing them. Boxplots can show outliers, symmetry/asymmetry, ranges, location, and spread. Furthermore, by lining up groups of boxplots side by side in a graph, we can study how such features compare among the different groups. However, boxplot graphs are not meaningful to study unless there are at least five data points per group. If there are less than five data points, use the point graph plot of the raw data for study.

For our example, there are five observations per group, so the boxplot graph is only marginally helpful. The statements to create Figure 2.2 are

```
> # data list-form will be reused
> sizelist <- split(size, grp)
> centers <- boxplot(sizelist, style.bxp = "att",
    medpch = "o", ylab = "Prostate Volume (cm3)")
> # Add group means
> points(centers, unlist(lapply(sizelist, mean,
    na.rm = T)), pch = "+")
```

Variability across the groups appears inconsistent according to the sizes of the interquartile range boxes. The NC group distribution display designates the highest and lowest values as potentially extreme. (Keep in mind though, the small sample sizes.) The median is portrayed by the open-circle character within the box, and the mean is portrayed by the plus sign (+) symbol. Both measures of central tendency suggest the same pattern of separations among the groups: CC less than NC, AE, and E1, which are in turn less than E2.

---

[7] These disadvantages of point graphs are clearly pronounced as sample sizes grow larger than, say, 10 data points per group.

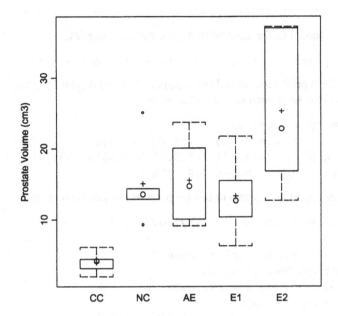

Figure 2.2. Boxplot graph.

We can refer to actual summary values such as quantiles, means, etc., within and across each of the groups by constructing a table of descriptive measures. A function descript is defined in the Appendix and a call to it produces the table:

```
> descript(sizelist)
      Min 25%ile Median 75%ile    Max   Mean Std Dev Std Err
CC  1.975  3.125  4.175  4.433  6.154  3.972  1.559  0.697
NC  9.301 12.840 13.531 14.336 25.102 15.022  5.954  2.663
AE  9.132 10.070 14.691 20.077 23.698 15.534  6.302  2.818
E1  6.313 10.356 12.651 15.464 21.708 13.298  5.772  2.582
E2 12.639 16.791 22.808 36.996 37.200 25.287 11.372  5.086
```

We end the exploratory part of the session with

```
> detach("psize.df")
```

2.3.2 Formal Analysis

Formal analysis of the multiple groups is done with one-way analysis-of-variance (ANOVA). Prostate size is modeled as a function of the groups as represented by the grp factor. In this analysis we will use "treatment" contrasts so that coefficients in the model reflect differences from the CC group. You can set this contrast type by using the options function:

```
> options(contrasts = c("contr.treatment", "contr.poly"))
```

Now the model fit for the ANOVA uses the aov function:

```
> aov.psize <- aov(size ~ grp, data = psize.df)
```

The default print method and the summary method display classical elements of the ANOVA table such as sums of squares, etc.:

```
> summary(aov.psize)
          Df Sum of Sq  Mean Sq   F Value        Pr(F)
     grp   4  1149.474 287.3686  5.980889  0.002469779
Residuals 20   960.956  48.0478
```

Some specific elements to use for purposes of this section are

```
> summ.aov.psize <- summary(aov.psize)
> # MSE
> mse <- summ.aov.psize$"Mean Sq"[[2]]
> # MSE degrees of freedom
> df.mse <- summ.aov.psize$Df[[2]]
> # Global F p-value
> gFpval <- summ.aov.psize$"Pr(F)"[[1]]
```

The global F-test on the grp factor yields

```
> round(gFpval, 3)
[1] 0.002
```

which not surprisingly suggests evidence of some significant differences among the five treatment groups.

Follow-up comparisons can be made with the multicomp function, which offers a variety of multiple comparison procedures (MCP). Typically in preclinical experiments there is more concern with missing potentially real effects (Type II errors) rather than statistical artifacts of declared significant differences (Type I errors). Any statistically significant effects will virtually always lead to studies (either in the same animal species or a different one) to evaluate the repeatability of the results. Thus we proceed with Fisher's Protected Least Significant Difference (LSD) Procedure (Fisher, 1935), which is generally regarded as the most liberal MCP.

A function display.mcp (based on the multicomp function) is defined in the Appendix to create a custom display of the MCP results. Running it on the fit object produces:

```
> display.mcp(aov.psize)
```

```
Fisher's Protected LSD results
 All possible pairwise comparisons
 95% Confidence Intervals & 5% Significance Level
       estimate stderr    lower    upper  P-value Signif?
CC-NC  -11.050   4.384  -20.194   -1.905   0.020     ***
CC-AE  -11.561   4.384  -20.706   -2.416   0.016     ***
CC-E1   -9.326   4.384  -18.471   -0.181   0.046     ***
CC-E2  -21.314   4.384  -30.459  -12.170 < 0.001     ***
NC-AE   -0.512   4.384   -9.656    8.633   0.908
NC-E1    1.724   4.384   -7.421   10.868   0.698
NC-E2  -10.265   4.384  -19.410   -1.120   0.030     ***
AE-E1    2.235   4.384   -6.910   11.380   0.616
AE-E2   -9.753   4.384  -18.898   -0.608   0.038     ***
E1-E2  -11.988   4.384  -21.133   -2.844   0.013     ***
```

A succinct display of these results can be achieved with the graphical method of Andrews et al. (1980). The purpose of what we will call here an Error Bar Graph is to visually convey which groups were found significantly different by the MCP. The following sequence of statements creates Figure 2.3:

```
> coefs.aov.psize <- coef(aov.psize)
> grpmeans <- coefs.aov.psize[1] +
    c(0, coefs.aov.psize[-1])
> names(grpmeans) <- levels(psize.df$grp)
> n <- 5 # sample size per group
> g <- length(grpmeans) # number of groups
> errorbarlength <- qt(0.975, df.mse) *
    sqrt(2*mse/n)/2
> error.bar(1:length(grpmeans), grpmeans,
    lower = errorbarlength, ylab = "Prostate Volume (cm3)",
    gap = F, xlab = "", pch = 16, xlim = c(0.5, g + 0.5),
    axes = F)
> axis(1, at = 1:g, label = names(grpmeans))
> axis(2)
> box()
```

The length of the error bars are chosen such that two groups whose bars do not overlap are found to be significantly different from each other. Conversely, if the error bars of two groups do overlap, they are not significantly different. In Figure 2.3 we see that compared to the castrated controls (CC) group, all of the other four groups have a significantly higher mean prostate size response, although the separation is not clear and requires careful study and/or reference back to the MCP results table. Similarly, the E2 group is found to have a significantly higher mean response than the NC, AE, and E1 groups.

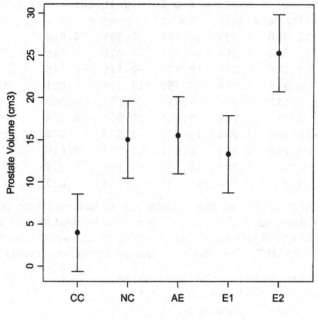

Figure 2.3. Error Bar Graph.

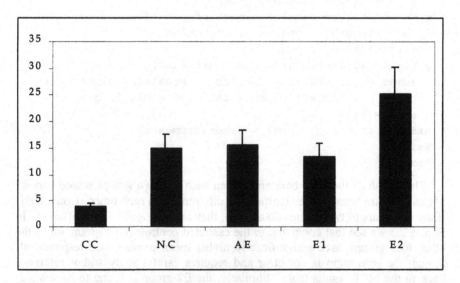

Figure 2.4. Typical representation of treatment group comparisons in the scientific literature.

The error bar graph method provides an eagle-eye, more informative, view that the often-used bar chart kinds of graphs in the scientific literature do not. Table 2.2 discusses the improvements, and Figure 2.4 shows a likely example of displaying the same analysis in the scientific literature.[8]

Table 2.2. Comparison of the error bar graph with common practice in scientific literature.

Error bar graph	Common scientific literature graph
Uses a plotting character to represent group means	Bar chart with bar heights representing group means.
Error bars on both sides.	Bar chart with solid bars that obliterate the display of the error bar on the lower side of the mean. This forces the viewer to try to picture what bars on the lower side look like in order to compare groups.
The length of the error bar is a multiple of the standard error determined from the analysis, such that if the bars from any two means do not overlap, their groups are significantly different.	Lengths of bars are either the observed standard error or standard deviation for each individual group, so there is no formal connection with the formal statistical analysis performed. Annotation with asterisks and footnotes are needed to convey which groups were found significantly different in the statistical analysis.
y-axis limits are chosen to minimize white space.	Bars start from zero.

One additional table that is useful is the mean, standard errors, and individual confidence intervals for each treatment group. This can be accomplished with the following commands:

```
> icilength <- qt(0.975, df.mse) * sqrt(mse/n)
> ilowerci <- grpmeans - icilength
> iupperci <- grpmeans + icilength
> indvgrps.psize <- cbind(n = rep(5, g), Mean = grpmeans,
    "Std Err" = rep(sqrt(mse/n), g), LowerCL = ilowerci,
    UpperCL = iupperci)
> display.indv <- function() {
    cat("Individual Groups Summary Table\n",
      "95% Confidence Intervals\n\n")
    round(indvgrps.psize, 3)
}
```

so that running the created function display.indv displays

---

[8] The last comparison item in Table 2.2 is not an issue for this example because the lower error bars for the CC group extend close to 0.

28   B. Pikounis

```
> display.indv()
Individual Groups Summary Table
 95% Confidence Interval

     n   Mean  Std Err  LowerCL  UpperCL
CC  5   3.972     3.1   -2.494   10.439
NC  5  15.022     3.1    8.556   21.488
AE  5  15.534     3.1    9.067   22.000
E1  5  13.298     3.1    6.832   19.765
E2  5  25.287     3.1   18.820   31.753
```

2.3.3 Diagnostics

Any statistical test or confidence interval, or the type of formal analysis where a
p-value is computed and significance is assessed, rests on an underlying model
for the data. This model always comes with assumptions that unfortunately can
never be verified to be exactly true. We know that a data distribution is never
exactly Gaussian (normal), nor are variances among groups exactly equal. So
what? Well, the degree of validity for probabilistic inferences (also known as
p-values) depends on the degree to which the underlying data approximate theoretical distributional forms. A p-value used to blindly determine whether a difference is significant or not may be grossly inaccurate. On the other hand, an
otherwise significant difference may be obscured by a wild outlier data point or
unequal variability among groups. Fortunately, empirical evidence and statistical research has shown that assumptions only need be reasonably satisfied or
approximated to ensure reasonably accurate probabilistic inferences on comparisons of multiple groups in a one-factor experimental design layout of equal or
near-equal sample sizes such as we have here (see, e.g., Fleiss (1986, pp. 59–
60); Scheffé (1959)).

In this section we focus on graphical methods for checking error assumptions of normality and equal group variances, since relying solely on tests to decide whether or not an assumption is met has the usual potential drawbacks as
alluded to above. Furthermore, tests do not inform us why there is a departure
from normality or equivariance for the data, or how reasonable the assumption
might appear to be. Put another way, graphical methods allow for evaluation of
how and to what degree a group of data departs from a specific theoretical distribution.

Figure 2.5 displays two plots for evaluation of the normality and equal group
variances assumptions. It was created with the following commands:

```
> r.psize <- resid(aov.psize)
> par(mfrow=c(1, 2), pty = "s")
> attach(psize.df) # original data set
> qqnorm(r.psize, xlab = "Standard Normal Quantile",
    ylab = "Residual", pch = 1)
```

```
> qqline(r.psize)
> # Equivariance graph
> sar.psize <- sqrt(abs(r.psize))
> grporigord <- factor(as.character(grp),
    levels = unique(as.character(grp)))
> sarmeans <- sapply(split(sar.psize, grporigord), mean)
> ordidx <- order(sarmeans)
> ordidx2 <- order(rep(sarmeans, each = 5))
> plot(jitter(as.numeric(grporigord)), sar.psize[ordidx2],
    ylab = "SqrtAbsResidual", xlab = "",
    xlim = c(0.5, 5.5), axes = F)
> axis(1, at = 1:5, label = levels(grporigord)[ordidx])
> axis(2)
> box()
> lines(1:5, sarmeans[ordidx], type = "l")
> par(mfrow = c(1, 1))
> detach("psize.df")
```

In the left panel is a quantile–quantile plot that suggests heavy tails and skewness in the high end of the data distribution. The right panel is based on a concept from John Tukey (see Cleveland (1993)), where the square-roots of the absolute values of the residuals are plotted against the groups, with a superposed line of the group means of the values. (Note that the groups are ordered in terms of these means.) If the group variances were equal, we would expect to see an approximately flat line, which is clearly not the case here.

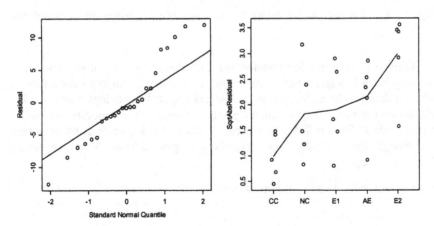

Figure 2.5. Graphs to assess normality and equivariance assumptions.

Transformation of the data using the logarithmic scale prior to data analysis is a useful technique to consider. The corresponding primary fit object is

```
> aov.log.psize <- aov(log(size) ~ grp, data = psize.df)
```

and the prior code to do the formal analysis and the diagnostics is applicable with some adjustments for working with the data in the log-transformed scale (see the Appendix for complete code listings). Alternatively, we could have also added the log-transformed variable to the data frame with

```
> psize.df$logsize <- log(psize.df$size)
```

Figure 2.6 demonstrates that the normality and equivariance assumptions are more reasonably met in the logarithmic scale although some distortions in the tails are still present in the left panel.[9]

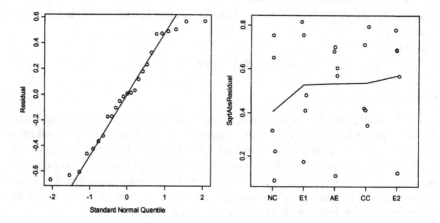

Figure 2.6. Graphs to assess normality and equivariance assumptions in logarithmic scale.

Analysis of data in a log-transformed scale corresponds to an analysis of ratios or percent changes when comparisons are involved. An expression like e^d, where d denotes the difference in the natural log scale, provides ratios, so that $100(e^d - 1)$ provides expression as percent change. If we modify the MCP-related code in Section 2.3.2 to express results of the log-scale analysis in percent change scale (see the Appendix), the analogous display to that created by display.mcp is

```
> display.mcp.log(aov.log.psize)
```

---

[9] Recall that the original size response was derived as a summary statistic, namely the mean over repeated measurements. Taking the logs first and then the average might be a preferable alternative for analyzing the data in the log scale. For these data, the generated values, by taking logs of the averages, are very similar so we stay with the simpler approach here.

```
Fisher's Protected LSD results
All possible pairwise comparisons
95% Confidence Intervals & 5% Significance Level
Analysis in Natural Log scale
```

	estimate	stderr	lower	upper	%Chg	SE %Chg	%Chg lower	%Chg upper	P-value	Signif?
CC-NC	-1.344	0.272	-1.911	-0.776	-73.910	7.101	-85.212	-53.971	< 0.001	***
CC-AE	-1.364	0.272	-1.931	-0.796	-74.427	6.960	-85.505	-54.884	< 0.001	***
CC-E1	-1.196	0.272	-1.763	-0.628	-69.753	8.232	-82.855	-46.636	< 0.001	***
CC-E2	-1.831	0.272	-2.399	-1.264	-83.980	4.360	-90.919	-71.736	< 0.001	***
NC-AE	-0.020	0.272	-0.588	0.548	-1.983	26.677	-44.443	72.926	0.942	
NC-E1	0.148	0.272	-0.420	0.716	15.935	31.553	-34.287	104.537	0.593	
NC-E2	-0.488	0.272	-1.055	0.080	-38.596	16.712	-65.195	8.332	0.088	
AE-E1	0.168	0.272	-0.400	0.736	18.280	32.192	-32.957	108.676	0.544	
AE-E2	-0.468	0.272	-1.035	0.100	-37.354	17.050	-64.491	10.524	0.101	
E1-E2	-0.636	0.272	-1.203	-0.068	-47.036	14.415	-69.979	-6.558	0.030	***

Note that four columns have been added. The fifth column is back-transformation to percent change where a negative value denotes a decrease; the sixth column contains the approximated standard error using the delta method (e.g. Agresti (1990, Ch. 12)). The percent change confidence intervals end-points in columns 7 and 8 use back-transformation of the lower and upper values in columns 3 and 4 and thus are asymmetric.

Similarly we obtain geometric means for the individual group summary statistics. See the Appendix for code analogous to display.indv in Section 2.3.2.

```
Individual Groups Summary Table
95% Confidence Intervals
Back-Transformed from Analysis in Log Scale
```

	n	GeoMean	SE GeoMean	LowerCL	UpperCL
CC	5	3.710	0.714	2.483	5.543
NC	5	14.220	2.737	9.518	21.245
AE	5	14.508	2.792	9.711	21.675
E1	5	12.266	2.361	8.210	18.325
E2	5	23.159	4.457	15.501	34.599

Figure 2.7 shows the error bar graph with log scaling on the y-axis and expression in the back-transformed scale. See the Appendix for the code that includes how the left- and right-hand side y-axes labels were constructed.

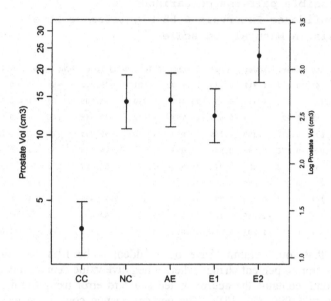

Figure 2.7. Error bar graph from analysis in log scale.

The outputs from the analysis in the log scale show significant statistical separation of four groups from the CC group, just as was found in the original scale case, although the *p*-values are much smaller and thus more definitive. However, only the E2 group was found significantly higher than the E1 group in the log-scale analysis, whereas the original scale analysis also indicated significant differences between E2 and each of NC and AE. Perhaps these additional significant difference findings were influenced by the apparent unequal variances in the original scale for prostate volume (cm^3) as we discussed with Figure 2.5.

Closer examination of the pairwise *p*-values, confidence intervals, and the error bar graphs between the two analysis scales shows not too much difference in a qualitative sense, however. Both exhibit a grouping pattern that makes scientific sense. Larger sample sizes would likely clarify the apparent groupings by increasing precision and power, and by downweighting undue influence of individual points like the two high values in the E2 group.

2.4 Planning the Next Study

Typically it is the case in the preclinical area that a subsequent study will be done, and therefore the findings of the analyzed data could be used to estimate power or sample sizes to help with the design of the new study.

Let us say that another study of the treatment effects on prostate size is planned and a one-factor layout will be used. It is desired to determine adequate sample size for the objectives of the study. The most helpful quantitative value that comes out of the previous study is an estimate of variability. More specifically we have s, a standard deviation estimate/conjecture from data of previous experiments that are similar to the one currently being planned. There are five other quantities to consider:

- α. The significance level of a test, typically set to 0.05 or 5%. This refers to the probability that a significant difference is claimed but in truth no such difference exists, so the common convention is to keep this at a low figure like 5%.
- $1 - \beta$. The power of a test. Power is defined as the probability of significantly detecting a true difference among the groups. A general rule of thumb for determining sample sizes is that the power should be at least 0.80, i.e., 80%.
- g. The number of groups in the study that is being planned.
- N. The total sample size, i.e., the total number of observations or experimental subjects. We assume that $N = gn$, where n denotes the common sample size per group.
- δ. A summary index of hypothesized differences among the groups in the study. This is related to the noncentrality parameter of the underlying reference distribution.

Most often it is difficult or uncomfortable for researchers to make a statement of expected or desired treatment effects in terms of magnitude and configuration. When a number of groups are studied, there are an infinite number of configurations in how they can be different and by what amounts they are different. Since we need a value for δ to make calculations tractable, a useful alternative is to base it on the hardest type of configuration to detect differences among group means. Such a scenario occurs when all means are equal except for two out of the g groups. Algebraically, this minimizes the noncentrality parameter under the nonnull parameter space. (Note that if $g = 2$, the groups are only different in one way, so for that case we only need concentrate on the amount of the difference.[10]) See, for example, Fleiss (1986, App. A.2, pp. 371–376) for a general discussion of specifying the noncentrality parameter and ways to determine a sample size for one-factor studies.

The above discussion presumes that we base our sample-size evaluations on the F-statistic for the global null hypothesis of no differences among the g groups. Below is a function that requires the user to specify a minimum scientifically meaningful difference (MSMD) between two out of the g groups under the hardest type of configuration to detect this scenario. Sample size is computed based on a single value or an array range of values that can be entered for the MSMD.

---

[10] S-PLUS has some built-in sample size and power functions for one- and two-sample cases.

Let us suppose the next study will have three groups, CC, E1, and E2, and we wish to detect at least a 50% change. To determine the required sample size for each group, we can use the function samplesizecalc defined in the Appendix:

```
> sampsizecalc(s = rmse.log, alpha = 0.05, power = 0.80,
    pctchg = T, g = 3, msmd = c(50, 75, 100, 150))
  msmd  n  N
1  50  23 69
2  75  13 39
3  100  9 27
4  150  6 18
```

Note that the estimate of root MSE is used from the analysis in the log scale, and that we set the pctchg=T option. This simple data-frame tabulation and/or an accompanying graph of MSMD versus N serves well to educate or remind the researcher how larger underlying differences require smaller sample sizes.

The function can certainly be adapted and improved in several ways. It uses a repeat / break structure to just keep the minimum value of n that satisfies the required power specification, so most likely an alternative using implicit looping structures of S-PLUS will be more efficient computationally. One might deem the hardest-to-detect approach or the use for the global F-test as too conservative or not of primary interest, so modifications to specify the noncentrality variable ncp would be in order. Indeed, if one can get specific configurations from the experimenter, that is most preferable. It is also often the case that the sample size available is fixed beforehand due to various economic constraints, and so the focus might be on power curves as a function of difference magnitudes. Any of these scenarios are easily programmable with S-PLUS.

It is always best kept in mind that statistical techniques used for determining power cannot be expected to be rigorously exact, since they depend on conjectures, like the estimates derived from previous studies. Empirical evidence has shown, however, that in general the techniques do very well in getting power estimates that are "in the neighborhood" of the true values. This message is a helpful standard attachment to any presentation of results to researchers.

2.5 Concluding Remarks

The goals stated at the outset of this chapter are motivated by experience as a statistician in the preclinical phases of drug research and development at a pharmaceutical company. One-factor designs are prevalent and there is way too much reliance on p-values to carry out comparative objectives with the data and subsequently make interpretations and conclusions. S-PLUS provides capabilities to maximize the extraction of information and value from data through an approach that relies heavily on good graphs and the expression of magnitudes of effects.

Of course there are details in the approach that a data analyst would wish to improve or adapt for a given study. Perhaps specific contrasts other than pairwise comparisons are of interest, or there are other diagnostic techniques to assess assumptions. The handling of unequal sample sizes and/or missing values would require careful attention to programming and interpretation. Other transformation functions might be more appropriate; for instance, a cube-root transformation on the prostate volume (cm^3) study data discussed here. Robust or resistant fits were not presented but should be considered an essential part of analysis. All the tools to accomplish these tasks are readily available in S-PLUS either as already existing functions or by straightforward programming.

2.6 References

Agresti, A.A. (1990). *Categorical Data Analysis*. Wiley, New York.

Andrews, H.P., Snee, R.D., and Sarner, M.H. (1980). Graphical display of means. *American Statistician* **34**, 195–199.

Chambers, J.M., Cleveland, W.S., Kleiner, B., and Tukey, P.A. (1983). *Graphical Methods for Data Analysis*. Wadsworth, Pacific Grove, CA.

Cleveland, W.S. (1993). *Visualizing Data*. Hobart Press, Summit, NJ.

Fisher, R.A. (1935). *The Design of Experiments*. Oliver & Boyd, London.

Fleiss, J. (1986). *The Design and Analysis of Clinical Experiments*. Wiley, New York.

Rhodes, L., Ding, V., Kemp, R., Khan, M.S., Nakhla, A.M., Pikounis, B., Rosner, B., Saunders, H.M., and Feeney, B. (2000). Estradiol causes a dose dependent stimulation of prostate growth in castrate beagle dogs. *The Prostate* **44**, 8–18.

Scheffé, H. (1959). *The Analysis of Variance*. Wiley, New York.

Tukey, J.W. (1977). *Exploratory Data Analysis*. Addison-Wesley, Reading, MA.

2.A.   Appendix

2.A.1   Descriptive Table of Summary Statistics (Section 2.3.1)

```
descript <- function(x) {
  comps <- sapply(x, function(x)
  {
    c(min(x), quantile(x, c(0.25, 0.5, 0.75)), max(x),
      mean(x), sqrt(var(x)), sqrt(var(x)/length(x)))
  }
  )
  dimnames(comps)[[1]] <- c("Min", "25%ile", "Median",
    "75%ile", "Max", "Mean", "Std Dev", "Std Err")
  round(t(comps), 3)
}
```

2.A.2   Multiple Comparisons (Section 2.3.2)

```
display.mcp <- function(fit) {
  mcp <- multicomp(fit, error = "cwe", method = "lsd")
  mcptbl <- data.frame(mcp$table)
  Pval <- 2 * (1 - pt(abs(mcptbl$estimate/mcptbl$stderr),
    df.mse))
  Pvalfmt <- ifelse(Pval <= 0.0005, "< 0.001",
    format(round(Pval, 3), nsmall = 3))
  snf <- ifelse(Pval < 0.05, "***", "   ")
  cat("Fisher's Protected LSD results\n",
    "All possible pairwise comparisons\n",
    "95% Confidence Intervals & 5% Significance Level\n\n")
  cbind(round(mcptbl, 3), "P-value" = Pvalfmt,
    "Signif?" = snf)
}
```

2.A.3   Diagnostics for Log-Scale (Section 2.3.3)

2.A.3.1   Q-Q and Equivariance Plots

```
# Log-scale code & functions analogous to
# original-scale evaluations
# Recall in text:
# options(contrasts = c("contr.treatment", "contr.poly"))
# aov.log.psize <- aov(log(size) ~ grp, data = psize.df)
# psize.df$logsize <- log(psize.df$size)
```

```
# Q-Q graph construction for Assessing Normality

r.log.psize <- resid(aov.log.psize)

attach(psize.df)  # original data set
par(mfrow = c(1,2), pty = "s")
qqnorm(r.log.psize, xlab = "Standard Normal Quantile",
  ylab = "Residual", pch = 1)
qqline(r.log.psize)

# Equivariance graph construction

sar.log.psize <- sqrt(abs(r.log.psize))
sarlogmeans <- sapply(split(sar.log.psize, grporigord),
  mean)
ordidx.log <- order(sarlogmeans)
ordidx2.log <- order(rep(sarlogmeans, each = 5))
plot(jitter(as.numeric(grporigord)),
  sar.log.psize[ordidx2.log], ylab = "SqrtAbsResidual",
  xlab = "", xlim = c(0.5, 5.5), axes = F)
axis(1, at = 1:5, label = levels(grporigord)[ordidx.log])
axis(2)
box()
lines(1:5, sarlogmeans[ordidx.log], type = "l")

par(mfrow = c(1, 1))
detach("psize.df")
```

2.A.3.2   *Formal Analysis*

```
# Classical ANOVA table display
summ.aov.log.psize <- summary(aov.log.psize)
# Components of ANOVA for later-use
mse.log <- summ.aov.log.psize$"Mean Sq"[[2]]
rmse.log <- sqrt(mse.log)
df.mse.log <- summ.aov.log.psize$Df[[2]]
gFpval.log <- summ.aov.log.psize$"Pr(F)"[[1]]
```

2.A.3.3   Display Multiple Comparisons Results

```
# Display of multiple comparison results associated with
# the ANOVA in the log-scale.  Note that it is essentially
# an annotated version of the display.mcp function in the
# text.  A better representation for general use would be
# to combine the two functions into one that has a
# log-scale analysis optional boolean argument.

display.mcp.log <- function(fit) {
  mcp.log <- multicomp(fit, error = "cwe", method = "lsd")
  mcptbl.log <- data.frame(mcp.log$table)
  Pval.log <- 2 * (1 - pt(abs(mcptbl.log$estimate/
    mcptbl.log$stderr), df.mse.log))
  Pvalfmt.log <- ifelse(Pval.log <= 0.0005, "< 0.001",
    format(round(Pval.log, 3), nsmall = 3))
  snf.log <- ifelse(Pval.log < 0.05, "***", "   ")
  mcptbl.log$pctchg <- 100 * (exp(mcptbl.log$estimate) - 1)
  mcptbl.log$se.pctchg <- 100 * exp(mcptbl.log$estimate) *
    mcptbl.log$stderr
  mcptbl.log$"%chglower" <- 100 * (exp(mcptbl.log$lower) -
    1)
  mcptbl.log$"%chgupper" <- 100 * (exp(mcptbl.log$upper) -
    1)
  names(mcptbl.log)[5:8] <- c("%Chg", "SE %Chg",
    "%Chg lower", "%Chg upper")
  cat("Fisher's Protected LSD results\n",
    "All possible pairwise comparisons\n",
    "95% Confidence Intervals & 5% Significance Level\n",
    "Analysis in Natural Log scale\n\n")
  cbind(round(mcptbl.log, 3), "P-value" = Pvalfmt.log,
    "Signif?" = snf.log)
}
# Function execution
display.mcp.log(aov.log.psize)

# Individual Groups Summary code for analysis in log scale

coefs.aov.log.psize <- coef(aov.log.psize)

grpmeans.log <- coefs.aov.log.psize[1] +
  c(0, coefs.aov.log.psize[-1])
names(grpmeans.log) <- levels(psize.df$grp)

icilength.log <- qt(0.975, df.mse.log) * sqrt(mse.log/n)
```

```
ilowerci.log <- grpmeans.log - icilength.log
iupperci.log <- grpmeans.log + icilength.log

indvgrps.psize.log <- cbind(n = rep(5, g),
  GeoMean = exp(grpmeans.log),
  "SE GeoMean" = exp(grpmeans.log) * sqrt(mse.log/n),
  LowerCL = exp(ilowerci.log), UpperCL = exp(iupperci.log))

display.indv.log <- function() {
  cat("Individual Groups Summary Table\n",
    "95% Confidence Intervals\n",
    "Back-Transformed from Analysis in Log Scale\n\n")
  round(indvgrps.psize.log,3)
}


# Function execution
display.indv.log()
```

2.A.3.4   Error Bar Graph

```
# This essentially mimics the version in the text for
# the analysis in the original scale.  The main annotations
# are in the axis() statements to denote logarithmic
# tick marks.

errorbarlength.log <- (qt(0.975, df.mse.log) *
  sqrt((2 * mse.log)/n))/2

error.bar(1:length(grpmeans.log), grpmeans.log,
  lower = errorbarlength.log,
  ylab = paste("Prostate Vol (cm3)", sep = ""), gap = F,
  xlab = "", pch = 16, xlim = c(0.5, g + 0.5), axes = F)
axis(1, at = 1:5, label = names(grpmeans.log))
axis(2, at = log(c(1, seq(5, 30, by = 5))),
  labels=paste(c(1, seq(5, 30, by = 5))), srt = 90)
axis(4, cex = 0.7)
mtext("Log Prostate Vol (cm3)", side = 4, cex = 0.7,
  line = 2)
box()
```

2.A.4   Sample Size Calculations (Section 2.4)

```
sampsizecalc <- function(s, g, msmd, pctchg = F,
  alpha = 0.05, power = 0.80, nmax = 100) {
#####
# Function to compute sample size
# nmax = upper bound on sample size iterations
# pctchg = are differences based on
#          log scale analysis?
#####
  nsolution <- vector("numeric", length = length(msmd))
  for(i in seq(along = msmd)) {
    n <- 2
    numdf <- g - 1
    msmdi <- msmd[i]
    if(pctchg) {msmdi <- log(msmdi/100 + 1)}
    repeat{
      dendf <- g*(n-1)
      ncp <- ( n * (msmdi)^2 ) / ( 2 * s^2 )
      pwr <- 1 - pf(qf(1 - alpha, numdf, dendf),
        numdf, dendf, ncp)
      n <- n + 1
      if (pwr > power | n > nmax) {
        nfinal <- n - 1
        break
      }
    }
    nsolution[i] <- nfinal
  }
  data.frame(msmd, n = nsolution, N = g*nsolution)
}
```

Part 3:

Pre-Clinical Safety Assessment

3

Analysis of Animal Carcinogenicity Data

Hongshik Ahn
State University of New York, Stony Brook, NY, USA

Ralph L. Kodell
Food and Drug Administration, Jefferson, AR, USA

3.1 Introduction

Animal carcinogenicity experiments are employed to test the carcinogenic potential of drugs and other chemical substances used by humans. Such bioassays are conducted in animals at doses that are generally well above human exposure levels, in order to detect carcinogenicity with relatively small numbers of animals. Animals are divided into several groups by randomization and treated with a test compound at different dose levels. A typical carcinogenicity study involves a control and 2 to 3 dose groups of 50 or more animals, usually rats or mice. Typically, a chemical is administered at a constant daily dose rate for a major portion of the lifetime of the test animal, for example, for 2 years. Sometimes, scheduled interim sacrifices are performed during the experiment. At the end of the study, all surviving animals are sacrificed and subjected to necropsy. For each animal in a given dose group, the age at death and the presence or absence of specific tumor types are recorded. Groups of animals are compared with respect to tumor development.

Many methods have been proposed for analyzing tumor incidence data from animal bioassays. For nonlethal tumors, Mantel–Haenszel-type tests have been applied to survival experiment data to test for differences in tumor prevalence (Hoel and Walburg (1972), Peto (1974), Gart (1975), Lagakos (1982)). For rapidly lethal tumors, Hoel and Walburg (1972) recommended a log-rank test for comparing the rate of death with tumor across different doses. However, most tumors are neither strictly nonlethal nor rapidly lethal. Further, the estimation of incidence rates for certain diseases can be affected by competing causes of death. Hence, many analyses require cause-of-death information (Kodell and Nelson (1980), Kodell et.al. (1982), Dinse and Lagakos (1982), Turnbull and Mitchell (1984)).

Peto (1974) and Peto et.al. (1980) proposed a method for analyzing tumor data in which tumors are observed in both the fatal and incidental contexts. This method is popular because it does not require large numbers of animals to be sacrificed at multiple timepoints, in order to observe the prevalence of occult tumors. This method assumes that pathologists can determine if a tumor affected an animal's risk of death. In practice, however, pathologists often claim that accurate determinations of the cause of death are impossible, and classification errors can produce biases. In many studies, cause of death is not assigned at all. Recently Ahn et.al. (2000) developed a new statistical approach for estimating numbers of fatal tumors in the absence of cause-of-death information. For experiments lacking cause-of-death data, tumor lethality is attributed through constrained nonparametric maximum likelihood estimation of the probability distributions of time-to-onset-of and time-to-death-from specific tumors of interest. The numbers of fatal tumors imputed by this method can be used to implement Peto's IARC (International Agency for Research on Cancer) test instead of using pathologist-assigned cause-of-death information.

Without cause-of-death information or simplifying assumptions, interim sacrifices of groups of animals are necessary in order for tumor incidence rates for occult tumors to be identifiable from bioassay data (McKnight and Crowley (1984)). Various statistical tests have been developed for experiments in which serial interim sacrifices are performed (McKnight and Crowley (1984), Portier and Dinse (1987), Malani and Van Ryzin (1988), Ahn and Kodell (1995), Kodell and Ahn (1996), Kodell and Ahn (1997)), or additional assumptions are given (Bailer and Portier (1988), Bieler and Williams (1993), Dinse (1993)). Among these, Bailer and Portier proposed the Poly-3 trend test, which made an adjustment to the Cochran–Armitage trend test (Cochran (1954), Armitage (1955)) for detecting a linear trend across dose groups in the overall proportions of animals with tumor. Bieler and Williams (1993) made a further adjustment to the Cochran–Armitage test.

However, most of the animal carcinogenicity studies are designed with a single terminal sacrifice. For data with no interim sacrifices, Dinse (1991) and Lindsey and Ryan (1994) proposed parametric statistical tests for dose-related trends. Dinse's test is based on the assumption of a constant difference between the death rates of animals with and without tumors, while Lindsey and Ryan's test assumes a constant ratio for those death rates. Kodell et.al. (1997) proposed a nonparametric age-adjusted trend test for a single terminal sacrifice. They assume the constant proportionality of tumor prevalence for live and dead animals.

Kodell and Nelson (1980) presented the results of an experiment with mice in which the effect of Benzidine dihydrochloride on the induction of liver tumors was investigated. The experiment was conducted at the National Center for Toxicological Research (NCTR) to study strain and sex differences with regard to chemically induced liver tumors in mice.

Male and female mice from two strains (F1 and F2) were used. The data from dose groups 60 ppm, 120 ppm, 200 ppm, and 400 ppm were reported in Kodell and Nelson (1980). In this chapter, the data from the F2 strain mice are chosen for illustration. As seen in Table 3.1, the sample sizes of the data are larger than those in typical animal carcinogenicity experiments. The counts of sacrificed animals, along with cause-of-death information assigned by pathologists in each interval are shown in Table 3.1.

Among the various methods discussed in this section, Peto's IARC test has been commonly used to evaluate tumorigenicity data for drugs and other chemicals. Recently the NTP (National Toxicology Program) adopted the Poly-3 test for dose-related trends with respect to the tumor incidence rate across several dose groups. In this chapter, we will illustrate Peto's IARC cause-of-death test, the Cochran–Armitage test, and the Poly-3 trend test (Bieler and Williams (1993)) using S-Plus. These tests will be conducted on the given data to test for a dose-related trend in tumor incidence across dose groups for each sex of mice. Since we are interested only in detecting a positive trend, only one-sided tests will be considered. As in other statistical areas, programs for analyzing data from a preclinical study like the Benzidine study can be easily implemented in S-Plus. The programs are provided in the Appendix.

3.2   Notation

Consider an experiment with g treatment groups, a control and $g - 1$ dose groups. Let l_i be the dose level of the ith group. Suppose N animals are initially placed on experiment, and N_i animals are assigned randomly to the ith treatment group. Let the animals in the ith treatment group be followed over time for the development of tumors. We assume that all animals come from the same population and are born tumor-free on day zero of the experiment. Divide the time span into m intervals such that the jth interval is $I_j = (t_{j-1}, t_j]$, $j = 1, \ldots, m$, where $t_0 = 0$ and t_m denotes the time at which the terminal sacrifice is scheduled. All the tumors are assumed to be irreversible.

3.3   Analysis of Data with Known Cause of Death

Peto (1974) and Peto et al. (1980) proposed a method for analyzing tumor data in which tumors are observed in both the fatal and incidental contexts. They recommended that pathologists assign a context of observation to each observed tumor.

Tumors that do not alter an animal's risk of death and are observed only as the result of a death from an unrelated cause are classified as incidental,

Table 3.1. Frequency data of the Benzidine Dihydrochloride experiment; F2-strain mice.

Sex	Dose[a]	j[b]	Death with fatal tumor[c]	Death with incidental tumor[c]	Death without tumor	Sacrificed with tumor	Sacrificed without tumor
Male	60	1	0	0	8	0	68
		2	0	0	3	3	41
		3	3	0	3	4	34
	120	1	0	0	3	0	46
		2	1	0	1	8	39
		3	3	0	2	12	27
	200	1	0	0	8	0	47
		2	1	0	4	6	31
		3	0	1	3	7	10
	400	1	0	0	3	1	23
		2	1	1	3	4	17
		3	5	1	0	6	6
Female	60	1	0	0	2	0	70
		2	0	0	1	10	38
		3	7	0	4	15	20
	120	1	0	0	7	2	44
		2	3	3	4	15	26
		3	15	1	2	20	1
	200	1	0	1	2	4	43
		2	14	3	1	23	12
		3	11	0	2	3	0
	400	1	0	0	3	8	14
		2	12	5	4	13	1
		3	9	0	1	1	0

[a]Dose in ppm.
[b]Time intervals 1 to 3 represent, respectively, 0 to 9.37, 9.37 to 14.07, and 14.07 to 18.70 months.
[c]Assigned by pathologists.

whereas tumors that affect mortality by either directly causing death or indirectly increasing the risk of death from other causes are classified as fatal. The analysis of data on occult tumors using cause of death or context of observation is performed separately for incidental tumors and fatal tumors.

Table 3.2. Tumor prevalence data for incidental tumors in interval I_j.

	Dose group			
	1	2	$\cdots$	g	Total
With tumors	y_{1j}	y_{2j}	$\cdots$	y_{gj}	$y_{.j}$
Without tumors	$n_{1j} - y_{1j}$	$n_{2j} - y_{2j}$	$\cdots$	$n_{gj} - y_{gj}$	$n_{.j} - y_{.j}$
Deaths	n_{1j}	n_{2j}	$\cdots$	n_{gj}	$n_{.j}$

One part of the test is for incidental tumors. Consider the animals which did not have the specific tumor before death and tumor-bearing animals which did not die of that tumor. Let n_{ij} be the number of animals in group i dying during interval I_j from causes unrelated to the presence of the tumor of interest, and let y_{ij} be the number of these animals in which the tumor was observed in the incidental context, for $i = 1, \ldots, g$. The totals of the n_{ij} and y_{ij} across the groups are $n_{.j} = \sum_{i=1}^{g} n_{ij}$ and $y_{.j} = \sum_{i=1}^{g} y_{ij}$, respectively. For each interval I_j, the tumor prevalence data may be summarized in a $2 \times g$ table as in Table 3.2. All tumors found in sacrificed animals are classified as incidental.

The expected number of tumors in the ith group for the jth interval is $E_{ij} = y_{.j} K_{ij}$, where $K_{ij} = n_{ij}/n_{.j}$. Thus, the observed and expected numbers of tumors in the ith group over the entire experiment are $O_i = \sum_{j=1}^{m} y_{ij}$ and $E_i = \sum_{j=1}^{m} E_{ij}$, respectively, for $i = 1, \ldots, g$.

Define

$$D_i = O_i - E_i = \sum_{j=1}^{m} (y_{ij} - E_{ij}) \tag{3.1}$$

$$V_{ri} = \sum_{j=1}^{m} \kappa_j K_{rj} (\delta_{ri} - K_{ij})$$

where $\kappa_j = y_{.j}(n_{.j} - y_{.j})/(n_{.j} - 1)$ and δ_{ri} is defined as 1 if $r = i$ and 0 otherwise. Let $D_a = (D_1, \ldots, D_g)'$ and let V_a be the $g \times g$ matrix with (r, i) entry V_{ri}.

The other part of the test is for tumors which are the cause of death. The method used is very similar to that used for the incidental tumors, except that each tumor–death time defines an interval. Table 3.3 is a contingency table for interval I_j. Let m_{ij} be the number of animals in group i surviving at the beginning of the interval, and let x_{ij} be the number of these animals dying of the tumor in that interval.

A vector D_b of differences of observed and expected values using the data is calculated in the same way as for the incidental tumors, and the corresponding covariance matrix V_b is computed. Although any life-table method may be used to implement the fatal-tumor part of Peto's test, the method just described is the log-rank test, which is especially powerful against proportional hazards alternatives.

Table 3.3. Tumor causing death in interval I_j.

	Dose group				
	1	2	$\cdots$	g	Total
With tumors	x_{1j}	x_{2j}	$\cdots$	x_{gj}	$x._j$
Surviving	m_{1j}	m_{2j}	$\cdots$	m_{gj}	$m._j$

The analysis of data on occult tumors using contexts of observation is based on the vector $D = D_a + D_b$, with covariance matrix $V = V_a + V_b$. Then

$$X_H = D'V^-D \qquad (3.2)$$

can serve as a test for heterogeneity among the g groups, where V^- is a generalized inverse of V. In our S-PLUS program, the inverse of matrix V is taken using the function solve if the matrix is of full rank. If V is singular, a generalized inverse is taken using the singular value decomposition since the inverse matrix cannot be obtained. See the function peto.test (3.A.1, page 63) for further detail. If there is no difference among the groups, then X_H is asymptotically distributed as a χ^2-distribution with $g - 1$ degrees of freedom. Also, a trend test can be considered by using

$$Z_R = (l'D)^2/(l'Vl) \qquad (3.3)$$

or

$$Z_R = l'D/\sqrt{l'Vl}$$

where $l = (l_1, \ldots, l_g)'$. If there is no dose-related trend, Z_R is asymptotically distributed as a standard normal. A test for departure from a monotonic dose–response relationship can be based on

$$X_M = X_H - X_R$$

which has a χ^2-distribution with $g - 2$ degrees of freedom under the null hypothesis that the dose–response relationship is linear. The program given in Section 3.A.1 calculates (3.2) and (3.3).

As introduced in Section 3.1, the Benzidine dihydrochloride data were analyzed using a Peto's IARC cause-of-death test program written in S-PLUS. As shown in Section 3.A.1, the program consists of three functions. The function peto.test receives the data, dose levels of the groups, number of groups, number of intervals and terminal sacrifice time as input, and returns the χ^2-values and the p-values of the heterogeneity test given in (3.2) and the z-values and the p-values of the dose-related trend test given in (3.3). The input data consist of five variables (columns), which are group ID, interval number, survival time, tumor indicator (1 if tumor present and 0 otherwise), cause of death (1 for fatal, 2 for incidental, and 3 for no tumor). The data used in this analysis were obtained from Kodell and Nelson (1980).

Table 3.4. Tumor prevalence data for incidental tumors in interval I_j. Benzidine Dihydrochloride experiment; F2-strain mice.

Sex	Int.	With Tumors					Without Tumors					Total				
		g1	g2	g3	g4	Total	g1	g2	g3	g4	Total	g1	g2	g3	g4	Total
M	1	0	0	0	1	1	76	49	55	26	206	76	49	55	27	207
	2	3	8	6	5	22	44	40	35	20	139	47	48	41	25	161
	3	4	12	8	7	31	37	29	13	6	85	41	41	21	13	116
F	1	0	2	5	8	15	72	51	45	17	185	72	53	50	25	200
	2	10	18	26	18	72	39	30	13	5	87	49	48	39	23	159
	3	15	21	3	1	40	24	3	2	1	30	39	24	5	2	70

Table 3.5. Tumor causing death in interval I_j. Benzidine Dihydrochloride experiment; F2-strain male mice.

Int.	With Tumors					Without Tumors					Total				
	g1	g2	g3	g4	Total	g1	g2	g3	g4	Total	g1	g2	g3	g4	Total
1	0	1	0	0	1	167	141	118	71	497	167	142	118	71	498
2	0	0	0	1	1	91	92	63	43	289	91	92	63	44	290
3	0	0	1	0	1	89	92	61	42	284	89	92	62	42	285
4	0	1	0	0	1	88	90	59	40	277	88	91	59	40	278
5	0	0	0	1	1	44	43	21	17	125	44	43	21	18	126
6	1	0	0	0	1	43	43	20	17	123	44	43	20	17	124
7	0	0	0	1	1	40	43	18	16	117	40	43	18	17	118
8	1	0	0	0	1	39	43	17	16	115	40	43	17	16	116
9	0	0	0	1	1	39	43	17	14	113	39	43	17	15	114
10	0	1	0	0	1	39	41	17	14	111	39	42	17	14	112
11	0	0	0	1	1	39	41	17	13	110	39	41	17	14	111
12	0	1	0	0	1	39	40	17	13	109	39	41	17	13	110
13	1	0	0	0	1	38	40	17	13	108	39	40	17	13	109
14	0	0	0	1	1	38	39	17	12	106	38	39	17	13	107
15	0	0	0	0	0	38	39	17	12	106	38	39	17	12	106

The unit of the survival time in this particular dataset is month. The table of tumor prevalence data for incidental tumors in each interval is produced by the function table.incidental.tumors. The intervals are determined by the sacrifice times. Table 3.4 shows the prevalence data for both males and females. These are the observed numbers for Table 3.2.

The function table.fatal.tumors produces the table of deaths caused by tumor in each interval. As mentioned earlier in this Section, the intervals are determined by each tumor death for the fatal tumors. The observed numbers for Table 3.3 are summarized in Table 3.5 for males and Tables 3.6 and 3.7 for females.

Table 3.6. Tumor causing death in interval I_j. Benzidine Dihydrochloride experiment; F2-strain female mice (part I).

	With Tumors					Without Tumors					Total				
Int.	g1	g2	g3	g4	Total	g1	g2	g3	g4	Total	g1	g2	g3	g4	Total
1	0	1	0	0	1	167	142	119	71	499	167	143	119	71	500
2	0	0	1	0	1	95	87	67	43	292	95	87	68	43	293
3	0	0	0	2	2	95	85	66	41	287	95	85	66	43	289
4	0	0	1	0	1	95	85	65	41	286	95	85	66	41	287
5	0	0	0	2	2	95	85	65	39	284	95	85	65	41	286
6	0	0	1	0	1	95	85	64	39	283	95	85	65	39	284
7	0	0	0	1	1	95	85	63	38	281	95	85	63	39	282
8	0	0	0	1	1	95	85	63	36	279	95	85	63	37	280
9	0	0	0	3	3	95	85	63	33	276	95	85	63	36	279
10	0	0	1	0	1	95	85	62	31	273	95	85	63	31	274
11	0	0	1	0	1	95	85	61	31	272	95	85	62	31	273
12	0	0	0	2	2	95	84	61	29	269	95	84	61	31	271
13	0	0	2	0	2	95	84	59	29	267	95	84	61	29	269
14	0	0	1	0	1	95	84	58	29	266	95	84	59	29	267
15	0	0	1	0	1	95	84	57	29	265	95	84	58	29	266
16	0	1	0	1	2	95	81	57	28	261	95	82	57	29	263
17	0	0	2	0	2	95	81	55	28	259	95	81	57	28	261
18	0	0	2	0	2	94	81	53	28	256	94	81	55	28	258
19	0	1	1	0	2	94	80	52	27	253	94	81	53	27	255
20	0	0	1	0	1	94	39	32	11	176	94	39	33	11	177
21	0	0	0	1	1	46	39	15	10	110	46	39	15	11	111
22	0	1	0	0	1	46	38	15	10	109	46	39	15	10	110
23	0	1	0	0	1	46	37	15	10	108	46	38	15	10	109
24	0	0	0	1	1	46	37	15	9	107	46	37	15	10	108
25	0	1	0	0	1	46	36	15	9	106	46	37	15	9	107
26	0	0	1	1	2	46	36	14	8	104	46	36	15	9	106
27	0	0	0	1	1	46	36	14	7	103	46	36	14	8	104
28	0	1	0	0	1	45	35	14	7	101	45	36	14	7	102
29	0	1	2	0	3	45	33	12	7	97	45	34	14	7	100
30	0	0	0	1	1	45	33	12	6	96	45	33	12	7	97

The following is a run of the Peto test program followed by an output:

```
> ngroup <- 4
> nint <- 3
> ntime <- 561
> dat[,3] <- ceiling(dat[,3]*30)
> peto.test(dat, dose, ngroup, nint, ntime)
                 Hetero(Incid) Hetero(Fatal) Hetero(Combo)
     Statistic   15.691292326   10.62255886   2.212110e+01
       p-value   0.001311791    0.01395215    6.155362e-05
```

Table 3.7. Tumor causing death in interval I_j. Benzidine Dihydrochloride experiment; F2-strain female mice (part II).

Int.	With Tumors					Without Tumors					Total				
	g1	g2	g3	g4	Total	g1	g2	g3	g4	Total	g1	g2	g3	g4	Total
31	0	0	2	0	2	45	32	10	6	93	45	32	12	6	95
32	1	0	0	0	1	44	32	10	6	92	45	32	10	6	93
33	0	0	1	0	1	42	32	9	6	89	42	32	10	6	90
34	0	1	0	0	1	42	31	9	6	88	42	32	9	6	89
35	1	0	0	0	1	41	31	9	6	87	42	31	9	6	88
36	0	1	0	0	1	41	30	9	6	86	41	31	9	6	87
37	1	1	0	0	2	40	29	9	6	84	41	30	9	6	86
38	0	2	0	0	2	40	27	9	5	81	40	29	9	5	83
39	0	0	0	1	1	40	27	9	4	80	40	27	9	5	81
40	0	1	0	0	1	40	26	9	4	79	40	27	9	4	80
41	0	0	1	0	1	40	26	8	4	78	40	26	9	4	79
42	0	0	0	1	1	40	26	8	3	77	40	26	8	4	78
43	0	0	1	0	1	40	26	7	3	76	40	26	8	3	77
44	0	0	0	1	1	40	26	6	2	74	40	26	6	3	75
45	1	0	0	0	1	38	26	6	2	72	39	26	6	2	73
46	0	0	1	0	1	38	26	5	2	71	38	26	6	2	72
47	1	1	0	0	2	37	25	5	2	69	38	26	5	2	71
48	0	2	1	0	3	37	23	4	2	66	37	25	5	2	69
49	0	0	0	1	1	37	23	4	1	65	37	23	4	2	66
50	0	1	0	0	1	37	22	4	1	64	37	23	4	1	65
51	1	0	0	0	1	36	21	3	1	61	37	21	3	1	62
52	1	0	0	0	1	35	21	3	1	60	36	21	3	1	61
53	0	0	0	0	0	35	21	3	1	60	35	21	3	1	60

	Trend(Incid)	Trend(Fatal)	Trend(Combo)
Statistic	3.5719829361	2.798487359	4.493313e+00
p-value	0.0001771443	0.002567129	3.506182e-06

The argument dat is the input dataset, dose is a vector of the dose metric for different groups, ngroup is the number of groups, nint is the number of sacrificing intervals used for the incidental tumors, and ntime is the final sacrifice time. The above output is for the male mice of the Benzidine Dihydrochloride data. In this example, we defined deaths on the same day as deaths at the same time. Because the unit of survival time in this dataset is month, the time was converted to days according to the above.

The summary table of tumor prevalence data for incidental tumors for male mice can be obtained as follows:

```
> ngroup <- 4
> nint <- 3
```

```
> table.incidental.tumors(dat, ngroup, nint)
             y_gp1 y_gp2 y_gp3 y_gp4 y_total
  interval 1     0     0     0     1       1
  interval 2     3     8     6     5      22
  interval 3     4    12     8     7      31


           n-y_gp1 n-y_gp2 n-y_gp3 n-y_gp4 n-y_total
  interval 1    76      49      55      26       206
  interval 2    44      40      35      20       139
  interval 3    37      29      13       6        85


           n_gp1 n_gp2 n_gp3 n_gp4 n_total
  interval 1   76    49    55    27     207
  interval 2   47    48    41    25     161
  interval 3   41    41    21    13     116
```

The function table.fatal.tumors produces the summary table of deaths caused by tumor with the same format as above.

3.4   Analysis of Data with Unknown Cause of Death

Cochran (1954) and Armitage (1955) introduced a trend test (the Cochran–Armitage test) for detecting a linear trend across dose groups in the overall proportions of animals with the tumor. This test needs an assumption of equal risk of getting the tumor among the dose groups over the duration of the study. By pooling all the time intervals, Table 3.2 can be modified to Table 3.8. The expected number of tumors in the ith group is $E_i = y.K_i$, where $K_i = N_i/N$. Defining D_i as in (3.1), the test statistic for a possible monotonic trend with dose is based on

$$X = \sum_{i=1}^{g} l_i D_i$$

and the variance is estimated by

$$V = \{y.(N - y.)/[N(N-1)]\} \sum_{i=1}^{g} N_i (l_i - l)^2$$

where $l = (\sum_{i=1}^{g} N_i l_i)/N$. The Cochran–Armitage test is

$$Z = X/\sqrt{V} \qquad (3.4)$$

where Z is asymptotically distributed as a standard normal variate under the null hypothesis of equal tumor incidence rates among the groups. Two-tailed tests may be based on Z^2, which is approximately χ^2-distributed

Table 3.8. Tumor data for the intervals combined in each group.

	Dose group				Total
	1	2	$\cdots$	g	
With tumors	y_1	y_2	$\cdots$	y_g	$y.$
Without tumors	$N_1 - y_1$	$N_2 - y_2$	$\cdots$	$N_g - y_g$	$N - y.$
Deaths	N_1	N_2	$\cdots$	N_g	N

[htp]

Table 3.9. Benzidine Dihydrochloride data for the intervals combined in each group for the Cochran–Armitage and Poly-3 tests.

		Dose group				Total
		1	2	3	4	
Male	with tumors	10	24	15	19	68
	without tumors	157	118	103	52	430
	Deaths	167	142	118	71	498
Female	with tumors	32	59	59	48	198
	without tumors	135	84	60	23	302
	Deaths	167	143	119	71	500

with one degree of freedom. The program given in Section 3.A.2 performs a one-sided test only.

The Benzidine data were analyzed using the Cochran–Armitage test. The S-PLUS program for this test is given in Section 3.A.2. The function cochran.armitage.test conducts the test. This function receives the data, dose levels of the groups, and number of the groups, and returns the z-value given in (3.3) and the p-value of the test. The user reads the data and calls the functions table.tumors for the summary table and cochran.armitage.test for the Cochran–Armitage test.

The summary data for the intervals combined in each group (the counts for Table 3.8) are given in Table 3.9. The input data consist of three variables (columns) which are group ID, survival time, and tumor indicator (1 if tumors present and 0 otherwise). The following is a sample run of the Cochran–Armitage test program with an output:

```
> ngroup <- 4
> cochran.armitage.test(dat,dose,ngroup)
      z-value        p-value
      3.845758   6.009009e-05
```

The argument dat is the input dataset, dose is a vector of the dose metric for each group, and ngroup is the number of groups. The above output is for the male mice of the Benzidine Dihydrochloride data.

The summary table for the male mice group is produced as follows:

```
> table.tumors(dat,ngroup)
              gp 1   gp 2   gp 3  total
  #w tumors     10     24     15     68
#w/o tumors    157    118    103    430
    #deaths    167    142    118    498
```

If the mortality patterns are similar across dose groups, then the Cochran-Armitage test will be appropriate, but it may not be valid if the mortality rates differ across dose groups. Bailer and Portier (1988) proposed the Poly-3 trend test, which made an adjustment of the Cochran–Armitage test by modifying the value of N to reflect decreased survival. Define the number at risk as the sum of N_i weights

$$r_i = \sum_{k=1}^{N_i} \omega_{ik}, \tag{3.5}$$

where ω_{ik} is the weight for the kth animal in the ith dose group. Note that the above test becomes the Cochran–Armitage test if the weights, ω_{ik}, are all equal to 1. Bailer and Portier (1988) defined the weights as $\omega_{ik} = 1$ if the kth animal in the ith dose group dies with the tumor, and $\omega_{ik} = (t_{ik}/t_{max})^3$ if not, where t_{max} is the maximum survival time across all groups. This weighting gives proportionally less weight to a tumor-free animal that dies at time t_{ik}. The third power comes from the observation that tumor incidence often seems to be a low-order polynomial in time (Portier et.al. (1986)). Bailer and Portier suggested a Poly-k test if the shape of the tumor incidence function is expected to follow time to some power k. Gart et.al. (1979) suggested the truncated trend test which defines the weights as $\omega_{ik} = 1$ if the age at death for the kth animal in the ith group exceeds the time of the first death with tumor present and $\omega_{ik} = 0$ if not. Another modification to the Cochran–Armitage test, which employs Bailer and Portier's adjusted quantal response rates and the delta method for variance estimation was introduced by Bieler and Williams (1993). Recently, it was adopted by the NTP. In our data analysis, the Poly-3 test of Bieler and Williams will be used.

The Poly-3 test program written in S-Plus is given in Section 3.A.3. The Benzidine Dihydrochloride data were analyzed using this program. The test is conducted in a function named poly3.test. This function receives the data, dose levels of the groups, number of the groups, and terminal sacrifice time, and returns the z-value and the p-value of the test. This function has one more input parameter than the Cochran–Armitage test because the survival time is used for calculating the weights given in (3.4).

As in the Cochran–Armitage test, the user reads the data and calls the functions table.tumors (given in Section 3.A.2) and poly3.test.

Table 3.10. Test results for Peto's cause-of-death test, the Cochran–Armitage test and the Poly-3 trend test for the Benzidine Dihydrochloride data for *male* mice given in Table 3.1. All tests are one-sided.

				Peto	C–A[a]	Poly-3
Male	Hetero-geneity	Incidental	χ^2-value[b]	15.69		
			p-value	0.0013		
		Fatal	χ^2-value	10.62		
			p-value	0.0140		
		Combined	χ^2-value	22.12		
			p-value	0.00006		
	Trend[c]	Incidental	Z-value	3.57		
			p-value	0.0002		
		Fatal	Z-value	2.80		
			p-value	0.0026		
		Combined	Z-value	4.49	3.85	4.25
			p-value	4×10^{-6}	0.00006	0.00001

[a]Cochran–Armitage test.
[b]Degrees of freedom: $g - 1 = 3$.
[c]Dose metric: Concentration in ppm.

The same input data as in the Cochran–Armitage test are used for the Poly-3 test. The following is a sample run of the Poly-3 test program:

```
> ngroup <- 4
> ntime <- 18.70
> poly3.test(dat,dose,ngroup,ntime)
        z-value        p-value
      4.251558    1.061444e-05
```

The argument ntime is the terminal-sacrifice time, and the other arguments are the same as those in function cochran.armitage.test. The above output is for the male mice of the Benzidine Dihydrochloride data.

The test results from the three tests illustrated in this section are summarized in Tables 3.10 and 3.11. S-PLUS plots of the time-to-tumor-onset survival functions obtained using the constrained nonparametric maximum likelihood method by Ahn et al. (2000) are given in Figure 3.2. The code for creating the figure is provided in Section 3.A.4. The input dataset consists of three columns which include times (column 1), tumor onset survival rates of male mice (column 2), and tumor onset survival rates of female mice (column 3).

Table 3.11. As Table 3.10, for *female* mice data.

				Peto	C–A[a]	Poly-3
Female	Hetero-geneity	Incidental	χ^2-value[b]	56.66		
			p-value	≈ 0		
		Fatal	χ^2-value	79.53		
			p-value	≈ 0		
		Combined	χ^2-value	131.16		
			p-value	≈ 0		
	Trend[c]	Incidental	Z-value	7.14		
			p-value	≈ 0		
		Fatal	Z-value	8.63		
			p-value	≈ 0		
		Combined	Z-value	11.09	7.04	7.66
			p-value	≈ 0	≈ 0	≈ 0

[a]Cochran–Armitage test.
[b]Degrees of freedom: $g - 1 = 3$.
[c]Dose metric: Concentration in ppm.

3.5 Summary

Various methods have been proposed for the analysis of tumor incidence data from animal bioassays. Among those methods, some of the widely used statistical tests were illustrated using S-PLUS.

The Cochran–Armitage test, Poly-3 test, and Peto's IARC test gave similar results. All the tests showed a highly significant dose-related trend. The tests gave p-values of almost zero for female mice. This implies that the tumor incidence rate increases as the level of dose increases. These results are similar to the results obtained in Kodell and Ahn (1996). Figure 3.2 indicates that the tumor onset survival time decreases as the dose level increases. It supports the test results obtained in this chapter.

The statistical methods for preclinical studies can be easily implemented in S-PLUS in a very efficient way. The programs written in S-PLUS can be very concise because S-PLUS provides many built-in functions compared to the programming languages such as Fortran or C. As we found from the programs given in the Appendix, a short S-PLUS program can contain many complicated computational procedures.

One drawback is that the calculation time of S-PLUS programs is longer than that of Fortran or C. Fortran or C are more appropriate than statistical languages such as S-PLUS or SAS for an intensive computation such as Monte Carlo simulation study or bootstrap. However, S-PLUS is a very convenient tool for a real data analysis.

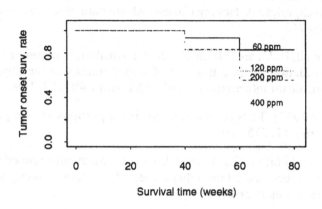

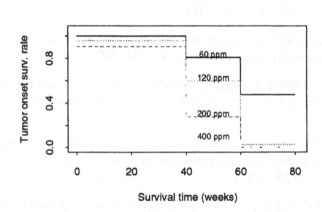

Figure 3.1. Time-to-tumor-onset survival functions obtained from the method of Ahn et al. (2000) for the Benzidine data.

Acknowledgement

Hongshik Ahn's work was supported by NIH grant 1 R29 CA77289-03.

3.6   References

Ahn, H., and Kodell, R. L. (1995). Estimation and testing of tumor incidence rates in experiments lacking cause-of-death data. *Biometrical Journal* **37**, 745–763.

Ahn, H., Kodell, R. L., and Moon, H. (2000). Attribution of tumor lethality and estimation of time to onset of occult tumors in the absence of cause-of-death information. *Applied Statistics* **49**, 157–169.

Armitage, P. (1955). Tests for linear trends in proportions and frequencies. *Biometrics* **11**, 375–386.

Bailer, A. J., and Portier, C. J. (1988). Effects of treatment-induced mortality and tumor-induced mortality on tests for carcinogenicity in small samples. *Biometrics* **14**, 417–431.

Bieler, G. S., and Williams, R. L. (1993). Ratio estimates, the delta method, and quantal response tests for increased carcinogenicity. *Biometrics* **49**, 793–801.

Cochran, W. G. (1954). Some methods for strengthening the common χ^2-tests. *Biometrics* **10**, 417–451.

Dinse, G. E. (1991). Constant risk differences in the analysis of animal tumorigenicity data. *Biometrics* **47**, 681–700.

Dinse, G. E. (1993). Evaluating constraints that allow survival-adjusted incidence analyses in single-sacrifice studies. *Biometrics* **49**, 399–407.

Dinse, G. E., and Lagakos, S. W. (1982). Nonparametric estimation of lifetime and disease onset distributions from incomplete observations. *Biometrics* **38**, 921–932.

Gart, J. J. (1975). Letter to the editor. *British Journal of Cancer* **31**, 696–697.

Gart, J. J., Chu, K. C., and Tarone, R. E. (1979). Statistical issues in interpretation of chronic bioassay tests for carcinogenicity. *Journal of the National Cancer Institute* **62**, 957–974.

Hoel, D. G., and Walburg, H. E. (1972). Statistical analysis of survival experiments. *Journal of the National Cancer Institute* **49**, 361–372.

Kodell, R. L., Ahn, H., Pearce, B. A., and Turturro, A. (1997). Age-adjusted trend test for the tumor incidence rate. *Drug Information Journal* **31**, 471–487.

Kodell, R. L., and Ahn, H. (1996). Nonparametric trend test for the cumulative tumor incidence rate. *Communications in Statistics - Theory and Methods* **25**, 1677–1692.

Kodell, R. L., and Ahn, H. (1997). An age-adjusted trend test for the tumor incidence rate. *Biometrics* **53**, 1467–1474.

Kodell, R. L., and Nelson, C. J. (1980). An illness-death model for the study of the carcinogenic process using survival/sacrifice data. *Biometrics* **36**, 267–277.

Kodell, R. L., Shaw, G. W., and Johnson, A. M. (1982). Nonparametric joint estimators for disease resistance and survival functions in survival/sacrifice experiments. *Biometrics* **38**, 43–58.

Lagakos, S. W. (1982). An evaluation of some two-sample tests used to analyze animal carcinogenicity experiments. *Utilitas Mathematica* **21B**, 239–260.

Lindsey, J. C., and Ryan, L. M. (1994). A comparison of continuous- and discrete- time three-state models for rodent tumorigenicity experiments. *Environmental Health Perspectives* **102, (Suppl. 1)**, 9–17.

Malani, H. M., and Van Ryzin, J. (1988). Comparison of two treatments in animal carcinogenicity experiments. *Journal of the American Statistical Association* **83**, 1171–1177.

McKnight, B., and Crowley, J. (1984). Tests for differences in tumor incidence based on animal carcinogenesis experiments. *Journal of the American Statistical Association* **79**, 639–648.

Peto, R. (1974). Guidelines on the analysis of tumour rates and death rates in experimental animals. *British Journal of Cancer* **29**, 101–105.

Peto, R., Pike, M. C., Day, N. E., Gray, R. G., Lee, P. N., Parish, S., Peto, J., Richards, S., and Wahrendorf, J. (1980). Guidelines for simple, sensitive significance tests for carcinogenic effects in long-term animal experiments. *IARC Monographs* **Supplement 2**, 311–426. Annex to: Long-term and Short-term Screening Assays for Carcinogens: a Critical Appraisal.

Portier, C., Hedges, J., and Hoel, D. G. (1986). Age-specific models of mortality and tumor onset for historical control animals in the national toxicology program's carcinogenicity experiments. *Cancer Research* **46**, 4372–4378.

Portier, C. J., and Dinse, G. E. (1987). Semiparametric analysis of tumor incidence rates in survival/sacrifice experiments. *Biometrics* **43**, 107–114.

Turnbull, B. W., and Mitchell, T. J. (1984). Nonparametric estimation of the distribution of time to onset for specific diseases in survival/sacrifice experiments. *Biometrics* **40**, 41–50.

3.A Appendix

3.A.1   Peto IARC Test

```
# ---
# peto.test:  Peto IARC Test
#    Parameters:
#       dat: data matrix.
#             column 1: group id
#             column 2: interval number
#             column 3: survival time
#             column 4: tumor indicator
#                       (1: with tumor; 0: w/o tumor)
#             column 5: cause of death
#                       (1: fatal; 2: incidental; 3: no tumor)
#       dose: vector of dose metric.
#             Length must be equal to the number of groups.
#       ngroup: maximum survival time.
#       nint: number of sacrificing intervals.
#       ntime: maximum survival time.
# ---
peto.test <- function(dat, dose, ngroup, nint, ntime)
{
  # Initialization of arrays
  y <- matrix(0, ngroup, nint)
  n <- matrix(0, ngroup, nint)
  gr.size <- rep(0, ngroup)
  group <- dat[, 1]
  int <- dat[, 2]
  time <- dat[, 3]
  tumor <- dat[, 4]
  cod <- dat[, 5]
  x.h <- matrix(0, 2, 3)
  z.r <- matrix(0, 2, 3)
  #*****************#
  # Incidental part #
  #*****************#
  # Formulation of Table 3.2 (Tumor prevalence data)
  # Intervals are based on the sacrifice times.
```

```
for(i in 1:ngroup) {
  gr.size[i] <- sum(group == i)
  for(j in 1:nint) {
    y[i, j] <- sum(tumor == 1 & cod !=
      1 & group == i & int == j)
    n[i, j] <- sum(cod != 1 & group ==
      i & int == j)
  }
}
# end for i
# Initialization of the arrays
k <- matrix(0, ngroup, nint)
e <- matrix(0, ngroup, nint)
delta <- diag(1, nrow = ngroup)
v.incid <- matrix(0, ngroup, ngroup)
ytotal <- apply(y, 2, sum)
ntotal <- apply(n, 2, sum)
kappa <- (ytotal * (ntotal - ytotal))/(ntotal - 1)
for(i in 1:ngroup) {
  k[i, ] <- n[i, ]/ntotal
  e[i, ] <- ytotal * k[i, ]
}
# end for i
# D_a = O - E in Equation (1)
ototal <- apply(y, 1, sum)
etotal <- apply(e, 1, sum)
dtotal <- ototal - etotal
for(r in 1:ngroup) {
  for(i in 1:ngroup) {
    temp <- 0
    for(j in 1:nint)
      temp <- temp + kappa[j] *
        k[r, j] * (delta[
        r, i] - k[i, j])
    v.incid[r, i] <- temp
  }
}
# end for r
d.incid <- dtotal
inv.svdd <- rep(0, ngroup)
# X_H = D_a' inv(V_a) D_a in Equation (2)
# If V_a is nonsingular, obtain inv(V_a) by solve(V_a).
# If V_a is singular, obtain inv(V_a) by singular
# value decomposition.
# Find Z_R in Equation (3)
```

```
eval <- eigen(v.incid)$values
if(eval[ngroup] > 10^(-7)) {
  vinv <- solve(v.incid)
}
else {
  temp <- svd(v.incid)
  svdd <- temp$d
  svdu <- temp$u
  svdv <- temp$v
  dim.svdd <- sum(svdd > 10^(-7))
  for(i in 1:dim.svdd) {
    inv.svdd[i] <- 1/svdd[i]
  }
  for(i in (dim.svdd + 1):ngroup) {
    inv.svdd[i] <- 0
  }
  dd <- diag(inv.svdd)
  vinv <- svdv %*% dd %*% t(svdu)
}
# end if
# z- and p-values for heterogeneity test and trend test
x.h[1, 1] <- t(d.incid) %*% vinv %*% d.incid
x.h[2, 1] <- 1 - pchisq(x.h[1, 1], ngroup - 1)
z.r[1, 1] <- crossprod(dose, d.incid)/sqrt(t(dose) %*%
  v.incid %*% dose)
z.r[2, 1] <- 1 - pnorm(z.r[1, 1])
#*************#
# Fatal part #
#*************#
# Formulation of Table 3.3 (Tumor causing death in
# interval I_j)
# Intervals are based on time to death with tumor.
dwhole <- time[cod == 1]
dwhole <- c(sort(dwhole), ntime)
nfatal <- 1
for(i in 2:length(dwhole)) {
  if(dwhole[i] > dwhole[i - 1]) {
    nfatal <- nfatal + 1
    dwhole[nfatal] <- dwhole[i]
  }
}
# end for i
d <- rep(0, nfatal)
d <- dwhole[1:nfatal]
m <- matrix(0, ngroup, nfatal)
```

```
x <- matrix(0, ngroup, nfatal)
if(nfatal == 1) {
  m[, 1] <- gr.size
  for(i in 1:ngroup)
    x[i, 1] <- sum(group == i & cod ==
      1)
}
else {
  m[, 1] <- gr.size
  for(i in 1:ngroup) {
    x[i, 1] <- sum(group == i & time <=
      d[1] & cod == 1)
    for(j in 2:nfatal) {
      m[i, j] <- gr.size[i] - sum(
        group == i & time <=
        d[j - 1])
      x[i, ] <- sum(group == i &
        time > d[j - 1] &
        time <= d[j] & cod ==
        1)
    }
  }
}
# end if
# Initialization of arrays
k <- matrix(0, ngroup, nfatal)
e <- matrix(0, ngroup, nfatal)
delta <- diag(1, nrow = ngroup)
v.fatal <- matrix(0, ngroup, ngroup)
# D_b = 0 - E in equation (1)
xtotal <- apply(x, 2, sum)
mtotal <- apply(m, 2, sum)
kappa <- (xtotal * (mtotal - xtotal))/(mtotal - 1)
for(i in 1:ngroup) {
  k[i, ] <- m[i, ]/mtotal
  e[i, ] <- xtotal * k[i, ]
}
# end for i
ototal <- apply(x, 1, sum)
etotal <- apply(e, 1, sum)
dtotal <- ototal - etotal
for(r in 1:ngroup) {
  for(i in 1:ngroup) {
    temp <- 0
    for(j in 1:nfatal)
```

```
        temp <- temp + kappa[j] *
          k[r, j] * (delta[
          r, i] - k[i, j])
      v.fatal[r, i] <- temp
    }
}
# end for r
d.fatal <- dtotal
# X_H = D_b' inv(V_b) D_b in Equation (2)
# Calculated similarly as in incidental part
# Find Z_R in Equation (3)
eval <- eigen(v.fatal)$values
if(eval[ngroup] > 10^(-7)) {
  vinv <- solve(v.fatal)
}
else {
  temp <- svd(v.fatal)
  svdd <- temp$d
  svdu <- temp$u
  svdv <- temp$v
  dim.svdd <- sum(svdd > 10^(-7))
  for(i in 1:dim.svdd) {
    inv.svdd[i] <- 1/svdd[i]
  }
  for(i in (dim.svdd + 1):ngroup) {
    inv.svdd[i] <- 0
  }
  dd <- diag(inv.svdd)
  vinv <- svdv %*% dd %*% t(svdu)
}
# end if
x.h[1, 2] <- t(d.fatal) %*% vinv %*% d.fatal
x.h[2, 2] <- 1 - pchisq(x.h[1, 2], ngroup - 1)
z.r[1, 2] <- crossprod(dose, d.fatal)/sqrt(t(dose) %*%
  v.fatal %*% dose)
z.r[2, 2] <- 1 - pnorm(z.r[1, 2])
# Combine incidental and fatal parts
# Find X_H = D' inv(V) D and Z_R in (2) and (3)
d.combo <- d.incid + d.fatal
v.combo <- v.incid + v.fatal
eval <- eigen(v.combo)$values
if(eval[ngroup] > 10^(-7)) {
  vinv <- solve(v.combo)
}
else {
```

```
    temp <- svd(v.combo)
    svdd <- temp$d
    svdu <- temp$u
    svdv <- temp$v
    dim.svdd <- sum(svdd > 10^(-7))
    for(i in 1:dim.svdd) {
      inv.svdd[i] <- 1/svdd[i]
    }
    for(i in (dim.svdd + 1):ngroup) {
      inv.svdd[i] <- 0
    }
    dd <- diag(inv.svdd)
    vinv <- svdv %*% dd %*% t(svdu)
  }
# end if
# Find the z-values and p-values
x.h[1, 3] <- t(d.combo) %*% vinv %*% d.combo
x.h[2, 3] <- 1 - pchisq(x.h[1, 3], ngroup - 1)
z.r[1, 3] <- crossprod(dose, d.combo)/sqrt(t(dose) %*%
  v.combo %*% dose)
z.r[2, 3] <- 1 - pnorm(z.r[1, 3])
# Output: z-values and p-values of heterogeneity test and
# trend test for incidental, fatal, and combined
peto.stat <- cbind(x.h, z.r)
dimnames(peto.stat) <- list(c("Statistic", "p-value"),
  c("Hetero(Incid)", "Hetero(Fatal)",
  "Hetero(Combo)", "Trend(Incid)",
  "Trend(Fatal)", "Trend(Combo)"))
return(peto.stat)
}
```

Summary Table for Incidental Part

```
# ---
# table.incidental.tumors:  Summary table for incidental part
#                           (see Table 3.2)
#
#   Parameters:
#     dat: data matrix.  the same as in function "peto.test."
#     ngroup: number of groups.
#     nint: number of intervals in each group.
# ---
table.incidental.tumors <- function(dat, ngroup, nint)
{
  y <- matrix(0, ngroup, nint)
  n <- matrix(0, ngroup, nint)
  group <- dat[, 1]
  int <- dat[, 2]
  time <- dat[, 3]
  tumor <- dat[, 4]
  cod <- dat[, 5]
  for(i in 1:ngroup) {
    for(j in 1:nint) {
      y[i, j] <- length(time[tumor == 1 &
        cod != 1 & group == i & int ==
        j])
      n[i, j] <- length(time[cod != 1 &
        group == i & int == j])
    }
  }
  # end for i
  ytotal <- apply(y, 2, sum)
  ntotal <- apply(n, 2, sum)
  count <- cbind(cbind(t(y), ytotal), cbind(t(n - y),
    ntotal - ytotal), cbind(t(n), ntotal))
  return(count)
}
```

Summary Table for Fatal Part

```
# ---
# table.fatal.tumors:  Summary table for fatal part
#                      (see Table 3.3)
#
#   Parameters:
#     dat: data matrix.  the same as in function "peto.test."
#     ngroup: number of groups.
#     nint: number of intervals in each group.
# ---
table.fatal.tumors <- function(dat, ngroup, ntime)
{
  group <- dat[, 1]
  time <- dat[, 3]
  cod <- dat[, 5]
  gr.size <- rep(0, ngroup)
  dwhole <- time[cod == 1]
  dwhole <- c(sort(dwhole), ntime)
  nfatal <- 1
  for(i in 1:ngroup)
    gr.size[i] <- sum(group == i)
  for(i in 2:length(dwhole)) {
    if(dwhole[i] > dwhole[i - 1]) {
      nfatal <- nfatal + 1
      dwhole[nfatal] <- dwhole[i]
    }
  }
  # end for i
  d <- rep(0, nfatal)
  d <- dwhole[1:nfatal]
  m <- matrix(0, ngroup, nfatal)
  x <- matrix(0, ngroup, nfatal)
  if(nfatal == 1) {
    m[, 1] <- gr.size
    for(i in 1:ngroup)
      x[i, 1] <- sum(group == i & cod ==
        1)
  }
  else {
    m[, 1] <- gr.size
    for(i in 1:ngroup) {
      x[i, 1] <- sum(group == i & time <=
        d[1] & cod == 1)
      for(j in 2:nfatal) {
```

```
          m[i, j] <- gr.size[i] - sum(
            group == i & (time <=
            d[j - 1]))
          x[i, j] <- sum(group == i &
            time > d[j - 1] &
            time <= d[j] & cod ==
            1)
        }
    }
  }
  # end if
  xtotal <- apply(x, 2, sum)
  mtotal <- apply(m, 2, sum)
  count <- cbind(cbind(t(x), xtotal), cbind(t(m - x),
    mtotal - xtotal), cbind(t(m), mtotal))
  return(count)
}
```

3.A.2   Cochran–Armitage Trend Test

```
# ---
# cochran.armitage.test:   Cochran-Armitage Trend Test
#
#    Parameters:
#      dat: data matrix.
#           column 1: group id
#           column 2: survival time
#           column 3: tumor indicator
#                     (1: with tumor; 0: w/o tumor)
#      dose: vector of dose metric.
#            Length must be equal to the number of groups.
#      ngroup: number of groups.
# ---
cochran.armitage.test <- function(dat, dose, ngroup)
{
  # Initialization of arrays
  y <- rep(0, ngroup)
  n <- rep(0, ngroup)
  group <- dat[, 1]
  time <- dat[, 2]
  tumory <- dat[, 3] == 1
  # Find y and N in Table 7
  for(i in 1:ngroup) {
```

```
    y[i] <- length(time[tumory & group == i])
    n[i] <- length(time[group == i])
}
# end for i
# Find the Cochran-Armitage test statistic
ysum <- sum(y)
nsum <- sum(n)
k <- n/nsum
e <- ysum * k
differ <- y - e
x <- sum(dose * differ)
dmean <- sum(n * dose)/nsum
temp <- sum(n * (dose - dmean)^2)
v <- (ysum * (nsum - ysum))/(nsum * (nsum - 1)) *
   temp
zca <- x/sqrt(v)
# Output: z-value and p-value
lb <- c("        z-value", "        p-value")
ca.stat <- c(zca, 1 - pnorm(zca))
names(ca.stat) <- lb
return(ca.stat)
}
```

Summary Table for the Cochran–Armitage and the Poly-3 Test

```
# ---
# table.tumors: Summary Table for Cochran-Armitage test and
#               Poly-3 test
#
#   Parameters:
#     dat: data matrix.
#       column 1: group id
#           column 2: survival time
#           column 3: tumor indicator
#                     (1: with tumor; 0: w/o tumor)
#     ngroup: number of groups.
# ---
table.tumors <- function(dat, ngroup)
{
  y <- rep(0, ngroup)
  n <- rep(0, ngroup)
  group <- dat[, 1]
  time <- dat[, 2]
  tumory <- dat[, 3] == 1
  for(i in 1:ngroup) {
    y[i] <- length(time[tumory & group == i])
    n[i] <- length(time[group == i])
  }
  # end for i
  ysum <- sum(y)
  nsum <- sum(n)
  count <- rbind(c(y, ysum), c(n - y, nsum - ysum),
    c(n, nsum))
  dimnames(count) <- list(c("  #w tumors",
    "#w/o tumors", "    #deaths"), c(paste("  gp",
    c(1:ngroup)), "total"))
  return(count)
}
```

3.A.3   Poly-3 Trend Test

```
# ---
# poly3.test:  Poly-3 (Bieler-Williams) Trend Test
#
#    Parameters:
#       dat: data matrix.
#            column 1: group id
#            column 2: survival time
#            column 3: tumor indicator
#                      (1: with tumor; 0: w/o tumor)
#       dose: vector of dose metric.
#             Length must be equal to the number of groups.
#       ngroup: number of groups.
#       ntime: maximum survival time.
# ---
poly3.test <- function(dat, dose, ngroup, ntime)
{
  # Initialization of arrays
  nn <- nrow(dat)
  y <- rep(0, ngroup)
  n <- rep(0, ngroup)
  n.wgt <- rep(0, ngroup)
  ci <- rep(0, ngroup)
  group <- dat[, 1]
  time <- dat[, 2]
  tumor <- dat[, 3]
  tumory <- tumor == 1
  tumorn <- tumor == 0
  # Time-at-risk weight associated with the (i,j)th animal
  yij <- tumor
  temp <- (time/ntime)^3
  temp[tumory] <- 1
  weight <- temp
  # Sum of the risk weights in group i, and y / N in Table 7
  for(i in 1:ngroup) {
    y[i] <- length(time[tumory & group == i])
    temp <- time[tumorn & group == i]/ntime
    n.wgt[i] <- y[i] + sum(temp^3)
    n[i] <- length(time[group == i])
  }
  # end for i
  ysum <- sum(y)
  nsum <- sum(n.wgt)
  # Calculate the Poly-3 test statistic
```

```
ai <- n.wgt^2/n
ppi <- y/n.wgt
pp <- ysum/nsum
zij <- yij - pp * weight
for(i in 1:ngroup) {
  zbar <- mean(zij[group == i])
  temp <- zij - zbar
  ci[i] <- sum((temp[group == i])^2)
}
# end for i
cc <- sum(ci)/(nn - ngroup)
ai.ppi.d <- sum(ai * ppi * dose)
ai.d <- sum(ai * dose)
ai.ppi <- sum(ai * ppi)
asum <- sum(ai)
ai.dd <- sum(ai * dose^2)
numerator <- ai.ppi.d - (ai.d * ai.ppi)/asum
denominator <- cc * (ai.dd - ai.d^2/asum)
# Output: z-value and p-value
z.poly3 <- numerator/sqrt(denominator)
lb <- c("      z-value", "       p-value")
poly3.stat <- c(z.poly3, 1 - pnorm(z.poly3))
names(poly3.stat) <- lb
return(poly3.stat)
}
```

3.A.4   Figure of Time-to-Tumor-Onset Survival Functions

The code to generate Figure 3.1, page 57.

```
# Benzidine, Tumor Onset Survival Function
par(mfrow = c(2, 1), oma = c(0, 4, 0, 4))
mtext(line = 3, cex = 1, outer = T, "")
dat <- matrix(scan("survbenz.dat"), ncol = 3, byrow = T)
surv.time <- cbind(dat[1:6, 1], dat[7:12, 1], dat[13:18,
  1], dat[19:24, 1])
tumor.rate.male <- cbind(dat[1:6, 2], dat[7:12, 2], dat[
  13:18, 2], dat[19:24, 2])
tumor.rate.female <- cbind(dat[1:6, 3], dat[7:12, 3], dat[
  13:18, 3], dat[19:24, 3])
par(mar = c(8, 5, 5, 2))
matplot(surv.time, tumor.rate.male, xlab =
  "Survival time (week)", ylab =
  "Tumor onset surv. rate", xlim = range(surv.time,
  0, 81), ylim = range(tumor.rate.male, 0, 1.05), main
  = "Male", type = "l", cex = 1)
text(70, 0.859, "60 ppm", cex =
  0.8)
text(70, 0.667 (S-Plus precision problem), "120 ppm", cex =
  0.8)
text(70, 0.582, "200 ppm", cex =
  0.8)
text(70, 0.353, "400 ppm", cex =
  0.8)
matplot(surv.time, tumor.rate.female, xlab =
  "Survival time (week)", ylab =
  "Tumor onset surv. rate", xlim = range(surv.time,
  0, 81), ylim = range(tumor.rate.male, 0, 1.05), main
  = "Female", type = "l", cex = 1)
text(50, 0.838, "60 ppm", cex =
  0.8)
text(50, 0.627, "120 ppm", cex = 0.8)
text(50, 0.308, "200 ppm", cex = 0.8)
text(50, 0.1, "400 ppm", cex =
  0.8)
rm(dat, surv.time, tumor.rate.male, tumor.rate.female)
```

4

Analysis of Toxicokinetic and Pharmacokinetic Data from Animal Studies

Wherly P. Hoffman, Michael A. Heathman,
James Z. Chou, and Darrel L. Allen
Lilly Research Laboratories, Indianapolis, IN, USA

4.1 Introduction

4.1.1 General Background

In the development process of human pharmaceutical therapeutics, toxicology studies play an important role in establishing the safety profiles of compounds. A battery of in vitro and in vivo studies is conducted for this purpose. Animals commonly used in toxicology studies include rats, mice, dogs, and monkeys. Interpretation of toxicological studies often hinges on the actual exposure of the animals to the compound. The amount of the compound administered to an animal is not necessarily the same as the amount reaching the intended site of action. Knowledge of the systemic exposure of an animal to a compound gives us an indication of the amount of the compound that is responsible for the actions. The exposure information allows an assessment of the linear correlation with pharmacodynamic measurements. For example, if heart rate was the biological/toxicological endpoint of interest, then increases in the area under the plasma concentration-time curve (AUC) could be compared to increases in the heart rate to determine a correlation between the pharmacodynamic and pharmacokinetic measurements. If there is a good correlation, then the area under the plasma concentration-time curve can be used to predict heart rate changes. If there is not a good correlation, this would suggest that the compound may be sequestered in tissues and, thus, have biological activity beyond the exposure measured in the blood. Alternatively, it could indicate that the dose range used was not wide enough to cause a range of responses in the heart rate.

Studies designed to provide information on systemic exposure are carried out either as separate pharmacokinetic studies or as companions to toxicity studies. These exposure studies, known as toxicokinetic studies, are pharmacokinetic studies with a focus on the toxicity assessment. In the International Conference on Harmonization (ICH) harmonized tripartite guideline (1995), it states that:

The primary objective of toxicokinetics is:

- To describe the systemic exposure achieved in animals and its relationship to dose level and the time course of the toxicity study.

Secondary objectives are:

- To relate the exposure achieved in toxicity studies to toxicological findings and contribute to the assessment of the relevance of these findings to clinical safety.
- To support the choice of species and treatment regimen in nonclinical toxicity studies.
- To provide information which, in conjunction with the toxicity findings, contributes to the design of subsequent nonclinical toxicity studies.

4.1.2   General Designs

For standard large animal toxicology studies utilizing dogs or monkeys, three to four animals per sex are typically included in each treatment group. Larger sample sizes may be used in rodent studies. It is stated in the ICH guideline that, in general, toxicokinetic data may be generated from the main study animals in large animal studies, while satellite animals may be required in rodent studies for evaluation of systemic exposure. The number of animals used in the toxicity studies should be the minimum needed to generate adequate toxicokinetic data. In practice, when feasible, blood samples are obtained from each animal at all selected time points to establish a plasma concentration-time curve for the evaluation of compound exposure to an animal. This is known as complete sampling. However, when drawing multiple blood samples in a time course from the same animal, blood collection may interfere with the evaluation of the toxicity of the compound. Therefore, sparse sampling designs may be useful and necessary (Pai et al., 1996). Two types of sparse sampling designs commonly used are destructive sampling design and composite sampling design. In a destructive sampling design study, animals are sacrificed after one blood sample is collected, while in a composite sampling design, blood samples are collected at a subset of designed time points from each animal. In a composite sampling design, two or more animals are assigned to each bleed session with preselected time points and the combined time points from all sessions in a treatment group make up the complete set of time points of the study design. The sparse sampling designs result in major challenges in the statistical evaluation of the compound exposure. Statistical approaches applied to studies using composite sampling can be applied to destructive sampling and complete sampling as special cases.

4.1.3   Scope of the Chapter

The primary purpose of this chapter is to describe the statistical methods applicable to exposure studies using composite sampling designs. Parameters for describing compound exposures in toxicokinetic and pharmacokinetic studies are AUC in ng/(mL × h), C_{max} in ng/mL, T_{max} in hours, and *Half-life* in hours as defined in Table 4.1.

Table 4.1. Parameters for describing compound exposures.

Parameter	Description
AUC	Area under the plasma concentration-time curve
C_{max}	Maximum concentration
T_{max}	Time of maximum concentration
Half-life	Time required for plasma concentration to be reduced by 50%

Plasma concentrations from blood samples taken over time are used to estimate these parameters. Although the parameters C_{max}, T_{max}, and *Half-life* are all very useful for assessing the exposure, the most important one is usually the AUC. Two AUC estimates are often discussed in a study. One is the AUC from the beginning to the end of the sampling period in a study. The other is the AUC for the sampling period plus the complete elimination phase obtained by extrapolating beyond the end of the sampling period in a study. The focus of this chapter is on the estimation of the AUC for the sampling period.

Approaches to model the concentration–time relationship include parametric and nonparametric (Vandenhende et al., 1998; Bailer and Ruberg, 1996; Yeh, 1990; Bailer, 1988). Since the primary objective of toxicokinetics is the interpretation of toxicity tests and not on characterizing the basic pharmacokinetic parameters of the substance studied, as stated in the ICH guideline, we'll discuss compound exposure based on AUC estimation using the linear trapezoidal rule. Although the design and analysis for composite sampling have not been discussed as much as for destructive sampling and complete sampling, they are frequently needed in practice for toxicokinetic or pharmacokinetic studies. When sparse sampling designs are used, typically only two or three animals are used at each time point and no formal statistical analyses are performed for evaluation of dose proportionality and changes between days or between sexes. Methods discussed in this chapter worked for the hypothetical study with a composite sampling design. The performance of these tests on data from rodent studies with sparse data merits further investigation. We will first generate data for a composite sampling design and then analyze the data using S-PLUS to evaluate the exposure of an animal to a compound.

In Section 4.2, we will describe the design of the study and how the plasma concentration data are generated. In Section 4.3, details on how data are processed to be ready for analysis are provided. In Section 4.4, an initial evaluation of the exposure data is entertained. Statistical analyses are performed and interpretations of results are provided. Conclusions are presented in Section 4.5.

4.2   Design and Source of Data

Four commonly used parametric models, describing the relationship between the plasma concentration and time, are discussed in a paper by Beatty and Piegorsch (1997). The four models are the one-compartment monoexponential open model associated with a single bolus injection, the one-compartment monoexponential model with zero-order absorption and first-order elimination associated with continuous or chronic exposure, the biexponential model with first-order absorption and elimination, and its generalization to the full biexponential model. The biexponential model with first-order absorption and elimination is the most commonly used one in pharmaceutical practice. The plasma concentration at time t for this model is described in (4.1):

$$c(t) = D\left(\frac{F}{V}\right)\frac{k_a}{(k_a - k_e)}\left(e^{-k_e t} - e^{-k_a t}\right) \tag{4.1}$$

where D is the dose, F is the fraction absorbed, V is the volume of distribution, and k_a and k_e are the absorption and elimination rate constants. An example of this model is shown in Figure 4.1 with $D = 3000$ μg, $F = 1.0$, $V = 50$ L, $k_a = 0.8$ h^{-1}, and $k_e = 0.2$ h^{-1}.

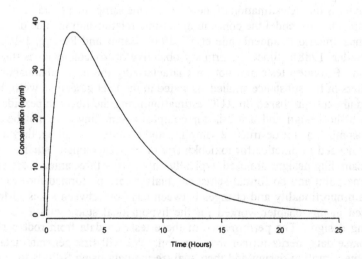

Figure 4.1. Plasma concentration–time curve.

Based on a hypothetical toxicokinetic study, plasma concentrations over a 24-hour study period are simulated for males and females for three dose levels. To include a time effect, the plasma concentrations are generated for Day 0 and Day 360. The three key hypotheses to evaluate are the presence of time effect, gender effect, and dose proportionality. The three dose groups (1, 3, and 10 mg/kg) each include 16 animals per sex. These 16 animals are divided among four bleed sessions each with four animals. Blood collection times are predose (0), 0.5, 1, 2, 3, 4, 5, 7, 10, 14, 18 and 24 hours post-dose. For each animal in a bleed session, blood samples are collected at three time points on each day using the same schedule. The composite design is summarized in Table 4.2. Plasma concentration data are simulated using (4.1). For each animal, a multivariate normal vector centered at the mean plasma concentration level with a covariance matrix is generated. The covariance matrix is assumed to have a first-order autoregressive (AR(1)) structure with a correlation coefficient of 0.3 and $CV = 10\%$. The parameter values are chosen based on previous studies of certain drugs (Table 4.3). The lower limit of quantitation of the assay is set at 10 ng/mL.

Table 4.2. Composite sampling design for each sampling day.

Dose	Animal numbers	Session 1 0^a, 3, 10 hours post-dose	Session 2 0.5, 4, 14 hours post-dose	Session 3 1, 5, 18 hours post-dose	Session 4 2, 7, 24 hours post-dose
1 mg/kg	Male: 1, 2, 3, 4 Female: 49, 50, 51, 52	x			
1 mg/kg	Male: 5, 6, 7, 8 Female: 53, 54, 55, 56		x		
1 mg/kg	Male: 9, 10, 11, 12 Female: 57, 58, 59, 60			x	
1 mg/kg	Male:13, 14, 15, 16 Female: 61, 62, 63, 64				x
3 mg/kg	Male: 17, 18, 19 ,20 Female: 65, 66, 67, 68	x			
3 mg/kg	Male: 21, 22, 23, 24 Female: 69, 70, 71, 72		x		
3 mg/kg	Male: 25, 26, 27, 28 Female: 73, 74, 75, 76			x	
3 mg/kg	Male: 29, 30, 31, 32 Female: 77, 78, 79, 80				x
10 mg/kg	Male: 33, 34, 35, 36 Female: 81, 82, 83, 84	x			
10 mg/kg	Male: 37, 38, 39, 40 Female: 85, 86, 87, 88		x		
10 mg/kg	Male: 41, 42, 43, 44 Female: 89, 90, 91, 92			x	
10 mg/kg	Male: 45, 46, 47, 48 Female: 93, 94, 95, 96				x

^{a} Predose.

Table 4.3. Parameter values used in simulation study.

Parameter	Value	
Fraction absorbed, F	1.0	
Absorption rate constant, k_a	Mean = 0.8 h^{-1}	$CV = 20\%$
Elimination rate constant, k_e	Mean = 0.2 h^{-1}	$CV = 20\%$
Volume of distribution, V	Mean = 50 L	$CV = 20\%$

A gender effect is introduced into the data through modification of the elimination rate constant, k_e. Females are assigned a parameter value twice that of males, 0.4 h^{-1} with $CV = 20\%$. The parameter values were chosen to illustrate a difference between the gender. Effect of repeated dosing is not generated for either sex on either day. Dose proportionality in AUC is simulated for both sexes on both days. AUC values can be calculated from the simulation parameter values for each combination of dose, gender and day of study:

$$\text{AUC}(0, \infty) = \int_0^\infty c(t)\, dt = D\left(\frac{F}{V}\right)\frac{1}{k_e} \qquad (4.2)$$

The resulting AUC values based on the simulation parameters are presented in Table 4.4. The difference between the AUC in males and females is successfully simulated.

Table 4.4. Mean AUC(0, ∞) in ng/(mL × h) calculated from simulation parameters.

Dose	Day 0		Day 360	
	Male	Female	Male	Female
1 mg/kg	474.29	207.45	377.20	220.39
3 mg/kg	1310.22	659.17	1162.95	707.42
10 mg/kg	3988.19	2322.46	4375.35	2035.16

Using the biexponential model, data are simulated in S-PLUS using the function sim.data (Appendix 4.A.1). Pharmacokinetic parameter values and variability, as well as dosing and sampling schemes, are passed to the function. Random animal weights are selected using a normal distribution, mean of 4 kg and standard deviation of 0.75 kg for this simulation. A limit of quantitation can be defined; all data below this limit are assigned the value NA. All data used in this chapter can be recreated using this code:

```
# Define pharmacokinetic parameters
f <- 1.0                    # Bioavailability parameter F
```

```
Mean.KA <- 0.80          # Mean value of absorption
                         #   rate constant, KA
Mean.KE <- 0.20          # Mean value of elimination
                         #   rate constant, KE
Mean.V <- 50.0           # Mean value of volume
                         #   of distribution, V
CV.KA <- 0.20            # CV for KA (inter-individual)
CV.KE <- 0.20            # CV for KE (inter-individual)
CV.V <- 0.20             # CV for V (inter-individual)
CV.Conc <- 0.10          # CV for concentration measurement
Rho <- 0.3               # Correlation factor for
                         #   covariance matrix
BQLcutoff <- 10.0        # Limit of quantitation
doses <- c(1, 3, 10)     # Dose groups in mg/kg
animals <- 4             # Animals per session

# Matrix to hold sampling schedules for each session
sampling <- matrix(nrow = 4,ncol = 3)
sampling[1,] <- c(0.0, 3.0, 10.0)  # times for session 1
sampling[2,] <- c(0.5, 4.0, 14.0)  # times for session 2
sampling[3,] <- c(1.0, 5.0, 18.0)  # times for session 3
sampling[4,] <- c(2.0, 7.0, 24.0)  # times for session 4

# Set random number seeds so simulations are reproducible
set.seed(214)
seeds <- floor(runif(4, min = 1, max = 1000))

# Generate simulated data for males at Day 0 and Day 360
male0 <- sim.data(f, Mean.KA, Mean.KE, Mean.V,
  CV.KA, CV.KE, CV.V, CV.Conc, BQLcutoff,
  doses, sampling, animals, Rho, seeds[1])
male360 <- sim.data(f, Mean.KA, Mean.KE, Mean.V,
  CV.KA, CV.KE, CV.V, CV.Conc, BQLcutoff,
  doses, sampling, animals, Rho, seeds[2])
N <- max(male0[,"ANIMAL"])

# Modify elimination rate constant and generate
# simulated data for females
Mean.KE <- 0.40
female0 <- sim.data(f, Mean.KA, Mean.KE, Mean.V,
  CV.KA, CV.KE, CV.V, CV.Conc, BQLcutoff,
  doses, sampling, animals, Rho, seeds[3])
female0[,"ANIMAL"] <- female0[,"ANIMAL"] + N
```

```
female360 <- sim.data(f, Mean.KA, Mean.KE, Mean.V,
  CV.KA, CV.KE, CV.V, CV.Conc, BQLcutoff,
  doses, sampling, animals, Rho, seeds[4])
female360[,"ANIMAL"] <- female360[,"ANIMAL"] + N

# Add Gender and Day information to simulated data
male0 <- cbind.data.frame(male0, SEX = 1, DAY = 0)
male360 <- cbind.data.frame(male360, SEX = 1, DAY = 360)
female0 <- cbind.data.frame(female0, SEX = 2, DAY = 0)
female360 <- cbind.data.frame(female360, SEX = 2,
  DAY = 360)

# Combine simulated data
simdata <- rbind(male0, male360, female0, female360)
```

4.3 Data Processing

Plasma concentration levels that are below the limit of quantitation, 10 ng/mL, are labeled as BQL. There are several approaches to treating BQLs in the statistical analysis (Rae et al., 1998; Bullingham et al., 1996). BQLs in the beginning and ending phases of the blood collection, in our opinion, should be considered differently from those that occur between the two phases. Before the initial treatment, the plasma concentration levels should be 0 because no drug has been delivered to the animals yet. Therefore, one would expect the concentrations to be BQLs. Toward the end of the sampling period, if there were no drug in the blood any more, one would expect the concentrations to be BQLs as well. Therefore, at the beginning phase or toward the end of the sampling period, if all plasma concentrations at a time point are BQLs, then they are set to 0. Assignments of other BQLs are not consistent among scientists. Some would ignore BQLs, some would assign BQLs to fixed values, and some would use all data including BQLs to model the concentration curve and estimate the parameters of interest by a likelihood approach. Four approaches of assigning BQLs to a fixed value are:

- assign them to 0;
- assign them to the limit of quantitation;
- assign them to the mid-value between 0 and the limit of quantitation;
- assign them to the value that is two-thirds of the limit of quantitation.

Other approaches to assigning BQLs as a function of the concentration and the percent of BQL can be structured and evaluated.

In the event that plasma concentrations from all animals at a given time point are all BQLs and that time point is neither at the beginning nor toward the end of the sampling period, one should examine the data closely to decide on the most appropriate treatment of those concentration levels. Since BQLs are, by definition, between 0 and the limit of quantitation, assignment of any BQLs to either 0 or the limit of quantitation will always result in a negative bias or positive bias, respectively, in the estimation of the AUC. In addition, the variance estimates of the AUC will be biased positively if 0 is assigned and negatively if the limit of quantitation is assigned. Assignments of BQLs to mid or two-thirds of the limit of quantitation are based on the concept of an expected value of the plasma concentration following a uniform or triangular distribution between 0 and the limit of quantitation. Therefore, one assignment may be a closer approximation to the actual concentration than the other, depending on the true but unknown distribution of the plasma concentration at that time point. BQL handling can be crucial to the evaluation of the exposure. Since there is no one approach agreed upon by all, readers should decide on the approach that is acceptable to the research team based on the questions to be answered. From the standpoint of statistical evaluation, once data are generated and BQL values are assigned, the statistical analysis procedures are the same. In this chapter, only the forth approach of assigning BQL values to two-thirds of the limit of quantitation is adopted as a demonstration.

In preparation for data analysis, the function BQL.Assign (Appendix 4.A.2) is used to assign a specific value, passed as a function argument, to observations below the limit of quantitation. If all observations at that sampling time are BQL the observation is set to a value of 0.0. The function returns a data frame identical to the original data, with BQL concentrations reassigned:

```
> BQLdata <- BQL.Assign(simdata, 6.67)
```

4.4   Statistical Analysis

The three primary questions to address here are:

- Is there any change in AUC from Day 0 to Day 360?
- Is there any difference in AUC between males and females?
- Do we have dose proportionality in AUC?

The three tests used are a one-sample or paired t-test for evaluating the time effect for each dose level and each sex, a two-sample t-test for evaluating the gender differences for each dose level on each day, and an F-test for evaluating dose proportionality for each sex on each day. Since the statistical tests are based on the variance and degrees of freedom of the estimate of the mean AUC, the estimation of the parameters are discussed before the statistical tests are performed.

In our simulated data, all time points with 100% BQLs are either at the beginning or toward the end of the sampling period. Therefore, the concentrations at time points with 100% BQLs are assigned to 0. Other BQL results reported here are based on the assignment of 6.67 ng/mL.

4.4.1   Estimation of AUC in the Sampling Period

For evaluation of the compound exposure, one has to obtain an estimate of the AUC and its associated variability. When the plasma concentration as a function of time, $c(t)$, is known, the area under the plasma concentration curve from time t_1 and t_2 hours post-dosing, $AUC(t_1, t_2)$, is defined as

$$AUC(t_1, t_2) = \int_{t_1}^{t_2} c(t)\,dt \qquad (4.3)$$

In practice, $c(t)$ is not known. One can assume that it follows certain parametric models based on previous experience and check for the fit of the assumed model against data. Due to the limited sample sizes and the sparse sampling in animal studies, it can be difficult to perform the check. A more common practice for the analysis of animal data is to use the nonparametric linear trapezoidal rule to empirically approximate the AUC by adding up all trapezoids. Since each animal is only sampled at a selected subset of time points, it is not feasible to obtain an AUC estimate for each animal. Only a partial AUC based on the concentrations at those selected time points can be calculated for each animal.

Let $AUC(t_0, t_k)$ be the estimated area under the plasma concentration curve from t_0 to t_k expressed as a linear combination of the observed concentrations, $y(t)$, at all time points. For an animal from a complete sampling design study, it is calculated as

$$AUC(t_0, t_k) = 0.5 \left[(t_1 - t_0)\,y(t_0) + \sum_{j=1}^{k} (t_{j+1} - t_{j-1})\,y(t_j) \right.$$

$$\left. + (t_k - t_{k-1})\,y(t_k) \right]$$

$$(4.4)$$

The AUC calculation is accomplished using the function Calc.AUC (Appendix 4.A.2). This function calculates a partial AUC for every observation. The return value is a data frame containing the partial AUC as an additional column:

```
> AUCdata <- Calc.AUC(BQLdata)
```

For ease of notation, AUC will be used to refer to the AUC from time 0 to the last sampling time point of the study, 24 hours for our simulated data. For animals in the composite sampling studies, the value of the observed concentration is missing at all but the selected sampling time points for each animal. Therefore, only a partial AUC can be obtained for each animal using (4.4). The mean partial AUC estimate of a bleed session is the average of the partial AUCs from all animals in the bleed session as presented in (4.5):

$$SessionMean(\text{AUC}) = \frac{\sum\limits_{j=1}^{n} partialAUC_j}{n} \quad (4.5)$$

where n is the number of animals included in the bleed session and partial AUC, is the partial AUC obtained from the jth animal in the bleed session. The AUC estimate for a dose group is computed by summing up the mean partial AUCs from all bleed sessions as in (4.6):

$$GrpMean(\text{AUC}) = \sum\limits_{m=1}^{s} Session_m Mean(\text{AUC}) \quad (4.6)$$

where s is the number of sessions in a group and $Session_m Mean(\text{AUC})$ is the mean AUC estimate for the mth bleed session.

4.4.2   Variability of AUC Estimates

For complete sampling designs, variability of an AUC estimate for a group can be calculated using the standard formula for standard errors. For destructive sampling designs, variability can be calculated by pooling the variability at each time point because all plasma concentration measurements are independent. However, for composite sampling designs where multiple blood samples are collected from each animal in a bleed session, the independence among plasma concentrations enjoyed by destructive sampling designs no longer holds. Therefore, to take into account the correlation of measurements within each animal, partial AUCs are calculated for each animal in each bleed session (Yeh, 1990). Variability for the estimate of the session mean AUC is calculated using the standard formula for standard errors, SE. Since different animals are used in different bleed sessions, the variability of the group mean AUC estimate on a specific day is simply the sum of the variability from all sessions in the dose group. This is

$$SE(Session\,Mean\,(\text{AUC}))$$

$$= \sqrt{\frac{\sum\limits_{j=1}^{n} (partial\,\text{AUC}_j - Session\,Mean\,(\text{AUC}))^2}{(n-1)n}} \quad (4.7)$$

and

$$SE(GrpMean(\text{AUC})) = \sqrt{\sum\limits_{m=1}^{s} (SE(Session_m Mean(\text{AUC})))^2} \quad (4.8)$$

The mean and standard error for each dose group on each day are calculated and summarized for males and females in Table 4.5 and plotted in Figure 4.2. The S-PLUS code for plotting this figure is included in Appendix 4.A.2. Visual examination of the data indicated the following. The mean AUC values are higher for males than females on both Day 0 and Day 360 for all dose groups. Overall, there does not appear to be a difference between Days 0 and 360 based on the mean AUC values. However, since each animal is sampled on both days, the evaluation of time effect should be based on the change in AUC from Day 0 to Day 360 in each animal. The increase in mean AUC values appears to be proportional to dose on both days for both sexes. A formal statistical analysis follows later in this section.

Table 4.5. Means and standard errors of AUC(0, 24) in ng/(mL × h) by linear trapezoidal rule.

Dose		Day 0		Day 360	
		Male	Female	Male	Female
1 mg/kg	Mean	451.06	186.88	354.12	200.92
	SE	46.44	14.39	25.13	19.88
3 mg/kg	Mean	1244.16	671.45	1132.20	679.19
	SE	140.66	57.21	129.77	67.85
10 mg/kg	Mean	4085.03	2456.97	4332.81	2023.04
	SE	433.89	238.59	399.22	159.53

Note: BQL values are replaced with 0 for time points with 100% BQLs and with 6.67 ng/mL for the rest.

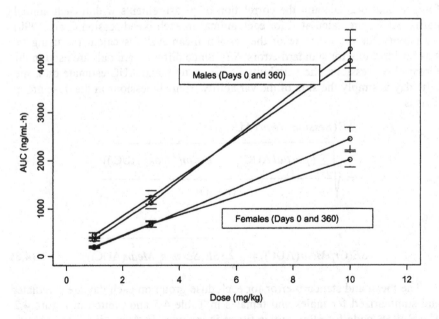

Figure 4.2. Plot of means and standard errors of AUC(0, 24) in ng/(mL × h) by linear trapezoidal rule.

4.4.3   Linear Trapezoidal Rule

The session means and standard errors are calculated, as shown in (4.5) and (4.7), using the function Session.Means (Appendix 4.A.2). This function returns session means and standard errors for all combinations of dose group, gender, day, and bleed session:

```
> sessiondata <- Session.Means(AUCdata)
```

The group means and standard errors reported in Table 4.5 are calculated, as described in (4.6) and (4.8), using the function Group.Means (Appendix 4.A.2). A data frame containing the session means and standard errors is passed into the function as an argument and the group means and standard errors are returned for all combinations of dose group, gender, and day:

```
> Group.Means(sessiondata)
```

4.4.4   Estimates of Degrees of Freedom

To perform statistical tests, we need to obtain the degrees of freedom associated with each estimated AUC. To account for heterogeneity of variances between comparison groups, the Satterthwaite procedure is adopted (Yeh, 1990) to approximate the degrees of freedom for each group. For example, the degrees of freedom associated with the estimate of the group mean AUC based on s bleed sessions is as follows:

$$DF = \frac{\left[\sum_{m=1}^{s} \left(SE(Session_m Mean(\text{AUC})) \right)^2 \right]^2}{\sum_{m=1}^{s} \left[\left(SE(Session_m Mean(\text{AUC})) \right)^4 / (n_m - 1) \right]} \qquad (4.9)$$

where $SE(Session_m Mean(\text{AUC}))$ is the standard error of the mth bleed session and n_m is the number of animals in that session. Note that if there is homogeneity and equal session size, n, then the degrees of freedom is $s(n - 1)$. The degrees of freedom as defined in (4.9) is calculated using the function calc.DF:

```
> calc.DF <- function(SE, N) {
    DF <- ((sum((SE)^2))^2)/sum((SE)^4/(N - 1))
    return(DF)
  }
```

4.4.5   Test for Change in AUC from Day 0 to Day 360

Since blood samples are collected from the same animal on Day 0 and Day 360, the partial AUC estimates are not independent. Therefore, the difference between the partial AUCs from the 2 days, partial DAUC, is calculated for each

88 W.P. Hoffman, M.A. Heathman, J.Z. Chou, and D.L. Allen

animal. The session mean is calculated using (4.5) with AUC replaced by DAUC and the group mean is calculated using (4.6) with AUC replaced by DAUC. The comparison is performed on the mean difference from all animals in each dose group for each sex using a two-sided t-test. The test statistic is

$$t = \frac{GrpMean(\text{DAUC})}{SE(GrpMean(\text{DAUC}))} \tag{4.10}$$

with degrees of freedom as calculated in (4.9).

Let t_{df} denote a random variable having Student's t-distribution with df degrees of freedom. The two-sided p-value associated with the test statistic is given by

$$p = 2 \times \Pr\left(t_{df} \geq |t|\right) \tag{4.11}$$

The results of testing for a change in AUC from Day 0 to Day 360 are reported in Table 4.6. The magnitudes of decreases and increases are small compared to the standard errors. No statistically significant changes are declared at the 0.05 significance level.

Mean changes in AUC from Day 0 to Day 360, as reported in Table 4.6, are calculated for all combinations of dose group and gender using the function Day.Test (Appendix 4.A.2).

```
> Day.Test(AUCdata)
```

Table 4.6. Summary of changes in AUC(0, 24) in ng/(mL × h) from Day 0 to Day 360.

Dose	Statistics	Male	Female
1 mg/kg	Diff	−96.94	14.03
	SE	64.63	25.16
	DF	9.79	7.80
	p-value	0.165	0.593
3 mg/kg	Diff	−111.95	7.74
	SE	190.22	86.20
	DF	7.26	8.90
	p-value	0.574	0.930
10 mg/kg	Diff	247.78	−433.93
	SE	479.54	266.23
	DF	9.28	6.48
	p-value	0.617	0.151

Note 1: Diff = mean of changes in partial AUC from Day 0 to Day 360.
Note 2: p-value is for a two-sided t-test.
Note 3: BQL values are replaced with 0 for time points with 100% BQLs and with 6.67 ng/mL for the rest.

4. Analysis of Toxicokinetic and Pharmacokinetic Data

4.4.6   Test for Difference in AUC Estimates
Between Males and Females

For the comparison of gender differences, the two-sample t-test (4.12) is used. The mean group AUC estimate of females is subtracted from that of males to provide a gender difference in AUC. The variability of the difference is obtained by summing the squared standard errors of the two independent sample estimates and the corresponding degrees of freedom are calculated by the Satterthwaite approximation using (4.13):

$$t = \frac{Grp_M Mean(\text{AUC}) - Grp_F Mean(\text{AUC})}{\sqrt{(SE(Grp_M Mean(\text{AUC})))^2 + (SE(Grp_F Mean(\text{AUC})))^2}} \qquad (4.12)$$

$$DF = \frac{DFnum}{DFden} \qquad (4.13)$$

where $DFnum$ and $DFden$ are defined below as a function of the standard errors of the group mean AUC estimates and the corresponding degrees of freedom calculated in (4.9):

$$DFnum = \left\{ \left[SE\left(Grp_M Mean\left(AUC\right)\right)\right]^2 + \left[SE\left(Grp_F Mean\left(AUC\right)\right)\right]^2 \right\}^2 \quad (4.14)$$

$$DFden = \frac{(SE(Grp_M Mean(\text{AUC})))^4}{DF_M(\text{AUC})} + \frac{(SE(Grp_F Mean(\text{AUC})))^4}{DF_F(\text{AUC})} \qquad (4.15)$$

The p-value associated with the two-sample t-statistic is obtained in the same manner as in (4.11) and the results are summarized in Table 4.7. A significant gender difference exists in all dose groups and on both sample collection days. Mean AUC estimates of males are consistently higher than females. The differences are similar on both days in the 1 mg/kg group and become larger in the 3 and 10 mg/kg groups. The standard errors of the mean gender differences of the three dose groups are heterogeneous.

The mean differences in AUC between males and females, as reported in Table 4.7, are calculated using the function Gender.Test (Appendix 4.A.2). This function uses session means and standard errors as input and returns two-sample t-test results for all combinations of dose group and day:

```
> Gender.Test(sessiondata)
```

Table 4.7. Summary of Gender Difference in AUC(0, 24) in ng/(mL × h).

Dose	Statistics	Day 0	Day 360
1 mg/kg	Diff	264.18	153.20
	SE	48.62	32.04
	DF	10.90	19.03
	p-value	< 0.001	< 0.001
3 mg/kg	Diff	572.71	453.01
	SE	151.85	146.44
	DF	7.03	6.04
	p-value	0.007	0.021
10 mg/kg	Diff	1628.06	2309.76
	SE	495.16	429.16
	DF	8.76	8.68
	p-value	0.010	0.001

Note 1: Diff =Group Mean AUC of males – group mean AUC of females.

Note 2: p-value is for a two-sided t-test.

Note 3: BQL values are replaced with 0 for time points with 100% BQLs and with 6.67 ng/mL for the rest.

4.4.7   Test for Dose Proportionality in AUC

For evaluation of the relationship between the dose administered to animals and drug exposure in animals, we would like to see that the exposure is a constant multiple of the dose for all dose levels, and, in general, would expect no exposure in animals if they have not been treated with any drug yet. This is equivalent to saying that the two-dimensional plot of the mean AUC versus dose should be a line passing through the origin (see Figure 4.2). This is an important fact in understanding how much of the drug is absorbed by the animal system. The statistical approach for evaluating the plausibility of dose proportionality is to test the equality of the adjusted mean AUCs of the three dose groups. The adjusted mean and *SE* of AUC estimates for each dose group is obtained by dividing the estimates (4.6) and (4.8) by the corresponding doses as in (4.14) and (4.15). If in fact the AUC were proportional to dose, then we would not expect to see any significant differences between the dose-adjusted group means for AUC:

$$AdjGrpMean(AUC) = \frac{GrpMean(\text{AUC})}{\text{Dose}} \tag{4.16}$$

$$AdjSE(GrpMean(\text{AUC})) = \frac{SE(GrpMean(\text{AUC}))}{\text{Dose}} \tag{4.17}$$

For a standard one-way design with I treatment groups and n_i subjects in each group, where $i = 1, 2, 3 \ldots I$, the F-statistic for testing for equality of the means is the ratio of the mean squares associated with treatment, MS_{TRT}, and the mean squares associated with the error, MS_{ERR}. MS_{TRT} is the ratio of the sum of squares associated with treatment, SS_{TRT}, and its degrees of freedom, df_{TRT}. Similarly, MS_{ERR} is the ratio of the sum of squares associated with the error term, SS_{ERR}, and its degrees of freedom, df_{ERR}. Denote the observation x_{ij} as the jth individual in the ith group, $\bar{x}_{i.}$ as the mean of the ith group, and $\bar{x}_{..}$ as the grand mean. Then SS_{TRT} and SS_{ERR} are

$$SS_{TRT} = \sum_i \sum_j (\bar{x}_{i.} - \bar{x}_{..})^2 \qquad (4.18)$$

$$SS_{ERR} = \sum_i \sum_j (x_{ij} - \bar{x}_{i.})^2 \qquad (4.19)$$

Since mean AUC is estimated from all 16 animals, each animal contributing to a partial AUC, we will use the degrees of freedom by the Satterthwaite method to approximate the second summations for the number of animals in SS_{TRT} and SS_{ERR}. Therefore, let $AdjGrp_iMean(\text{AUC})$ and df_i be the mean and the degrees of freedom for the ith group, and let $GrandMean$ be the average of the three adjusted group means. Then SS_{TRT} and SS_{ERR} are calculated by

$$SS_{TRT} = \sum_{i=1}^{3} (AdjGrp_iMean(\text{AUC}) - GrandMean)^2 (df_i + 1) \qquad (4.20)$$

$$SS_{ERR} = \sum_{i=1}^{3} (AdjSE(Grp_iMean(\text{AUC})))^2 (df_i)(df_i + 1) \qquad (4.21)$$

The F-ratio is approximated by an F-distribution with degrees of freedom (df_{TRT}, df_{ERR}) where $df_{TRT} = 2$ for 3 dose groups and $df_{ERR} = df_1 + df_2 + df_3$ as

$$F = \frac{SS_{TRT}/df_{TRT}}{SS_{ERR}/df_{ERR}} \qquad (4.22)$$

The summary of the F-tests is presented in Table 4.8. The adjusted and un-adjusted group mean AUCs and the difference between the adjusted means, the F-ratio, associated degrees of freedom, and the p-values are reported for each sex on each day. For males, mid- to low-dose adjusted AUC ratios are 0.92 and 1.07 on Day 0 and Day 360, respectively. The high- to low-dose adjusted AUC ratios are 0.91 and 1.07 on Day 0 and Day 360, respectively. For females, mid- to low-dose adjusted AUC ratios are 1.20 and 1.13 on Day 0 and Day 360, respectively. The high- to low-dose adjusted AUC ratios are 1.31 and 1.01 on Day 0 and Day 360, respectively. Since all p-values are greater than the nominal significance level of 0.05, dose proportionality is concluded for both sexes on both days.

The F-test for dose proportionality is performed using the function `Dose.Test` (Appendix 4.A.2). The session mean data is passed as a function argument and the return value contains all the results in Table 4.8:

```
> Dose.Test(sessiondata)
```

Table 4.8. Summary of dose proportionality in AUC(0, 24) in ng/(mL × h) for males and females on each day.

Gender	Day	Statistics	1 mg/kg	3 mg/kg	10 mg/kg	Diff. Ratio Mid /Low	High /Low	Degrees of freedom	F-ratio	p-value
Male	0	Mean	451.06	1244.16	4085.03					
		AdjMean	451.06	414.72	408.50	−36.34 (0.92)	42.56 (0.91)	(2, 20.00)	0.27	0.763
	360	Mean	354.12	1132.20	4332.81					
		AdjMean	354.12	377.40	433.28	23.29 (1.07)	79.17 (1.22)	(2, 21.23)	1.61	0.223
Female	0	Mean	186.88	671.45	2456.97					
		AdjMean	186.88	223.82	245.70	36.93 (1.20)	58.81 (1.31)	(2, 21.12)	2.49	0.107
	360	Mean	200.92	679.19	2023.04					
		AdjMean	200.92	226.40	202.30	25.48 (1.13)	1.39 (1.01)	(2, 23.95)	0.49	0.617

Note 1: AdjMean is the mean AUC divided by dose.
Note 2: Mid/Low Diff = *Group AdjMean* AUC of mid − *Group AdjMean* AUC of low.
    High/Low Diff = *Group AdjMean* AUC of high − *Group AdjMean* AUC of low.
Note 3: Degrees of freedom are approximated using the Satterthwaite method.
Note 4: BQL values are replaced with 0 for time points with 100% BQLs and with 6.67 ng/mL for the rest.

4.5   Conclusions

In toxicokinetic and pharmacokinetic studies, AUC is routinely used in assessing the systemic exposure of animals to a compound to estimate animal-to-human exposure multiples. Sparse sampling can be successfully used to estimate the AUC values in these studies. To evaluate a compound for a long period such as in this simulated 1-year study, AUC values at the beginning and end of the study are always compared to determine any alteration of systemic exposure due to repeated dosing such as accumulation of the compound or enzyme induction or inhibition. Sometimes, a mid-point in time is also included in a study to confirm the observed "trend" from the beginning to the end of the study. Assignment of BQL values is an important issue and should be carefully thought out before the assessment of the exposure. A sensitivity analysis may be performed to have a feel for the effects of different treatments for the BQL values. The evaluation of proportionality in AUC involves the substitution of the approximated degrees of freedom for session size in both the numerator and denominator of the F-ratio statistic; therefore, results based on the approximation depend on the quality of the approximation. Since multiple tests were performed in the evaluation of the effects, in general, a multiplicity adjustment should be considered to avoid an inflated false positive rate.

Based on the assignment of 6.67 ng/mL for concentrations at time points with less than 100% BQL values and 0 for time points with 100% BQL values, results in Table 4.4 suggest the following: no clear differences in AUC values exist between days; clear differences between males and females are present on both days; and proportionality in AUC is plausible on both days for both sexes. These observations are consistent with the effects designed into the simulated data. Detailed statistical analyses confirmed the observations.

4.6 References

Bailer, A.J. (1998). Testing for the equality of area under the curves when using destructive measurement techniques. *Journal of Pharmacokinetics and Biopharmaceutics* **16**, 303–309.

Bailer, A.J. and Ruberg, S.J. (1996). Randomization tests for assessing the equality of area under curves for studies using destructive sampling. *Journal of Applied Toxicology* **16**, 391–395.

Beatty, D.A. and Peigorsch, W.W. (1997). Optimal statistical design for toxicokinetic studies. *Statistical Methods in Medical Research* **6**, 359–376.

Bullingham, R., Monroe, S., Nicholls, A. and Hale, M. (1996). Pharmacokinetics and bioavailability of mycophenolate mofetil in healthy subjects after single-dose oral and intravenous administration. *Journal of Clinical Pharmacology* **36**, 315–324.

International Conference on Harmonization (1995). Guideline on the assessment of systemic exposure in toxicity studies. *Federal Register* **60**, 11264–11268.

Pai, S.M., Fettner, S.H , Hjian, G., Cayen, M.N., and Batra, V.K. (1996). Characterization of AUCs from sparsely sampled populations in toxicology studies. *Pharmaceutical Research* **13**, 1283–1290.

Piegorsch, W.W. and Bailer, A.J. (1989). Optimal design allocations for estimating area under curves for studies employing destructive sampling. *Journal of Pharmacokinetics and Biopharmaceutics* **17**, 493–507.

Rae, S., Raboud, J.M., Conway, B., Reiss, P., Vella, S., Cooper, D., Lange, J., Harris, M., Wainberg, M.A., Robinson, P., Myers, M., Hall, D., and Montaner, J.S.G. (1998). Estimates of the virological benefit of antiretroviral therapy are both assay- and analysis-dependent. *AIDS.* **12**, 2185–2192.

Vandenhende, F., Dewe, W., and Hoffman, W.P. (1998). A likelihood-based analysis of drug exposure in toxicokinetic studies. *The XIXth International Biometric Conference.* IBC98, p. 297.

Yeh, C. (1990). Estimation and significance tests of area under the curve derived from incomplete blood sampling. *American Statistical Association, 1990 Proceedings of the Biopharmaceutical Section*, pp. 74–81.

4.A. Appendix

4.A.1 S-PLUS Code for Pharmacokinetic Data Simulation

```
# Code to simulate plasma concentrations using
# biexponential pharmacokinetic model,
# 1-compartment oral absorption

sim.data <- function(f,Mean.KA, Mean.KE, Mean.V, CV.KA,
  CV.KE, CV.V, CV.Conc, BQLcutoff, doses, sampling,
  animals, Rho, seed)
{
# f           Bioavailability parameter F
# Mean.KA     Mean value of absorption rate constant, KA
# Mean.KE     Mean value of elimination rate constant, KE
# Mean.V      Mean value of volume of distribution, V
# CV.KA       CV for KA (inter-individual)
# CV.KE       CV for KE (inter-individual)
# CV.V        CV for V (inter-individual)
# CV.Conc     CV for final concentration measurment
# BQLcutoff   Lower limit of quantitation
# doses       Dose groups in mg/kg
# sampling    Array of sampling times for each session
# animals     Number of animals per session
# Rho         Correlation factor for covariance matrix
# seed        Seed for random generation of variability
  set.seed(seed)
# Number of sessions in simulated dataset
  sessions <- nrow(sampling)
# Number of samples per animal
  samples <- ncol(sampling)
  first <- T
  id <- 1  # initialize animal ID counter
# Loop through doses, sessions and animals per session
  for(dd in doses) {
    for(jj in c(1:sessions)) {
      for(ii in c(1:animals)) {
# Generate random weight for this animal
# Mean body weight of 4 kg, standard deviation of 0.75
```

```
        wt <- round(rnorm(1, mean = 4, sd = 0.75),
           digits = 4)
# calculate dose (in micrograms) based on weight
        dose <- round(wt * dd * 1000, digits = 2)
# variability in KA for this animal
        varKA <- rnorm(1, mean = 1, sd = CV.KA)
# variability in KE for this animal
        varKE <- rnorm(1, mean = 1, sd = CV.KE)
# variability in V for this animal
        varV <- rnorm(1, mean = 1, sd = CV.V)
        KA <- Mean.KA * varKA  # Apply variabilty to KA
        KE <- Mean.KE * varKE  # Apply variability to KE
        V <- Mean.V * varV  # Apply variability to V
# Get sampling times from sampling matrix
        tfds <- sampling[jj, ]
# Calculate concentrations at these times
# Equation assumes no accumulation from previous doses
        Mean.Conc <- (f * dose * KA)/(V * (KA - KE)) *
        (exp( - KE * tfds) - exp( - KA * tfds))
# Generate a row vector of standard deviations
# from Mean and CV for final measurement
        sigma <- Mean.Conc * CV.Conc
# Generat covariance matrix for independent observations
        SIGMA <- outer(sigma, sigma)
# Correlation matrix for a
# first order auto regressive (AR(1)) structure
        correl <- Rho^abs(row(SIGMA) - col(SIGMA))
# Generate covariance matrix with AR(1) structure
        SIGMA <- SIGMA * correl
# Perform Choleski decomposition
        y <- chol(SIGMA, pivot = T)
        pivots <- attr(y, "pivot")
        U <- y[pivots, pivots]
# Generate correlated normal vectors
# for plasma concentrations and
# apply variability to calculated values
        Z <- rnorm(samples, mean = 0, sd = 1)
        Conc <- Mean.Conc + t(U) %*% Z
# Combined data
        animdata <- cbind.data.frame(HOUR = tfds,
          CONC = Conc, GROUP = dd, SESSION = jj,
          ANIMAL = id)
        if(first) {
          outdata <- animdata
          first <- F
```

```
      }
      else {
        outdata <- rbind(outdata, animdata)
      }
      id <- id + 1  # Increment animal ID counter
    }
  }
}
# Select concentrations below limit of quantitation
# and assign value of NA (missing)
  sel <- outdata[, "CONC"] <= BQLcutoff
  outdata[sel, "CONC"] <- NA
  return(outdata)
}
```

4.A.2   S-PLUS Code for BQL Handling

```
# Define function to count missing values (BQL)
count.BQL <- function(x)
{
  sel <- is.na(x)
  len <- length(x[sel])
  return(len)
}


# Function to assign a value to BQL concentrations
# 0.0 is assigned to time points where
# all measurements are BQL
BQL.Assign <- function(tdata, value)
{
# Create matrix containing all unique combinations
#   of GROUP, SEX, DAY and HOUR
# BQLtot column contains count of BQL values for
#   each combination, calculated using count.BQL()
  categories <- list(tdata[, "HOUR"], tdata[, "DAY"],
    tdata[, "SEX"], tdata[, "GROUP"])
  BQLdata <- cbind(
    GROUP = as.vector(tapply(tdata[, "GROUP"], categories,
      unique)),
    SEX = as.vector(tapply(tdata[, "SEX"], categories,
      unique)),
    DAY = as.vector(tapply(tdata[, "DAY"], categories,
      unique)),
    HOUR = as.vector(tapply(tdata[, "HOUR"], categories,
```

```
        unique)),
      BQLtot = as.vector(tapply(tdata[, "CONC"], categories,
        count.BQL)))
# Combine BQL count data with full data set by
# matching GROUP, SEX, DAY and HOUR
  rdata <- merge(tdata, BQLdata,
    by = c("GROUP", "SEX", "DAY", "HOUR"))
# Assign value to all BQL concentrations
  sel <- (is.na(rdata[, "CONC"]))
  rdata[sel, "CONC"] <- value
# Assign value of 0.0 to concentrations where
# all measurements were BQL (BQLtot = 4)
  sel <- (rdata[, "BQLtot"] == 4)
  rdata[sel, "CONC"] <- 0
  return(rdata)
}
```

4.A.3   Function for Calculating Partial AUCs

```
Calc.AUC <- function(tdata)
{
# Add column to contain calculated
# partial AUC values, YPAUC
  tdata <- cbind(tdata, YPAUC = 0)
# Generate sorted list of available sampling times
  hours <- sort(unique(tdata[, "HOUR"]))
# Select data from very first sampling time
# and calculate partial AUC by multiplying
# the concentration by the difference between
# first and first and second sampling times
  sel <- tdata[, "HOUR"] == hours[1]
  tdata[sel, "YPAUC"] <- 0.5 *
    (hours[2] - hours[1]) * tdata[sel, "CONC"]
# Step through intermediate sampling times
# calculating partial AUC, by multiplying the
# concentration by the difference between
# subsequent and previous sampling times
  for(ii in 2:(length(hours) - 1)) {
    sel <- tdata[, "HOUR"] == hours[ii]
    tdata[sel, "YPAUC"] <- 0.5 *
      (hours[ii + 1] - hours[ii - 1]) * tdata[sel, "CONC"]
  }
# Select data from very last sampling time and
# calculate partial AUC, by multiplying the
```

```
# concentration by the difference between
# last and next-to-last sampling times
  sel <- tdata[, "HOUR"] == hours[length(hours)]
  tdata[sel, "YPAUC"] <- 0.5 *
    (hours[length(hours)] - hours[length(hours) - 1]) *
      tdata[sel, "CONC"]
  return(tdata)
}
```

4.A.4   S-PLUS Code for Calculating Means and Standard Errors

```
### Function for Calculating Session Means and SEs ###
Session.Means <- function(tdata)
{
# Create matrix containing all combinations of
# GROUP, SEX, DAY and SESSION
  categories <- list(tdata[, "SESSION"], tdata[, "DAY"],
    tdata[, "SEX"], tdata[, "GROUP"])
  sessiondata <- cbind(
    GROUP = as.vector(tapply(tdata[, "GROUP"], categories,
      unique)),
    SEX = as.vector(tapply(tdata[, "SEX"], categories,
      unique)),
    DAY = as.vector(tapply(tdata[, "DAY"], categories,
      unique)),
    SESSION = as.vector(tapply(tdata[, "SESSION"],
      categories, unique)),
    MEAN = 0, SE = 0, N = 0)
# Step through all combinations and
# calculate statistics
  for(ii in 1:nrow(sessiondata)) {
# Select all data for this combination
    sel <- (tdata[, "GROUP"] == sessiondata[ii, "GROUP"]) &
      (tdata[, "SEX"] == sessiondata[ii, "SEX"]) &
      (tdata[, "DAY"] == sessiondata[ii, "DAY"]) &
      (tdata[, "SESSION"] == sessiondata[ii, "SESSION"])
# N is the number of unique animal IDs
    sessiondata[ii, "N"] <- N <-
      length(unique(tdata[sel, "ANIMAL"]))
# Array to hold the sum of partial AUCs
# for each animal
    temp <- tapply(tdata[sel, "YPAUC"],
      tdata[sel, "ANIMAL"], sum)
# The session mean is the mean of the sum of
```

```
# partial AUCs for each animal
    sessiondata[ii, "MEAN"] <- mean(temp)
# Calculate standard error of session mean
    sessiondata[ii, "SE"] <- sqrt(var(temp)/N)
  }
  return(sessiondata)
}


# Function to sum standard errors
sum.SE <- function(x)
{
  SE <- sqrt(sum(x^2))
  return(SE)
}


#### Calculating group means and SEs for all ####
#### combinations of dose group, gender and day ####
Group.Means <- function(sessiondata)
{
# Create matrix containing all unique combinations
# of GROUP, DAY and SEX
# MEAN column is the group mean,
# calculated by summing the session means
# SE is the SE of the group mean,
# calculated using sum.SE()
  categories <- list(sessiondata[, "SEX"],
    sessiondata[, "DAY"], sessiondata[, "GROUP"])
  tabl <- cbind(
    DoseGroup = as.vector(tapply(sessiondata[, "GROUP"],
      categories, unique)),
    Gender = as.vector(tapply(sessiondata[, "SEX"],
      categories, unique)),
    DAY = as.vector(tapply(sessiondata[, "DAY"],
      categories, unique)),
    MEAN = as.vector(tapply(sessiondata[, "MEAN"],
      categories, sum)),
    SE = as.vector(tapply(sessiondata[, "SE"], categories,
      sum.SE)))
# Round MEAN and SE values to two decimal places for output
  tabl[, "MEAN"] <- round(tabl[, "MEAN"], digits = 2)
  tabl[, "SE"] <- round(tabl[, "SE"], digits = 2)
  return(tabl)
}


# Function to calculate degrees of freedom
```

```
calc.DF <- function(SE,N)
{
  DF <- ((sum((SE)^2))^2)/sum((SE)^4/(N - 1))
  return(DF)
}
```

4.A.5   Two-Sided T-Test for Change in AUC from Day 0 to Day 360

```
Day.Test <- function(tdata)
{
# Construct matrix for all combinations of
# GROUP, SEX and SESSION to hold difference
# data for Day 0 to Day 360
  categories <- list(tdata[, "SESSION"], tdata[, "SEX"],
    tdata[, "GROUP"])
  daydata <- cbind(
    GROUP = as.vector(tapply(tdata[, "GROUP"], categories,
      unique)),
    SEX = as.vector(tapply(tdata[, "SEX"], categories,
      unique)),
    SESSION = as.vector(tapply(tdata[, "SESSION"],
      categories, unique)),
    DIFF = 0, SE = 0, N = 0)
# Step through combinations of GROUP, SEX and DAY
  for(ii in 1:nrow(daydata)) {
# Select all data for Day 0 for this combination
    sel0 <- (tdata[, "GROUP"] == daydata[ii, "GROUP"]) &
      (tdata[, "SEX"] == daydata[ii, "SEX"]) &
      (tdata[, "SESSION"] == daydata[ii, "SESSION"]) &
      (tdata[, "DAY"] == 0)
# Select all data for Day 360 for this combination
    sel360 <- (tdata[, "GROUP"] == daydata[ii, "GROUP"]) &
      (tdata[, "SEX"] == daydata[ii, "SEX"]) &
      (tdata[, "SESSION"] == daydata[ii, "SESSION"]) &
      (tdata[, "DAY"] == 360)
# Array of differences from Day 0 to Day 360
    diff <- tdata[sel360, "YPAUC"] - tdata[sel0, "YPAUC"]
# Append differences to selected data
    diffdata <- cbind(tdata[sel0,  ], DIFF = diff)
# N is the number of unique animals in this combination
    daydata[ii, "N"] <- N <-
      length(unique(diffdata[, "ANIMAL"]))
# Array of differences for each animal
```

```
    temp <- tapply(diffdata[, "DIFF"],
      diffdata[, "ANIMAL"], sum)
# Calculate session mean of differences
    daydata[ii, "DIFF"] <- sum(temp)/N
# Calculate SE of session mean
    daydata[ii, "SE"] <- sqrt(var(temp)/N)
  }
# Construct matrix to hold statistics of Day 0
# to Day 360 differences by GROUP and SEX
# The Group Mean is calculated as
# the sum of session means,
# SE of the group mean is calculated using sum.SE()
  categories <- list(daydata[, "SEX"], daydata[, "GROUP"])
  tabl <- cbind(
    DoseGroup = as.vector(tapply(daydata[, "GROUP"],
      categories, unique)),
    Sex = as.vector(tapply(daydata[, "SEX"], categories,
      unique)),
    Difference = as.vector(tapply(daydata[, "DIFF"],
      categories, sum)),
    SE = as.vector(tapply(daydata[, "SE"], categories,
      sum.SE)),
    DF = 0, Pvalue = 0)
# Step through combinations and calculate
# Degrees of Freedom
  for(ii in 1:nrow(tabl)) {
    sel <- (daydata[, "GROUP"] == tabl[ii, "DoseGroup"]) &
      (daydata[, "SEX"] == tabl[ii, "Sex"])
    tabl[ii, "DF"] <- calc.DF(daydata[sel, "SE"],
      daydata[sel, "N"])
  }
# Calculate p-values for two-sided t-test
  tabl[, "PValue"] <- 2 * (1 -
    pt(abs(tabl[, "Difference"]/tabl[, "SE"]),
      tabl[, "DF"]))
# Round data for output
  tabl[, "Difference"] <- round(tabl[, "Difference"],
    digits = 2)
  tabl[, "SE"] <- round(tabl[, "SE"], digits = 2)
  tabl[, "DF"] <- round(tabl[, "DF"], digits = 2)
  tabl[, "PValue"] <- round(tabl[, "PValue"], digits = 3)
  return(tabl)
}
```

4.A.6   Two-Sample T-test for Difference in AUC
Between Males and Females

```
# Construct table to contain statistics of
# differences between males and females,
# by GROUP and DAY
  categories <- list(sessiondata[, "DAY"],
    sessiondata[, "GROUP"])
  tabl <- cbind(
    DoseGroup = as.vector(tapply(sessiondata[, "GROUP"],
      categories, unique)),
    Day = as.vector(tapply(sessiondata[, "DAY"],
      categories, unique)),
    Difference = 0, SE = 0, DF = 0, Pvalue = 0)
# Step through combinations of GROUP and DAY
  for(ii in 1:nrow(tabl)) {
# Select all data for males,
# for these GROUP and DAY values
    selm <-
      (sessiondata[, "GROUP"] == tabl[ii, "DoseGroup"]) &
      (sessiondata[, "DAY"] == tabl[ii, "Day"]) &
      (sessiondata[, "SEX"] == 1)
# Select data for females,
# for these GROUP and DAY values
    self <-
      (sessiondata[, "GROUP"] == tabl[ii, "DoseGroup"]) &
      (sessiondata[, "DAY"] == tabl[ii, "Day"]) &
      (sessiondata[, "SEX"] == 2)
# Sum difference in session mean values
#    to produce group mean difference
    tabl[ii, "Difference"] <- sum(
      sessiondata[selm, "MEAN"] -
      sessiondata[self, "MEAN"])
# Calculate SE of group mean from
#    SEs of session means
    tabl[ii, "SE"] <- sqrt(
      sum((sessiondata[selm, "SE"])^2) +
      sum((sessiondata[self, "SE"])^2))
# calculate N, degrees of freedom and
#    SE for males alone
    NM <- length(sessiondata[selm, "SE"])
    DFM <- calc.DF(sessiondata[selm, "SE"], NM)
    SEM <- sum.SE(sessiondata[selm, "SE"])
# calculate N, degrees of freedom and SE
```

```
#    for females alone
    NF <- length(sessiondata[self, "SE"])
    DFF <- calc.DF(sessiondata[self, "SE"], NF)
    SEF <- sum.SE(sessiondata[self, "SE"])
# Numerator of equation for degrees of freedom
    DFNUM <- (SEM^2 + SEF^2)^2
# Denominator of equation for degrees of freedom
    DFDEN <- ((SEM)^4/DFM) + ((SEF)^4/DFF)
# Calculate degrees of freedom
    tabl[ii, "DF"] <- DFNUM/DFDEN
# Calculate T-statistic
    TSTAT <- tabl[ii, "Difference"]/tabl[ii, "SE"]
# Calculate p-value for two-sided t-test
    tabl[ii, "PValue"] <- 2 * (1 -
      pt(abs(TSTAT), tabl[ii, "DF"]))
  }
# Round values for output
  tabl[, "Difference"] <- round(tabl[, "Difference"],
    digits = 2)
  tabl[, "SE"] <- round(tabl[, "SE"], digits = 2)
  tabl[, "DF"] <- round(tabl[, "DF"], digits = 2)
  tabl[, "PValue"] <- round(tabl[, "PValue"], digits = 3)
  return(tabl)
}
```

4.A.7   F-Test for Dose Proportionality in AUC

```
Dose.Test <- function(sessiondata)
{
# Construct matrix to hold statistics of
# dose groups, by SEX and DAY
  categories <- list(sessiondata[, "DAY"],
    sessiondata[, "SEX"])
  tabl <- cbind(
    Gender = as.vector(tapply(sessiondata[, "SEX"],
      categories, unique)),
    Day = as.vector(tapply(sessiondata[, "DAY"],
      categories, unique)),
    "Mean 1mg" = 0, "adjMean 1mg" = 0, "Mean 3mg" = 0,
    "adjMean 3mg" = 0, "Mean 10mg" = 0, "adjMean 10mg" = 0,
    "ML Diff" = 0, "ML Ratio" = 0, "HL Diff" = 0,
    "HL Ratio" = 0, DF = 0, "F-Ratio" = 0, Pvalue = 0)
# Construct matrix to hold temp statistics
# for each DoseGroup
```

```
  groups <- unique(sessiondata[, "GROUP"])
  dosedata <- cbind(DOSE = groups, MEAN = 0, ADJMEAN = 0,
    SE = 0, ADJSE = 0, DF = 0)
# Step through combinations of SEX and DAY
  for(ii in 1:nrow(tabl)) {
# Step through DoseGroups
    for(kk in 1:length(groups)) {
      dose <- dosedata[kk, "DOSE"]
# Select all data for this SEX, DAY and dose group
      sel <- (sessiondata[, "SEX"] == tabl[ii, "Gender"]) &
        (sessiondata[, "DAY"] == tabl[ii, "Day"]) &
        (sessiondata[, "GROUP"] == dose)
# Calculate group mean by summing session means
      dosedata[kk, "MEAN"] <- sum(sessiondata[sel, "MEAN"])
# Calculate adjusted mean by dividing by dose
      dosedata[kk, "ADJMEAN"] <- dosedata[kk, "MEAN"]/dose
# Calculate SE from SEs of session means
      dosedata[kk, "SE"] <- sum.SE(sessiondata[sel, "SE"])
# Divide by dose to produce adjusted SE
      dosedata[kk, "ADJSE"] <- dosedata[kk, "SE"]/dose
# Calculate degrees of freedom from SEs of session means
      N <- length(sessiondata[sel, "SE"])
      dosedata[kk, "DF"] <- calc.DF(sessiondata[sel, "SE"],
        N)
    }
# Copy means and adjusted means into table
    tabl[ii, "Mean 1mg"] <- dosedata[1, "MEAN"]
    tabl[ii, "adjMean 1mg"] <- dosedata[1, "ADJMEAN"]
    tabl[ii, "Mean 3mg"] <- dosedata[2, "MEAN"]
    tabl[ii, "adjMean 3mg"] <- dosedata[2, "ADJMEAN"]
    tabl[ii, "Mean 10mg"] <- dosedata[3, "MEAN"]
    tabl[ii, "adjMean 10mg"] <- dosedata[3, "ADJMEAN"]
# Calculate differences and ratios
#   from mid-to-low and high-to-low dose groups
    tabl[ii, "ML Diff"] <- dosedata[2, "ADJMEAN"] -
      dosedata[1, "ADJMEAN"]
    tabl[ii, "ML Ratio"] <- dosedata[2, "ADJMEAN"]/
      dosedata[1, "ADJMEAN"]
    tabl[ii, "HL Diff"] <- dosedata[3, "ADJMEAN"] -
      dosedata[1, "ADJMEAN"]
    tabl[ii, "HL Ratio"] <- dosedata[3, "ADJMEAN"]/
      dosedata[1, "ADJMEAN"]
# Calculate grand mean of dose groups
    GRAND.MEAN <- mean(dosedata[, "ADJMEAN"])
# Calculate sum of squares associated
```

```
#    with treatment and the
#    corresponding degrees of freedom
     SSTRT <- sum(((dosedata[, "ADJMEAN"] - GRAND.MEAN)^2) *
       (dosedata[, "DF"] + 1))
     dfTRT <- length(groups) - 1
# Calculate sum of squares associated with
#    standard error and the
#    corresponding degrees of freedom
     SSERR <- sum(((dosedata[, "ADJSE"])^2) *
       (dosedata[, "DF"]) * (dosedata[, "DF"] + 1))
     tabl[ii, "DF"] <- dfERR <- sum(dosedata[, "DF"])
# Calculate F-ratio
     tabl[ii, "F-Ratio"] <- FRatio <-
       (SSTRT/dfTRT)/(SSERR/dfERR)
# Calculate p-value for F-test
     tabl[ii, "PValue"] <- PVALUE <-
       (1 - pf(FRatio, dfTRT, dfERR))
   }
# Round values for output
   tabl[, "Mean 1mg"] <- round(tabl[, "Mean 1mg"],
     digits = 2)
   tabl[, "adjMean 1mg"] <- round(tabl[, "adjMean 1mg"],
     digits = 2)
   tabl[, "Mean 3mg"] <- round(tabl[, "Mean 3mg"],
     digits = 2)
   tabl[, "adjMean 3mg"] <- round(tabl[, "adjMean 3mg"],
     digits = 2)
   tabl[, "Mean 10mg"] <- round(tabl[, "Mean 10mg"],
     digits = 2)
   tabl[, "adjMean 10mg"] <- round(tabl[, "adjMean 10mg"],
     digits = 2)
   tabl[, "ML Diff"] <- round(tabl[, "ML Diff"], digits = 2)
   tabl[, "ML Ratio"] <- round(tabl[, "ML Ratio"],
     digits = 2)
   tabl[, "HL Diff"] <- round(tabl[, "HL Diff"], digits = 2)
   tabl[, "HL Ratio"] <- round(tabl[, "HL Ratio"],
     digits = 2)
   tabl[, "DF"] <- round(tabl[, "DF"], digits = 2)
   tabl[, "F-Ratio"] <- round(tabl[, "F-Ratio"], digits = 2)
   tabl[, "PValue"] <- round(tabl[, "PValue"], digits = 3)
   return(tabl)
}
```

4.A.8   S-PLUS Code for Figure 4.2

```
tabl <- Group.Means(sessiondata)

# Set X and Y Axis Limits
xlm <- c(0, 12)
ylm <- c(0, max(tabl[ ,"MEAN"]) + max(tabl[ ,"SE"]))
# Plot error bars for Means an SEs
error.bar(tabl[ ,"DoseGroup"], y = tabl[ ,"MEAN"],
  tabl[ ,"SE"], tabl[,"SE"], xlim = xlm, ylim = ylm,
  xlab = "Dose (mg/kg)", ylab = "AUC (ng/mL·h)")

# Draw lines to join error bars by Day and Gender
for(ii in unique(tabl[ ,"Gender"]))  {
  for(jj in unique(tabl[ ,"DAY"]))  {
    sel <- (tabl[ ,"Gender"] == ii) & (tabl[ ,"DAY"] == jj)
    lines(tabl[sel, "DoseGroup"], tabl[sel, "MEAN"],
      lwd = 2)
  }
}

# Add Legends
legend(3, 4000, "Males (Days 0 and 360)")
legend(5.5, 1000, "Females (Days 0 and 360)")
```

Part 4:

Phase I Studies

5

Analysis of Pharmacokinetic Data

Ha Nguyen
Merck Research Laboratories, Rahway, NJ, USA

Dhammika Amaratunga
The R.W. Johnson Pharmaceutical Research Institute, Raritan, NJ, USA

5.1 Introduction

Pharmacokinetics is the study of the time course of a drug administered to a biological organism, in particular, a human. A drug, following administration, is processed by the body in a series of stages: absorption, distribution, metabolism and elimination; these stages are often abbreviated by their initials, ADME. Since the nature, intensity, and duration of a drug's biologic effects depend on the amount of drug available to the body and, more specifically, to the amount of drug available at the target site in the body, pharmacokinetic studies play an integral role in a drug's development.

Pharmacokinetic (henceforth abbreviated to PK for convenience) studies involve profiling the concentration, $C(t)$, of drug present in the plasma (or other relevant biological fluid or tissue), over time, t. Two typical $C(t)$ versus t profiles are shown in Figure 5.1. Different formulations and different routes of administration of the same drug will exhibit different PK profiles. Findings from PK studies, interpreted in conjunction with efficacy, toxicity, and formulation considerations, serve to identify the most effective formulation and route of administration for a drug.

The remainder of this chapter is laid out as follows. Section 5.2 outlines how PK data are collected. Section 5.3 describes simple exploratory graphical techniques useful for a rapid examination of PK profile data. PK profiles may be analyzed using a model-independent approach by summarizing each subject's profile via one or more measures, the most common being the AUC, the area under the PK profile curve; this approach is described in Section 5.4. Where possible, the relationship between $C(t)$ versus t may be formally modeled using an appropriate nonlinear model; this approach to analyzing PK data is described in Section 5.5. The use of S-PLUS graphics in "population pharmacokinetics" (PPK for short), in which an attempt is made to relate features of the PK profiles to the subjects' demographic characteristics, is also outlined in the latter two sections. Some concluding remarks are made in Section 5.6. User-written S-PLUS functions mentioned in the text are collected in the Appendix of this chapter.

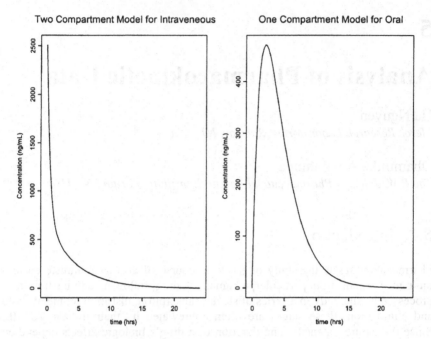

Figure 5.1. Examples of PK profiles for an intravenous and oral formulation.

5.2 Source of Data

We consider a single-dose PK study, i.e., one in which each of several (say n) subjects are administered a single dose of drug. Then the concentration y_{ij} of drug present in a blood sample is measured at several times t_{ij}, where $i = 1, ..., n$ indexes the subjects and $j = 1, ..., m_i$ indexes the times at which the blood samples were drawn for the ith project. We could write $y_{ij} = C(t_{ij}) + \varepsilon_{ij}$, where ε_{ij} reflects measurement error.

The PK data given in Appendix 5.A.6 are from a study involving a certain test drug code-named "M2000" (the data are slightly modified from the original). In this study, each of 12 healthy volunteers received 100 mg of drug by intravenous infusion. Plasma levels of M2000, in ng/ml, were measured at 14 time points: immediately after and 10, 15, 30, 60, and 90 min and 2, 4, 6, 8, 10, 12, 16, and 24 h after administration of the drug. The PK data were input into an S-PLUS data frame, m2000, and demographic data (i.e., the subjects' gender, age, height, and weight) were input into a separate S-PLUS data frame, m2000.demog; the first few records of these two data frames are shown below:

```
> m2000[1:4, ]
  subject time conc
1     271 0.00 2950
```

```
2         271 0.17 2164
3         271 0.25 1884
4         271 0.50 1366


> m2000.demog[1:4,]
  subject gender age.yr ht.cm wt.kg
1     271      F   69.0   157  63.7
2     272      M   66.8   165  66.5
3     273      F   67.7   163  57.1
4     274      M   73.7   167  81.3
```

Two S-PLUS library datasets, Quinidine (136 subjects measured at various time points) and Theoph (12 subjects measured at each of 11 time points), are from PK studies (type ?Quinidine and ?Theoph at the S-PLUS prompt for details of these datasets); the first two records of these datasets are

```
> Quinidine[1:2,]
Grouped Data: conc ~ time | Subject
  Subject time conc dose interval Age Height Weight
1       1    0    0  NA  249       NA  60     69    106
2       2    1    3  NA  249       NA  60     69    106


          Race Smoke Ethanol    Heart Creatinine glyco
1 Caucasian      no current Moderate       >= 50  0.41
2 Caucasian      no current Moderate       >= 50  0.41


> Theoph[1:2,]
Grouped Data: conc ~ Time | Subject
  Subject   Wt Dose Time conc
1       1 79.6 4.02 0.00 0.74
2       1 79.6 4.02 0.25 2.84
```

While in this chapter we shall use exclusively the m2000 data for illustration, the reader wishing to try out the S-PLUS commands given here can also invoke them with at most minor changes with these built-in datasets.

5.3   Exploratory Graphics

A natural first step when exploring PK data is to graph each subject's PK profile. S-PLUS's powerful graphics capabilities offers several ways to do this. One way is to plot each subject's PK profile in different panels having identical axes (Figure 5.2) using S-PLUS's Trellis Graphics function xyplot:

112   H. Nguyen and D. Amaratunga

```
> xyplot(conc ~ time | subject, data = m2000,
    xlab = "time (hrs)", ylab = "Concentration (ng/mL)",
    type = "o")
```

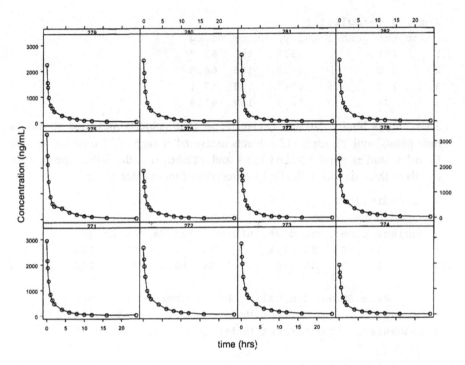

Figure 5.2. Concentration versus time plot for the individual subjects.

Another useful display is a plot of all the subjects' PK profiles overlaid on a single graph (Figure 5.3); this can be accomplished, again using xyplot, with the groups and panel.superpose features of Trellis:

```
> xyplot(conc ~ time, data = m2000, groups = subject,
    panel = panel.superpose, type = "l", col = 1:6,
    lty = 1, xlab = "time (hrs)",
    ylab = "Concentration (ng/mL)",
    scales = list(y = list(las = 0)))
```

These plots should be scrutinized for features such as:

1. What is the general pattern?
2. Are there any unusual individual data points?
3. Are any subjects substantially different from the others?

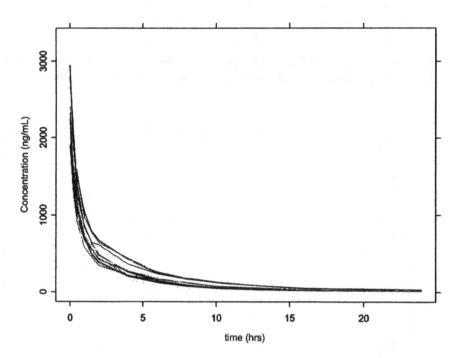

Figure **5.3.** Concentration versus time plot for the individual subjects (overlaid).

In this example we observe an immediate high concentration of the drug in the blood, then a rapid decline in concentration over the first two hours, followed by a more gradual decline, until at 24 hours the drug has been almost completely eliminated from the body; this type of profile is typical of an intravenous formulation following two-compartment model kinetics (Welling, 1986). The variability appears to decrease along with concentration level. No unusual individual data points or subjects are immediately apparent.

If the measurements for all the subjects were taken at the same times, as they are in this example, a summary profile may be graphed by calculating a suitable location estimate, such as a mean or median, for the concentrations at each time point, and then plotting these location estimates over time (Figure 5.4); medians are preferable to means as they are not influenced by a few outliers:

```
> m2000.medians <- tapply(m2000$conc, m2000$time, median)

> round(m2000.medians, 0)
   0 0.17 0.25  0.5    1  1.5    2    4    6    8   10   12   16   24
2406 1888 1616 1154  740  558  449  262  158  106   66   51   28   14

> m2000.time <- unique(m2000$time)
```

```
> xyplot(m2000.medians ~ m2000.time, type = "l",
    xlab = "time (hrs)",
    ylab = "Median concentration (ng/mL)",
    scales = list(y = list(las = 0)))
```

Note that summary profiles tend to appear smoother than individual profiles and could be misleading if used for selecting a model for individual PK profiles.

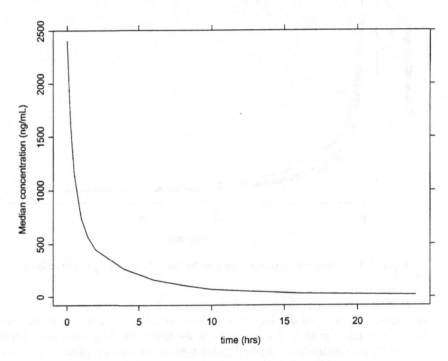

Figure **5.4.** Median concentration versus time plot.

It is helpful to add variability information to this plot. To do this, we first calculate the median absolute deviations about the median (MAD), as a resistant measure of variability, at each time point:

```
> m2000.mad <- tapply(m2000$conc, m2000$time, mad)

> round(m2000.mad, 0)
   0 0.17 0.25 0.5    1 1.5    2    4    6    8 10 12 16 24
 482  384  288 242 164 107 108 90 56 42 27 20 14  5
```

The MAD decreases with time, i.e., with concentration. Now we construct the plot (Figure 5.5), using S-PLUS's standard plotting routine (rather than Trellis):

```
> plot(m2000.time, m2000.medians,
      ylim = c(min(m2000.medians - m2000.mad),
        max(m2000.medians + m2000.mad)), type = "b",
      xlab = "time (hrs)",
      ylab = "Median Concentration (ng/mL)")

> error.bar(m2000.time, m2000.medians, m2000.mad, add = T,
      gap = F)
```

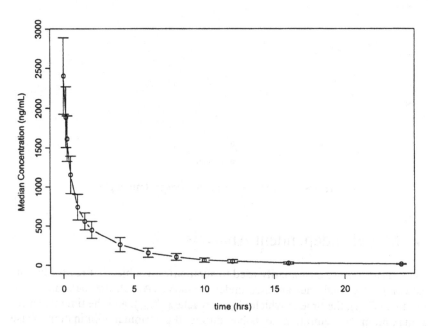

Figure 5.5. Median concentration versus time plot (with MAD bars).

Another graph useful in this regard is a spread-versus-level plot, in which the MAD is plotted against the median, with both on a log scale (Figure 5.6):

```
> plot(log(m2000.medians), log(m2000.mad),
      xlab="log(Median)", ylab="log(MAD)")
```

Figure 5.6 indicates that log(MAD) is roughly linearly related to log(median); i.e., that it is reasonable to assume that the error variability is proportional to a power of the concentration, i.e., that $V(\varepsilon_{ij}) \propto C(t_{ij})^{\xi}$ for some power ξ.

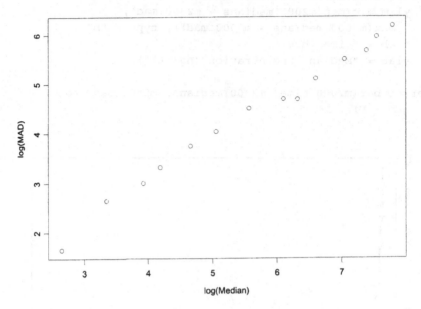

Figure 5.6.   MAD versus median concentration plot.

5.4   Model-Independent Analysis

Several measures are customarily used to summarize a subject's PK profile: the most commonly used ones are area under the curve (AUC), the maximum concentration (C_{max}), the time at which C_{max} is reached (T_{max}), and the time taken for the maximum concentration to halve during the terminal elimination phase ($T_{1/2}$). We describe below how each of these measures can be calculated using S-PLUS.

The most important of these measures is the AUC, the area under the PK profile curve, which reflects the total exposure of an individual subject to the drug. The AUC may be determined for the period under study ($AUC_{(0-T)}$) or, by extrapolation with or without formal modeling, for all time ($AUC_{(0-\infty)}$). If most of the drug has been eliminated by the last measurement, as in the m2000 data, these two values should be approximately equal.

In PK studies, $AUC_{(0-T)}$ is often calculated using the trapezoidal rule (Welling, 1986). The area, A_i, between t_{i-1} and t_i is calculated by linearly interpolating between t_{i-1} and t_i. Then the value of $AUC_{(0-T)}$ is obtained by adding up the $\{A_i\}$: $AUC_{(0-T)} = \Sigma A_i$. A simple function, f.auc, can be written to perform the necessary computations (see Appendix 5.A.1). For the m2000 data we have

```
> round(unlist(by(m2000, m2000$subject, f.auc)), 0)
  271   272   273   274   275   276   277   278   279   280   281   282
 4126  5675  5784  3169  5525  3193  3283  5161  3803  4086  2659  3558
```

The Lagrange method improves upon the trapezoidal rule by using cubic polynomial interpolation between t_{i-2} and t_{i+1} to calculate A_i (quadratic interpolation is used at the ends; Yeh and Kwan (1978) provide details). Employing a higher degree polynomial for interpolation allows the inherent curvature of a PK profile to be better tracked with only a modest cost in complexity. An S-PLUS function, f.lagrange, that implements the Lagrange method is given in Appendix 5.A.2. For the m2000 data, it gives for $AUC_{(0-T)}$:

```
> round(unlist(by(m2000, m2000$subject, f.lagrange)), 0)
  271   272   273   274   275   276   277   278   279   280   281   282
 3955  5489  5587  3041  5338  3053  3160  5050  3643  3952  2533  3422
```

The two sets of $AUC_{(0-T)}$'s for this study are in reasonable agreement. Figure 5.7 is the dotplot of the $AUC_{(0-T)}$'s, computed using the trapezoidal rule and the Lagrange method (see Appendix 5.A.3 for code to produce this plot).

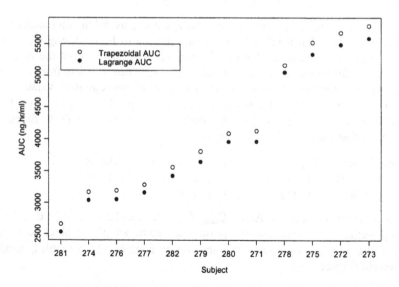

Figure 5.7. Dotplot of the $AUC_{(0-T)}$'s, computed using the trapezoidal rule and the Lagrange method.

A subject's $AUC_{(0-\infty)}$ can be calculated by adding $AUC_{(T-\infty)}$, obtained by extrapolating the data in the terminal elimination phase of the subject's PK profile, to $AUC_{(0-T)}$. Both the f.auc and f.lagrange functions, when called with the tail.auc=T option set, will calculate $AUC_{(0-\infty)}$'s:

```
> round(unlist(by(m2000, m2000$subject, f.auc, tail.auc = T)), 0)
   271   272   273   274   275   276   277   278   279   280   281   282
  4276  5970  6189  3305  5791  3351  3430  5711  4030  4244  2701  3730

> round(unlist(by(m2000, m2000$subject, f.lagrange,
    tail.auc = T)), 0)
   271   272   273   274   275   276   277   278   279   280   281   282
  4040  5644  5767  3100  5506  3122  3228  5292  3755  4047  2561  3495
```

Since C_{max} is just the maximum concentration in a subject's PK profile, it is easily determined using the S-PLUS function max:

```
> tapply(m2000$conc, m2000$subject, max)
   271   272   273   274   275   276   277   278   279   280   281   282
  2950  2661  2802  1944  3303  1882  1901  2316  2254  2409  2620  2404
```

The time, T_{max}, at which a subject reaches C_{max} can be calculated by writing a short S-PLUS function, f.tmax (see Appendix 5.A.4):

```
> unlist(by(m2000, m2000$subject, f.tmax))
 271 272 273 274 275 276 277 278 279 280 281 282
   0   0   0   0   0   0   0   0   0   0   0   0
```

T_{max} is obviously not informative in this study, but it is useful in other studies.

The time, $T_{1/2}$, that it takes for the concentration to halve during the exponential terminal elimination phase is determined as follows. First the data are examined to define the terminal elimination phase. Then, for data from this period only, a straight line is fitted to the logarithm of concentration versus time using least squares. If b_e is the resultant estimated slope, then $T_{1/2} = -\log(2)/b_e$ is an estimate of the subject's $T_{1/2}$. Appendix 5.A.5 gives an S-PLUS function, f.thalf, that performs this calculation:

```
> round(unlist(by(m2000, m2000$subject, f.thalf)), 1)
 271 272 273 274 275 276 277 278 279 280 281 282
 3.8 4.2 4.3 3.7 4.5   4 3.8 4.7 4.2   4 3.5 3.8
```

These various measures, AUC, C_{max}, T_{max}, etc. can be summarized and examined further. For example, let us now use some simple graphics to check whether the $AUC_{(0-T)}$'s are related in any way to demographic variables, gender and weight (Figure 5.8):

```
> m2000.auc.lagr <- unlist(by(m2000, m2000$subject,
    f.lagrange))
> par(mfrow = c(2, 2))
> # Divide AUC by 1000 so y-axis tick labels use less space
> boxplot(split(m2000.auc.lagr/1000, m2000.demog$gender),
    xlab = "Gender", ylab = "AUC (ug.hr/mL)")
```

```
> plot(m2000.demog$wt.kg, m2000.auc.lagr/1000,
    xlab = "Weight", ylab = "AUC (ug.hr/mL)")
> boxplot(split(m2000.demog$wt.kg, m2000.demog$gender),
    xlab = "Gender", ylab = "Weight")
> par(mfrow = c(1, 1))
```

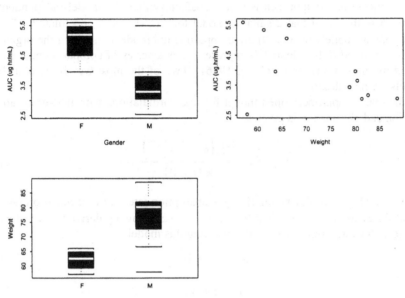

Figure 5.8. Plots of AUC versus gender, AUC versus weight, and weight versus gender.

Figure 5.8(a) indicates that females are exposed to more drugs than males (except for one male outlier). Figures 5.8(b) and 5.8(c) seem to offer an explanation in that it appears that heavier subjects tend to absorb less drugs, and males tend to be heavier (with two exceptions, one of whom corresponds to the outlier in Figure 5.8(a).

5.5 Model-Based Analysis

In the model-based analysis approach, it is assumed that the relationship between $C(t)$ and t can be represented by a mathematical model involving several (say p) undetermined parameters, denoted by the p-vector, θ. When we wish to show the model's dependence on θ, we may write $C(t)$ as $C(t, \theta)$ and $y_{ij} = C(t_{ij}) + \varepsilon_{ij}$ as $y_{ij} = C(t_{ij}, \theta_i) + \varepsilon_{ij}$. Observe that, although the general form, $C(t, \theta)$, of the functional relationship, is assumed to be the same for all subjects, the parameter, θ_i, that specifies the individual profile for the ith subject, is permitted to vary from subject to subject. This is a means of incorporating a common kinetic model for all subjects, and reflects the structure seen in the PK profile plots (Figures 5.2 and 5.3), where all the subjects have PK profiles of similar shape, but the precise form of the PK profiles varies from subject to subject. The parameters, θ_i, may be regarded as random variates with population mean θ and covariance matrix Λ.

The model-based approach is more satisfactory than the model-independent approach in the event that a suitable model for $C(t, \theta)$ can be postulated. The most popular models for $C(t, \theta)$ are compartmental models, in which the organism is presumed to be adequately represented as a series of compartments connected by linear transfers (Welling, 1986). Two of the most widely used such models are now described.

The one-compartment open model for oral formulations with first-order absorption and elimination is given by:

$$C(t, \theta) = \frac{\alpha\beta\left(e^{-\beta t} - e^{-\alpha t}\right)}{\kappa(\alpha - \beta)} \tag{5.1}$$

In this model, $\theta = (\alpha, \beta, \kappa)$ comprises the absorption rate, α, the elimination rate, β, and the clearance, κ ($\alpha > 0$, $\beta > 0$, $\kappa > 0$). The following derived parameters are useful for interpreting the results of fitting this model:

$$AUC_{(0-\infty)} = 1/\kappa \tag{5.2}$$

$$T_{1/2} = \ln(2)/\beta \tag{5.3}$$

$$T_{max} = \ln(\alpha/\beta)/(\alpha - \beta) \tag{5.4}$$

$$C_{max} = (\beta/\kappa)\exp(-\beta T_{max}) \tag{5.5}$$

The biexponential model for intravenously administered formulations following two-compartment model kinetics with complete absorption is given by

$$C(t, \theta) = Ae^{-\alpha t} + Be^{-\beta t} \tag{5.6}$$

In this model, $\theta = (A, B, \alpha, \beta)$ and $A > 0$, $B > 0$, $\alpha > \beta > 0$. The following derived parameters are useful for interpreting the results of fitting this model:

$$AUC_{(0-\infty)} = (A/\alpha) + (B/\beta) \tag{5.7}$$

$$k_{21} = [\alpha\beta(\alpha + \beta)]/(A\beta + B\alpha) \tag{5.8}$$

$$k_{el} = (\alpha\beta)/k_{21} \tag{5.9}$$

$$T_{1/2} = \ln(2)/\beta \tag{5.10}$$

where k_{21} denotes the transfer rate between the two compartments and k_{el} denotes the elimination rate.

These models may be parameterized differently, sometimes for enhanced interpretability, for imposing constraints, and/or for greater numerical stability in fitting. For example, writing the biexponential model as

$$C(t, \theta) = Ae^{-\exp(\alpha')t} + Be^{-\exp(\beta')t} \tag{5.11}$$

forces the estimates of $\alpha = \exp(\alpha')$ and $\beta = \exp(\beta')$ to be positive as required. The m2000 dataset, being from an intravenous formulation, was modeled using the biexponential model.

The measurement errors, ε_{ij}, are assumed to have zero mean and to be mutually independent. Their variance may be assumed either to be the same for all observations (i.e., $V(\varepsilon_{ij}) = \sigma^2$) or to be functionally related to the value, $C(t_{ij}, \theta_i)$, being measured (e.g., $V(\varepsilon_{ij}) = \sigma^2 C(t_{ij}, \theta_i)^\xi$), the appropriate choice being dependent on the nature of the measurement process. The random variates in the model, ε_{ij} and/or θ_i, are assumed to follow a Gaussian distribution where necessary (e.g., to apply likelihood theory).

Several ways of fitting such models, which fall under the category of nonlinear mixed effects models, have been proposed. The fitting procedures can be classified as two-stage methods and linearization methods. A comprehensive review of the topic is well beyond the scope of this chapter; instead the interested reader is referred to Davidian and Giltinan (1993, 1995) and Pinheiro and Bates (2000).

The remainder of Section 5.5 is devoted to outlining a few ways of fitting such models using the nlme3 library in S-PLUS. We begin by noting that S-PLUS greatly simplifies the fitting of the above two models by setting them up as Self Starting Functions, SSfol and SSbiexp respectively, thereby taking care of issues such as providing initial values for the iteration, parameterizing the model optimally, and other technical details.

5.5.1 Two-Stage Methods

As the name implies, this class of methods fits the model to the data in two stages: at the first stage, subject-specific estimates $\hat{?}_i$ of θ_i, are determined; at the second stage, these individual estimates $\hat{?}_i$ are combined to form an overall estimate θ.

It is simplest if homoscedasticity ($V(\varepsilon_{ij}) = \sigma^2$) can be assumed. Then, at Stage 1, the nonlinear model is fitted to each subject using ordinary least squares (Bates and Watts, 1988). This can be done either by using nls for each subject or, as we have done here for fitting the biexponential model to each of the m2000 PK profiles, by using nlsList, which automatically runs nls for all the subjects.

With the nlme3 library in S-PLUS, the m2000 data frame object must be converted into a groupedData object before the functions in the library can be used (Pinheiro and Bates, 2000). A groupedData object contains the data values stored as a data frame, a formula designating the response variable, the primary covariate and a grouping factor (for m2000, these are conc, time, and subject, respectively), and other optional features such as axis labels for plots.

```
> m2000.gr <- groupedData(conc ~ time | subject,
    data = m2000,
    labels = list(x = "Time", y = "Concentration"),
    units = list(x = "(hrs)", y = "(ng/mL)"))
```

Now we can invoke nlsList with SSbiexp, the self-starting function for the biexponential model.

```
> m2000.nlslist <- nlsList(conc ~
    SSbiexp(time, A1, lrc1, A2, lrc2), data = m2000.gr)
```

In this example, there was sufficient data for each subject and the data was of sufficiently good quality that convergence was achieved for all subjects. This may not always be the case, however (a potential drawback of the two-stage procedure). In this example, we collect the subject-specific estimates $\hat{\theta}_i$ of θ_i from Stage 1 in a dataset, m2000.nlslist.param:

```
m2000.nlslist.param <- coef(m2000.nlslist)
```

Then, at Stage 2, $\hat{\theta}$ is taken to be the average of the $\hat{\theta}_i$ as in Steimer et al. (1984):

```
> apply(m2000.nlslist.param, 2, mean)
       A1       lrc1        A2        lrc2
 1781.718 0.7752041 684.9784 -1.520266
```

This can also be done directly via the S-PLUS function fixed.effects:

```
> fixed.effects(m2000.nlslist)
       A1       lrc1        A2        lrc2
 1781.718 0.7752041 684.9784 -1.520266
```

Note that the parameters lrc1 and lrc2 of the function SSbiexp refer to α' and β'; hence, to report the estimates of α and β, the last two values have to be exponentiated:

```
> m2000.nlslist.fixed <- fixed.effects(m2000.nlslist)
> c(m2000.nlslist.fixed[c(1, 3)],
    exp(m2000.nlslist.fixed[c(2, 4)]))[c(1, 3, 2, 4)]
       A1       lrc1        A2        lrc2
 1781.718 2.171035 684.9784 0.2186538
```

In general, it is more realistic to assume that the measurement error variance is proportional to the quantity being measured, e.g., by assuming $V(\varepsilon_{ij}) = \sigma^2 C(t_{ij}, \theta_i)^\xi$. It is then more appropriate to apply generalized least squares for fitting the model, borrowing strength across subjects to improve efficiency (Davidian and Giltinan (1993, 1995) describe a fitting procedure).

5.5.2   Linearization Methods

These methods involve linearizing the model, $y_{ij} = C(t_{ij}, \theta_i) + \varepsilon_{ij}$, about an appropriate point, θ_0, via a Taylor series expansion: $y_{ij} = C(t_{ij}, \theta_0) + C'(t_{ij}, \theta_0)(\theta_i - \theta_0) + \varepsilon_{ij}$, and then applying maximum likelihood-based linear mixed effects model techniques. The process begins at a specified θ_0 which is then iteratively updated until convergence. This class of methods was introduced by Sheiner et at. (1972) and Beal and Sheiner (1982). A refinement of their approach, suggested by Lindstrom and Bates (1990), in which linearization is about the conditional mode of $\{\theta_i\}$, is implemented in S-PLUS as the function nlme (Pinheiro, 1998; Pinheiro and Bates, 2000).

Since we have an nlsList fit available, we can run nlme as follows:

```
> m2000.nlme <- nlme(m2000.nlslist)
```

The maximum likelihood estimates of the fixed effects are:

```
> fixed.effects(m2000.nlme)
       A1       lrc1       A2       lrc2
 1784.413 0.7735786 682.0468 -1.518759
```

Diagnostic displays can be used to identify potential problems with the model fitted. One such display is a plot of the standardized residuals (residuals divided by the estimated within-group standard deviation) against fitted values as shown in Figure 5.9.

```
> plot(m2000.nlme)
```

Both Figures 5.6 and 5.9 indicate the presence of heteroscedasticity of the form $V(\varepsilon_{ij}) = \sigma^2 C(t_{ij}, \theta_i)^\xi$, so we may improve inference with a fit that takes this into account.

```
> m2000.nlme.hetero <- update(m2000.nlme,
    weights = varPower())
```

Figure 5.10 shows the diagnostic plot for this model:

```
> plot(m2000.nlme.hetero)
```

The fixed effects estimates for this model are slightly different:

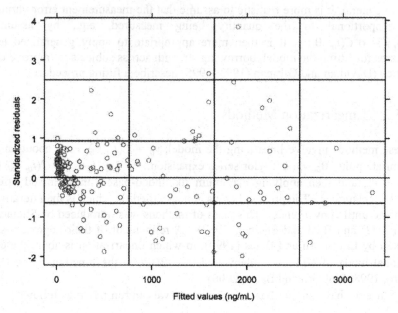

Figure 5.9. Diagnostic plot for homoscedastic model.

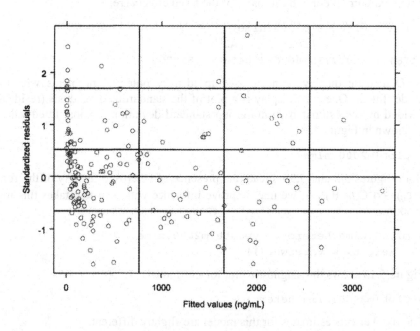

Figure 5.10. Diagnostic plot for heteroscedastic model.

```
> fixed.effects(m2000.nlme.hetero)
      A1        lrc1        A2        lrc2
 1801.224 0.7536314 661.7569 -1.550913
```

The output from summary(m2000.nlme.hetero) indicates that the variance function parameter, ξ, is estimated to be 0.26.

We can use penalized likelihood criteria, such as the Akaike's Information Criterion (AIC) or the Bayesian Information Criterion (BIC) to choose between competing models: the output from summary(m2000.nlme) indicates that AIC is 1767 and BIC is 1814 for the homoscedastic model while the output from summary(m2000.nlme.hetero) indicates that AIC is 1708 and BIC is 1758 for the heteroscedastic model; since the latter model has smaller values for AIC and BIC, it is the preferred model for this example.

It would be interesting to see how closely the PK profiles predicted by this model track the observed PK profiles. Figure 5.11 shows the observed data that we plotted in Figure 5.2, along with the individual predicted fit (solid line) and the predicted population fit (dotted line) overlaid. The resulting profiles are so close that it is hard to differentiate the three!

```
> plot(augPred(m2000.nlme.hetero, level=0:1),
    layout=c(3, 4), lty = c(1, 2))
```

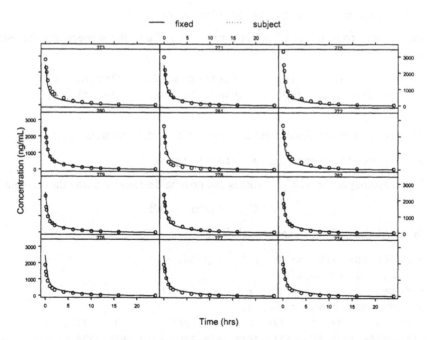

Figure 5.11. Concentration versus time for the individual subjects (with the individual and population fits overlaid).

For the m2000 data there are a sufficient number of observations for each subject, and the data are of sufficiently good quality that nlme converges. When nlme does not converge, you may have to tinker with the convergence criteria, the starting values, the number of iterations, the parameterization, or even the model specification to coax it to converge. Here is an example of how to approach the latter in the context of modeling PK data: suppose that historical data indicated that one of the PK parameters, say the absorption rate, varied only slightly across the subjects; in this case, one may try to rerun nlme treating the absorption rate constant as a fixed (although unknown) value; the resulting model, being simpler, has a better chance of convergence.

5.6   Examining the Effects of Covariates on PK Profiles

Either the individual parameters or, preferably, interpretable functions of the parameters, such as AUC and clearance, can be examined to check whether they are being affected by covariates, particularly demographic variables such as gender and weight.

For the m2000 data, the subject-specific parameter estimates can be obtained by first calculating the individual random effects:

```
> b <- random.effects(m2000.nlme.hetero)
```

then adding back the fixed effect estimates to obtain the subject-specific estimates:

```
> b <- b + matrix(rep(fixed.effects(m2000.nlme.hetero),
    length(unique(m2000$subject))), ncol = length(b),
    byrow = T)
```

They can also be obtained using the coefficients function:

```
> b <- coefficients(m2000.nlme.hetero)
```

The subject-specific $AUC_{(0-\infty)}$ values can now be determined using the formula

$$AUC_{(0-\infty)} = (A/\alpha) + (B/\beta) \tag{5.12}$$

for an $AUC_{(0-\infty)}$ for a biexponential model:

```
> m2000.auc.nlme <- b[ , 1]/exp(b[, 2]) +
    b[, 3]/exp(b[, 4])
> names(m2000.auc.nlme) <- row.names(b)
> round(m2000.auc.nlme, 0)
  276  277  274  279  278  282  280  281  272  273  271  275
 3072 3158 3055 3653 4940 3449 3934 2569 5500 5614 3996 5368
```

These values are in reasonable agreement with those obtained via the trapezoidal and Lagrange methods, especially the latter.

As in Section 5.4, we shall now investigate whether these subject-specific $AUC_{(0-\infty)}$'s obtained from the heteroscedastic model fit are related to the demographic data using graphical techniques. Having observed there that $AUC_{(0-\infty)}$ was related to both gender and weight, here we shall study them simultaneously in a single graph by using gender-specific plotting symbols in a plot of $AUC_{(0-\infty)}$ against weight (Figure 5.12):

```
> plot(m2000.demog$wt.kg, m2000.auc.nlme, type = "n",
    xlab = "Weight", ylab = "Model AUC")
> text(m2000.demog$wt.kg, m2000.auc.nlme,
    labels = as.character(m2000.demog$gender))
```

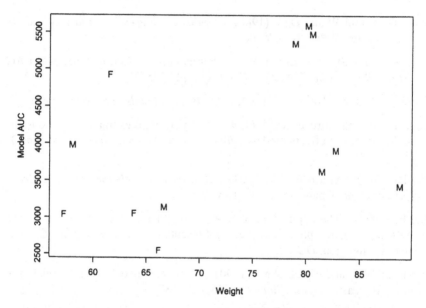

Figure 5.12. Area under the curve versus weight for the individual subjects.

We again observe that heavier subjects tend to absorb less drugs. Females tend to be lighter and therefore absorb greater quantities. One outlier becomes apparent: a man who, despite being relatively very light, does not absorb much drug.

5.7 Concluding Remarks

The interactive environment provided by S-PLUS is an excellent platform on which to study PK data. The availability of a powerful nonlinear mixed effects modeling function, nlme, along with standard PK models set up as self-starting functions, is an added benefit. Coupling the two (the interactive capability and the modeling capability) reduces some of the "black box" feel of mixed effects modeling. Our focus in this chapter was to describe how a reasonably standard PK analysis could be performed with S-PLUS. Ette (1998) takes this paradigm further and uses a battery of sophisticated S-PLUS tools for a more detailed analysis that is appropriate with larger datasets. The developers of S-PLUS, recognizing the effectiveness of their software for analyzing PK data, are, at the time of writing, developing a PK module that will make such analyses more accessible to even casual users of S-PLUS.

5.8   References

Bates, D.M. and Watts, D.G. (1988). *Nonlinear Regression Analysis and Its Applications.* Wiley, New York.

Beal, S.L. and Sheiner, L.B. (1982). Estimating population kinetics. *CRC Critical Reviews in Biomedical Engineering* **8**, 195–222.

Chambers, J.M. and Hastie T.J. (1992). *Statistical Models in S.* Wadsworth.

Davidian, M. and Giltinan, D.M. (1993). Some general estimation methods for nonlinear mixed effects models. *Journal of Biopharmaceutical Statistics* **3**, 23–55.

Davidian, M. and Giltinan, D.M. (1995). *Nonlinear Models for Repeated Measurement Data.* Chapman & Hall, New York.

Ette, E. (1998). The application of S-PLUS graphics, modeling and statistical tools in population pharmacokinetics. Presented at the *S-PLUS User Conference*, Washington, DC.

Lindstrom, M.J. and Bates, D.M. (1990). Nonlinear mixed effects models for repeated measures data. *Biometrics* **46**, 673–687.

Pinheiro, J.C. (1998). NLME: Software for mixed effects models. http://franz.stat.wisc.edu/pub/NLME.

Pinheiro, J.C. and Bates, D.M. (2000). *Mixed Effects Models in S and S-PLUS.* Springer-Verlag, New York.

Sheiner, L.B., Rosenberg, B., and Melmon, K.L. (1972). Modeling of individual pharmacokinetics data for computer aided drug dosing. *Computers and Biomedical Research* **5**, 441–459.

Steimer, J.L., Mallet, A., Golmard, J.L., and Boisvieux, J.F. (1984). Alternative approaches to estimation of population pharmacokinetic parameters: Comparison with the nonlinear mixed effects model. *Drug Metabolism Reviews* **15**, 265–292.

Venables, W.N. and Ripley, B.D. (1997). *Modern Applied Statistics with S-PLUS, 2nd Ed.* Springer–Verlag, New York.

Welling, P.G. (1986). *Pharmacokinetics: Processes and Mathematics.* American Chemical Society.

Yeh, K.C. and Kwan, K.C. (1978). A comparison of numerical integrating algorithms by trapezoidal, Lagrange, and spline approximation. *Journal of Pharmacokinetics and Biopharmaceutics* **6**, 79–89.

5.A. Appendix

5.A.1 Function to Calculate Area under the Curve Using the Trapezoid Rule

```
f.auc <- function(x, conc = "conc", time = "time",
  tail.auc = F, btail = n - 1)
{
###
### Calculates: AUC by the trapezoidal rule
###    for a single subject.
### Input: Data frame in which variable "time" has the
###    measurement times and variable "conc" has the
###    concentration measurements.
###
  n <- nrow(x)
  auc0 <- sum((x[-1, conc] + x[ - n, conc]) *
    (x[-1, time] - x[ - n, time]))/2
### Tail Area
  if(tail.auc) {
    if(missing(btail)) {
      nmax <- order(x[, conc])[n]
      nlast <- floor(n - (n - nmax)/3)
      if(nlast == n)
        btail <- n - 1
    }
    tmp <- lm(log(x[btail:n, conc]) ~
      x[btail:n, time])$coef
    auc1 <- exp(tmp[1] + (tmp[2] * x[n, time]))/( - tmp[2])
  }
  else auc1 <- 0
  return(as.numeric(auc0 + auc1))
}
```

5.A.2 Function to Calculate Area under the Curve Using the Lagrange Method

```
f.lagrange <- function(x, jconc = "conc", jtime = "time",
  tail.auc = F)
{
###
### Calculates: AUC by the Lagrange method.
### Input: Data frame in which variable "time" has the
###   measurement times and variable "conc" has the
###   concentration measurements.
###
  time <- x[!is.na(x[, jconc]), jtime]
  conc <- x[!is.na(x[, jconc]), jconc]
### Compute the mean concentration over time.
  conc <- tapply(conc, time, mean)
  time <- tapply(time, time, mean)
### Compute the AUC for each subinterval
### First interval
  auc1 <- f.subauc(c(1, 2, 3), time, conc)
### Second interval and more ...
  n <- length(conc)
  x <- cbind(1:(n - 3), 2:(n - 2), 3:(n - 1), 4:n)
  auc2 <- apply(x, 1, f.subauc, time, conc)
### Last interval
  auc3 <- f.subauc(c(n - 2, n - 1, n), time, conc)
### Tail Area
  auc4 <- 0
  if(tail.auc) {
    nmax <- order(conc)[n]
    nlast <- floor(n - (n - nmax)/3)
    if(nlast == n)
      nlast <- n - 1
    tmp <- lm(log(conc[nlast:n]) ~ time[nlast:n])$coef
    auc4 <- exp(tmp[1] + (tmp[2] * time[n]))/( - tmp[2])
  }
### Add them up
  auc <- auc1 + sum(auc2) + auc3 + auc4
  return(as.numeric(auc))
}
```

```
f.subauc <- function(index, time, conc)
{
###
### Compute the AUC for intervals specified by index.
###
### First or last interval
  if(length(index) == 3) {
    tmp <- solve(cbind(rep(1, length(index)), time[index],
      time[index]^2), conc[index])
    if(index[1] == 1) {
      loindex <- index[1]
      hiindex <- index[2]
    }
    else {
      loindex <- index[2]
      hiindex <- index[3]
    }
    auc <- tmp[1] * (time[hiindex] - time[loindex]) +
      tmp[2]/2 * (time[hiindex]^2 - time[loindex]^2) +
      tmp[3]/3 * (time[hiindex]^3 - time[loindex]^3)
    return(auc)
  }
  else if(length(index) == 4) {
    tmp <- solve(cbind(rep(1, length(index)), time[index],
      time[index]^2, time[index]^3), conc[index])
    loindex <- index[2]
    hiindex <- index[3]
    auc <- tmp[1] * (time[hiindex] - time[loindex]) +
      tmp[2]/2 * (time[hiindex]^2 - time[loindex]^2) +
      tmp[3]/3 * (time[hiindex]^3 - time[loindex]^3) +
      tmp[4]/4 * (time[hiindex]^4 - time[loindex]^4)
    return(auc)
  }
}
```

5.A.3   S-PLUS Code to Produce Figure 5.7

```
AUC.t <- unlist(by(m2000, m2000$subject, f.auc))
AUC.l <- unlist(by(m2000, m2000$subject, f.lagrange))
n <- length(AUC.t)
index <- order(AUC.t)
plot(1:n, sort(AUC.t), pch = 1, ylim = range(AUC.t, AUC.l),
  xlab = "Subject", ylab = "AUC (ng.hr/ml)", xaxt = "n")
axis(1, at = 1:n, labels = names(AUC.t)[index])
```

```
points(1:n, AUC.1[index], pch = 16)
legend(1, 5500, legend = c("Trapezoidal AUC",
  "Lagrange AUC"), marks = c(1, 16))
```

5.A.4 Function to Calculate T_{max}

```
f.tmax <- function(x)
{
###
### Calculates: TMAX for a single subject.
### Input: Data frame in which variable "time" has the
###    measurement times and variable "conc" has the
###    concentration measurements.
###
  x$time[x$conc == max(x$conc)]
}
```

5.A.5 Function to Calculate $T_{1/2}$

```
f.thalf <- function(x, jconc = "conc", jtime = "time",
  terminal.time)
{
###
### Calculates: THALF for a single subject.
### Input: Data frame in which variable "time" has the
###    measurement times and variable "conc" has the
###    concentration measurements.
###
  time <- x[!is.na(x[, jconc]), jtime]
  conc <- x[!is.na(x[, jconc]), jconc]
  if(missing(terminal.time)) {
    ttime <- unique(time)
    n <- length(ttime)
    start <- max(2, floor(n/3))
    terminal.time <- ttime[start]
  }
  conc <- conc[time > terminal.time]
  time <- time[time > terminal.time]
  half.life <-  - log(2)/lm(log(conc) ~ time)$coef[2]
  return(as.numeric(half.life))
}
```

5.A.6   Data

Table 5.1. Table of demographic data for m2000.

Subject	Gender	Age (yrs)	Height (cm)	Weight (kg)
271	F	69.0	157.0	63.7
272	M	66.8	165.0	66.5
273	F	67.7	163.0	57.1
274	M	73.7	167.0	81.3
275	F	68.0	149.0	61.4
276	M	73.6	189.0	88.6
277	M	67.8	169.0	82.6
278	F	73.2	165.0	66.0
279	M	66.5	178.0	80.4
280	M	65.6	173.5	80.0
281	M	66.8	170.0	57.9
282	M	73.5	174.0	78.8

Table 5.2. Table of concentration levels of M2000, in ng/ml, in plasma, measured at 14 time points: immediately after and 10, 15, 30, 60, and 90 min and 2, 4, 6, 8, 10, 12, 16, and 24 h after administration of the drug.

Sbj	0	0.17	0.25	0.5	1	1.5	2	4	6	8	10	12	16	24
271	2950	2164	1884	1366	844	625	478	256	147	88	58	48	26	13
272	2661	2200	1924	1525	962	770	642	433	256	176	119	92	46	24
273	2802	2285	2007	1493	1033	790	669	422	243	183	128	95	46	27
274	1944	1646	1473	1057	660	474	366	195	116	80	45	38	18	10
275	3303	2479	2132	1459	863	607	500	417	266	182	124	93	53	25
276	1882	1459	1244	891	574	452	341	218	150	97	62	46	22	12
277	1901	1665	1485	1079	700	492	386	207	123	75	50	37	20	11
278	2316	1857	1641	1246	856	637	611	361	235	166	122	94	51	32
279	2254	1550	1370	976	672	551	422	267	165	122	74	59	31	17
280	2409	1919	1607	1185	757	566	461	274	181	115	70	54	32	15
281	2620	1999	1625	1004	457	330	247	158	90	49	35	26	13	5
282	2404	1789	1561	1122	722	545	437	218	136	82	53	40	21	12

6

Graphical Presentation of Single Patient Results

Jürgen Bock

F. Hoffmann-La Roche AG, Basel, Switzerland

6.1 Introduction

This chapter illustrates the usefulness of graphical presentations of single patient results for small- or medium-sized clinical studies. Single patient results include single observations per patient, patient profiles, or derived values per patient. These graphs serve as exploratory tools. They allow us to check for outliers, the validity of distribution assumptions, the linearity of relationships, etc. It is not intended to give an introduction to graphical diagnostic tools. Only simple statistical models will be applied. Consequently, there will be no analysis of residuals for any model, only presentation of raw or transformed data.

The graphs were created using Trellis graphics, a powerful tool for generating multipanel plots. The S-PLUS functions included in the Trellis library let you display the same type of plot in separate panels for the subsets of the data frame split by factor levels. This is the graphical equivalent to a numerical subgroup analysis using the by function. It allows the investigation of the influence of factors on the observations. The graphics commands and the computations performed within the panels or subsets are defined by the panel functions.

For each general display function such as xyplot, dotplot, bwplot, etc., the corresponding default panel function panel.xyplot, panel.dotplot, panel.bwplot exists, but users may write their own panel functions. These usually include traditional graphics functions like points, lines, abline, but they are not restricted to graphics functions. Almost any analytical function that accepts a data frame as an argument can be included in panel function. This has the advantage that numerical and graphical analyses are run in the same session, i.e., they are always based on the same data frame. Computational results for the subsets together with the plots can then be displayed within the panels. This is demonstrated by an example on asthma where the effect of an allergen is measured by the area above a curve indicating the impairment of the lung function.

6.2   Point Plots

6.2.1   Dotplots–A Bioequivalence Study

Simple presentations of grouped observations can be created by dotplots or stripplots. They display the single observations of the subgroups as dots on parallel lines. In each panel one line is drawn for each level of the same factor. Generally, with Trellis graphics, the dependency on further factors is investigated by "conditioning" on their levels. The panels display the data of subgroups that are created by splitting the whole data frame by the levels of the other factors.

Even though dotplots are very basic they are extremely useful for getting a feel for spread, skewness, and outliers as well as the effects of the factors. They help to become familiar with the data. For larger studies it is better to use more sophisticated plots such as Box-Whiskers plots, quantile plots, or histograms to explore the distributional properties of the data. If the number of observations is small, these plots can however be misleading. For further details refer to the S-Plus manuals.

The following example of a bioequivalence study shows how to apply dotplots to explore treatment and period effects. Often bioequivalence studies are designed as cross-over studies where two formulations of a drug are administered to healthy volunteers in two sequential periods. In the first period the subjects are randomized to two groups getting one of the formulations. The treatments are then interchanged in the second period. Since the groups are identified by their treatment sequences, we call them sequence groups. To avoid carry-over effects, a sufficiently long washout period is placed between the two periods.

In this two-way cross-over study (Bührens and others (1991)) 12 subjects are treated with two formulations of Allopurinol, a reference formulation R, and a test formulation T. The objective is to demonstrate that the two formulations are bioequivalent, i.e., that the rate and extent of absorption, distribution, and elimination of the two drugs is similar.

The blood plasma levels of the drugs are assessed from blood samples taken according to a prespecified time schedule during the days with drug intake. The main kinetic parameter for the analysis is the area under the plasma level time curve (AUC). The data frame Bioequi contains the AUC-values for both treatment periods.

The first four records are

SUBJECT	AUC	period	treatment	lgAUC
1	3.881	1	T	0.5889
1	4.894	2	R	0.6897
2	4.835	1	T	0.6844
2	6.504	2	R	0.8132

Bioequivalence study: AUC

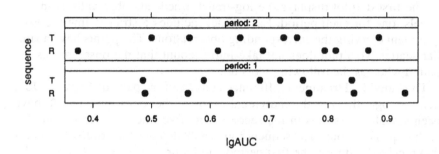

Bioequivalence study: Treatment differences

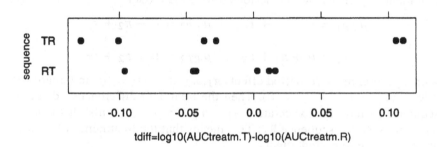

Bioequivalence study: Period differences period1 - period2

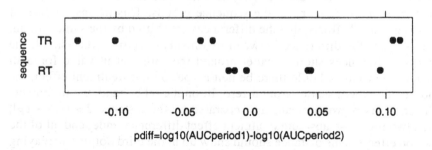

Figure 6.1. Dotplots for a bioequivalence study.

In bioequivalence trials, a logarithmic transformation is often applied to the AUC in order to change from the multiplicative models used in pharmacokinetics to linear models in the logarithmic scale.

In addition, a better approximation to the normal distribution is often achieved by the transformation. The column lgAUC provides the values log10(AUC). These values are used to generate the three dotplots arranged in tandem in Figure 6.1.

The first dotplot displays the log-transformed raw observations in two panels, one for each period. Each panel includes a straight line for each treatment carrying the corresponding observations. The points have a similar scatter. This plot does not take into account that the observations are paired whereas the following plots do.

The individual treatment differences tdiff of the pairs of log10(AUC), i.e., the logarithms of the individual ratios of the areas for T and R, have been visualized as dots in the second plot. Two lines are shown, one for each sequence group. The sequence group TR denotes the group of subjects that received T during the first period and R during the second period. For the group RT the sequence is reversed.

Let's assume that the population means of log10(AUC) indexed by the treatment sequences and the periods follow the model

$$\mu_{TR1} = \mu + \pi_1 + \tau_T \qquad \mu_{TR2} = \mu + \pi_2 + \tau_R$$

$$\mu_{RT1} = \mu + \pi_1 + \tau_R \qquad \mu_{RT2} = \mu + \pi_2 + \tau_T$$

with a general mean μ, period effects π_1, π_2 ($\pi_1 + \pi_2 = 0$), and treatment effects τ_T, τ_R ($\tau_T + \tau_R = 0$). Then the treatment differences tdiff in sequence TR have the expectation $\pi_1 - \pi_2 + \tau_T - \tau_R$, whilst this is $\pi_2 - \pi_1 + \tau_T - \tau_R$ for sequence RT. This means that the treatment differences are biased due to period effects.

The factor effect differences become better identifiable by generating the individual period differences pdiff of log10(AUC), i.e., the differences between period 1 and period 2. They have the expectations $\pi_1 - \pi_2 + \tau_T - \tau_R$ and $\pi_1 - \pi_2 + \tau_R - \tau_T$ in the sequence groups. This means, in case of existing period effects, that the differences are shifted by the same amount $\pi_1 - \pi_2$. If the difference between treatments is zero or very small, the period differences should scatter around the same mean value for both sequence groups. Should there be neither period nor treatment effect, the observations will scatter around zero. If, on the other hand, the treatments differ, the two group means are separated by the distance $2 * |\tau_T - \tau_R|$, i.e., two times the absolute treatment effect difference, independent of the period effects. This distance should show up in the third dotplot displaying the period differences. Although the points spread differently, the means in both sequence groups are close to zero. The variation is not as high as it may appear by looking at the plot. A log10 value of -0.1 corresponds to a ratio of 0.79, and a log10 value of 0.1 to a ratio of 1.26. Usually the interval [0.8, 1.25] is accepted as equivalence range for geometric means in case of mean bioequivalence. This is a very tight range for the individual ratios. The treatment effect difference is very small. The plots suggest that the formulations can be regarded as bioequivalent. Mean bioequivalence can be confirmed by the calculation of 90% confidence limits for the treatment mean ratio. Both limits are included in the interval [0.80, 1.25]. Here we are more interested in the plots.

The code for these plots can be found in the Appendix. The function `arrange.plots` is used to place several plot on one page. It can be found in the library.

6.2.2   Scatterplots–Looking at the Effect of IL12 in Asthmatics

Scatterplots of paired observations allow you to explore the relationship between variables. They are often used for checking distributional assumptions of statistical procedures as shown by the following example.

Example. In a small study (Bryan et.al. (2000)) 40 mild asthmatic patients have been treated with either a placebo or Interleukin12 (IL12). The primary objective of this study was to see whether IL12 reduces the impairment of the lung function that has been caused by an allergen. As a secondary objective it has been explored in a subset of 27 patients whether IL12 modifies the immune system reactions to allergens. Eosinophil granulocytes in sputum indicate the allergic status of asthmatic patients. Eosinophils are cells that can be colored with Eosin to make their counting easy. The cell composition of sputum varies greatly from sample to sample depending on the part of the lung it comes from. Therefore the Eosinophil counts have been related to the number of nonsquamous cells, i.e., percentages of Eosinophils are used for the analysis, but not the absolute counts. They have been assessed before the first treatment at week

1 and 1 day after the last treatment at week 4. Two patients have been discarded because of outlying observations. Twenty-five data records are provided in the data frame Eos that can be found in the S-PLUS library. The first four records are

PT	TRT	VIOLATOR	EOS1PCT	EOS4PCT
1	IL12		2.50	4.50
2	Placebo		3.20	1.50
4	IL12	PV	15.40	8.90
5	Placebo		49.67	39.00

where PT denotes the patient numbers, EOS1PCT, EOS4PCT the Eosinophil percentages at week 1 and 4, respectively, and TRT the treatment factor. The variable VIOLATOR indicates whether a patient was fully compliant with all procedures described in the protocol. If any criterion has been violated, e.g., the drug has not been taken as prescribed, the patient has dropped out of the study because of side effects or has missed an essential assessment or visit, then he is a protocol violator, denoted by "PV."

To investigate the dependency of the effect of IL12 on the baseline status and explore the sensitivity to protocol violations, scatterplots of EOS4PCT vs. EOS1PCT have been created for each treatment group. In the first two panels of the first plot the protocol violators are indicated by the string "PV," whereas in the last two panels the patients can be identified by their numbers. The plot is created by two calls of the display function xyplot. Each call generates a two-panel plot for the levels "IL12" and "Placebo" of the treatment factor TRT. The two plots have then been placed on one page using the function arrange.plots included in the library.

The detailed code can be found in the library. The panel functions are included in the Appendix. The function point(x,y) in the panel function plots the points (x=EOS1PCT, y=EOS1PCT) for the subsets of rows with TRT="IL12" in the left panel and the remaining points in the right panel. To add the values of a third column to the plot the same rows must be selected by using the subscripts argument of the panel function. The patient numbers PT, belonging to the same records as used for the panel, are printed at the positions of the displayed points by

```
> text(x, y, Eos$PT[subscripts], cex=0.6, adj=0.5)
```

The 45 degree lines through the origin are plotted by

```
> abline(c(0,1), lty=2, lwd=2)
```

to support the detection of changes in Eosinophil percentages. Regression lines have been fitted by the least squares method within each panel by

```
> abline(lm(y~x, na.action=na.omit), lwd=4)
```

The smoother loess is called by the panel function

```
> panel.loess(x, y, lwd=2)
```

A comparison between the smoothing curves generated by loess and the fitted regression lines indicates that a linear relationship may be assumed. The protocol violators can easily be identified by comparing the panels. The protocol violators are found within the bulk of the data. We can therefore conclude that the exclusion of protocol violators would not change the results substantially.

Usually variance homogeneity needs to be assumed for linear regression models. This assumption is often violated dealing with percentages. However in our case it is not clear whether the assumption has been violated since there are many factors which could have influenced the composition of the sputum samples and consequently the variability of the Eosinophil percentages. From Figure 6.2 one cannot see whether the variance is constant since the small values are clustered, and the number of larger values is too small to judge their variability.

For percentages, the log- or logit-transformation is often used to stabilize the variance. Since we have a bulk of smaller values, the log-transformation

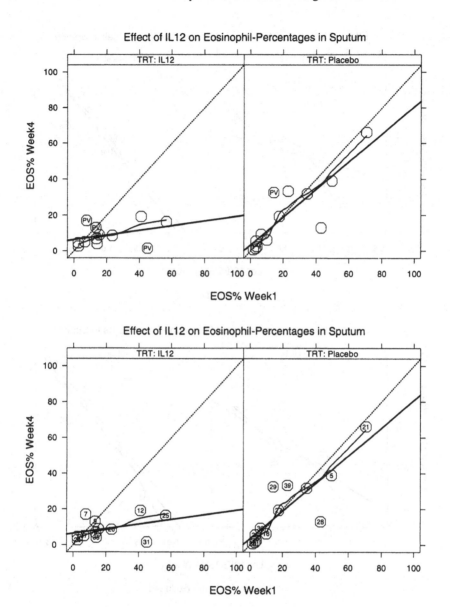

Figure 6.2. Scatterplot of sputum Eosinophils.

should widen this range. A second plot of the same type has been produced with the log-transformed percentages to explore their relationship (Figure 6.3). The upper panels show the scatterplots of all subjects, whilst the protocol violators have been excluded in the plot below. The protocol violators

142   J. Bock

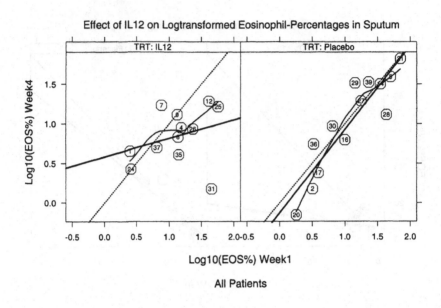

All Patients

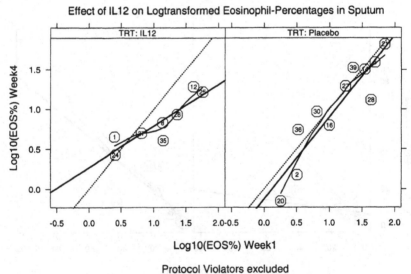

Protocol Violators excluded

Figure 6.3. Scatterplot of log-transformed sputum Eosinophils.

have gained more influence on the slope and variance estimate after applying the transformation. By excluding them we get a clearer picture of the
relationship for compliant patients.

The linearity assumption seems to hold more when the transformation is applied. The usual assumption of equal slopes for the analysis of covariance does not hold here. Therefore the treatment effect cannot be described by the distance between the lines. The value of the covariate must be chosen carefully to define a mean treatment effect, often set to the distance between the lines at the grand mean of the covariate. Since the treatment effect is not just additive, it may be better described by both parameters of the regression line instead of the distance at a fixed value of the covariate.

We must be aware of the fact that the number of observations in this example is small. Exploration of data helps to formulate reasonable hypotheses, but these must be proved in another study.

Assumption checking is not the only purpose of such presentations. Although the statistical analysis will preferably be based on the transformed data, it makes sense to provide the plot of the original values to the investigator. It is easier for him to judge the magnitude of the effects on the original scale. As the plots show, the treatment effect is non-additive. It depends on the baseline status. A larger effect of IL12 can be expected for more allergic patients. Nevertheless, the small number of severely allergic patients in this study leaves some doubts about the validity of this conclusion. The results must be confirmed in a sufficiently large study.

6.3   "Sheep Flock"–Displays

When profiles of patients are assessed during a study, individual profile curves can be displayed together in a single plot. When there are more than four to five curves it becomes difficult to discriminate between them. Nevertheless, it is often required to identify patients with profiles that differ essentially from others. Using patient as a factor in a Trellis graph makes it very easy to get individual displays. Looking at the whole group of panels, as you would do with a flock of sheep, lets you check whether the shape of the curves is similar and identify patients with strongly differing profiles (e.g., the black sheep!).

Example. The data frame Anti50 contains antibody assessments for patients treated with a 50 mg dose of a Tumor Necrosis Factor binding protein in a study in Multiple Sclerosis (Arnason et.al. (1999)). This treatment should change the immune status of patients. The patients are followed up over a maximum time period of 1 year. The first 4 records are

PT	DAY	ANTIBODY
103	1	0
103	29	60
103	57	20
103	85	100

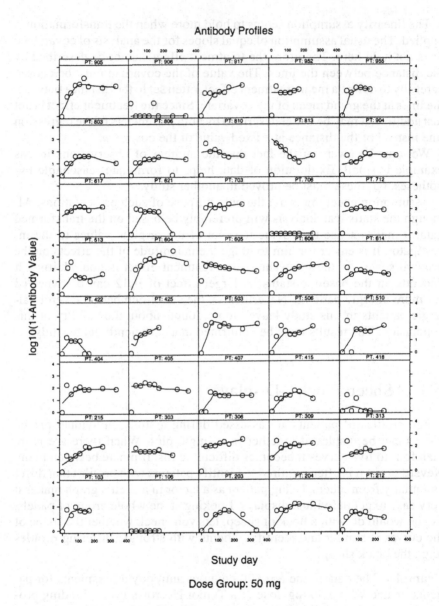

Figure 6.4. Antibody formation.

For the graphical display (Figure 6.4) the antibody values have been transformed using log $10(1+x)$. The 1 has been added to allow the inclusion of zero values.

The panel.loess function has been applied to make it easier to identify the shape of the profiles. For most of the patients the transformed antibody curves reached a plateau quickly. Surprisingly, some patients in the dose group have not developed antibodies.

6.4   Effect Curves and Areas

As demonstrated before, Trellis graphics is a powerful tool for exploring the data. Writing appropriate panel functions gives you much more power and flexibility than the default panel functions provide. Panel functions are not restricted to graphics commands. Here you can also include numerical functions that are applied to the same columns of the data frame and the same subgroup of data as plotted in the panel. If also the corresponding data from other columns of the data frame shall be displayed or used in the calculations then the subscripts argument needs to be included in the panel function. This lets you produce highly sophisticated plots, as will be illustrated by the following example:

Example. Allergen challenge studies in asthma patients are widely used to explore the dilatory and anti-inflammatory effects of potential asthmatic drugs. Before the first drug intake and at the end of the treatment period an allergen challenge is conducted. The patient's lungs are provoked by inhaling an allergen (e.g., grass pollen), the amount having been determined during screening. The effect on the lungs is observeded for several hours.

Thereby the lung function is measured by the FEV1 (i.e., the forced expiratory volume in 1 second). Usually a decline of FEV1 is seen during the first 2 hours after the challenge (early response). Most of the patients recover and their lung function returns to normal. In many patients the allergen triggers an inflammation leading to a new impairment of the lung function starting about 4 hours after the provocation (late response). The magnitude of the early and late response is often described by the area between the baseline FEV1 line parallel to the abscissa and the FEV1 time profile curve.

The data for this example come from the study mentioned in the example on Eosinophils in sputum (Bryan et al. (2000)). The first four records are

```
PT FEV1 FEV1BL       TIME              ASSMNT
 1 3.70   3.70  -0.13333333 WEEK 1. DAY -1.
 1 2.27   3.70   0.08333333 WEEK 1. DAY -1.
 1 2.27   3.70   0.16666667 WEEK 1. DAY -1.
 1 2.12   3.70   0.25000000 WEEK 1. DAY -1.
```

The data for two patients are provided by the data frame Fev1, where PT denotes the patient number. The next two columns contain the FEV1 values and the baseline FEV1, i.e., the value at time zero. The factor ASSMNT gives the assessment day, TIME denotes the time after drug intake.

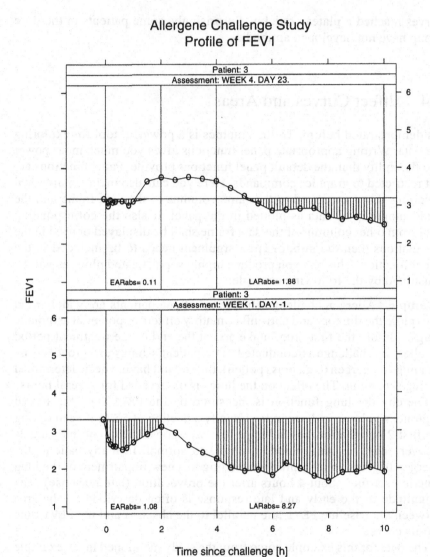

EARabs, LARabs: Area below baseline between 0-2h and 4-10h resp.

Figure 6.5. Early and late asthmatic response, patient 3.

The patients lung function has been followed up until 10 hours after inhaling the allergen. The early asthmatic response (EAR) is defined as the area calculated for the first 2 hours. The late asthmatic response (LAR) is measured by the area over the time interval 4 to 10 hours.

The FEV1 profile is plotted together with the baseline FEV1 value in Figure 6.5. The code to generate this plot can be found in the library of this book.

The areas describing the early and late response could be computed before applying a plotting procedure. If the dataset has been modified it is easy to forget updating the areas. To avoid that, we decided to calculate the areas from the actual values used for plotting. The panel function created for this purpose is attached in the Appendix. The statement

```
> abline(h=Fev1$FEV1BL[subscripts])
```

draws a horizontal line at the height given by the actual baseline FEV1 value. It is essential to use the subscripts argument here, because the baseline values differ from patient to patient, and the actual baseline value has to be selected. The panel function retrieves automatically the x, y coordinates of the points to be plotted. The baseline values are picked from a third column of the data frame, and the correct values have to be selected by using the same subscripts as for the records used for plotting. By the statements

```
> ivlar <- x>3.9 & x<10.2
> x.1 <- x[ivlar]
> y.1 <- y[ivlar]
> bl <- Fev1$FEV1BL[subscripts]
> bl.1 <- bl[ivlar]
```

the time points and the corresponding FEV1 values and baseline values for the late response time interval are selected. You can see that some FEV1 values exceed the baseline values, indicating an improvement of the lung function over the baseline status. Since only the area below the baseline values shall be computed, larger y-values have to be reset to baseline temporarily:

```
> y.1 <- ifelse(y.1>bl.1, bl.1, y.1)
> n.1 <- length(x.1)
```

gives the number of time points included, and

```
> abline(v=c(min(x.1), max(x.1)))
```

draws vertical lines displaying the actual time interval.

Then the area describing the late asthmatic response is calculated according to the trapezoid rule:

```
> lar <- round((max(x.1)-min(x.1))*
  mean(Fev1$FEV1BL[subscripts]) -
  sum((x.1[2:n.1]-x.1[1:(n.1-1)])*
  (y.1[1:(n.1-1)]+y.1[2:n.1])/2), 2)
```

and printed to the panel by

```
> text(6, 1, paste("LARabs=", lar), cex=0.7, col=3)
```

Finally, the polygon function hatches the area just computed. This allows checking whether the correct area has been calculated.

The same procedure is repeated for the early response. The page function is used to stamp the pages of the plot with the author's name and the date of creation

```
> page=function(n)
    text(0.9, -0.14, paste("Author", date()), cex=0.5)
```

Here "Author" can be replaced by the author's name.

By "conditioning" the patient numbers (i.e., defining patient as a factor) the individual profiles can be plotted in separate panels. The plot of patient number 3 has been included as an example. The lower panel shows the response to the allergen challenge before treatment with IL12. Both, the early and late asthmatic responses are strong, showing that the patient is highly sensitive to the allergen. The FEV1 profile after 4 weeks of treatment is presented in the upper panel. Both areas are reduced, indicating that the allergic status of the patient has changed essentially. Unfortunately, this favorable change could only be observed in a few patients. The variability of the effects was high. The hypothesis that IL12 would reduce the asthmatic response could not be demonstrated.

6.5   Summary

Small studies can only give meaningful results if the variability of the involved variates is sufficiently small compared to the size of the investigated effects. In this case it makes sense to display single subject results. With Trellis graphics one can easily produce multipanel plots for subsets of data belonging to the levels of factors. This allows to investigate the influence of factors and check the model assumptions. In addition, the freedom to write specific panel functions makes the system very flexible and allows to create highly sophisticated plots. For data exploration this is a big advantage for Trellis graphics compared to other software, where the subgroup plots have to be generated separately, before they can be placed into panels.

6.6   References

Arnason, B., Jacobs, G. et al. (1999). TNF neutralization in MS. Results of a randomized, placebo-controlled multicenter study. *Neurology* **53**, 457–465.

Bryan, S., O'Connor, B. et al. (2000). Effects of recombinant human interleukin-12 on eosinophils, airway hyperreactivity and late asthmatic response. *Lancet* **356**, 2149–53.

Bührens et al. (1991). Zum Nachweis der Bioäquivalenz von zwei Allopurinolpräparaten. *Arzneimittel-Forschung/Drug-Research* **41**(1), 250–253.

6.A Appendix

6.A.1   Dotplot Generation for the Bioequivalence Study

```
eqplt <- vector("list", 3)
eqplt[[3]] <- dotplot(treatment ~ lgAUC | period,
  Bioequi, strip = function(...)
strip.default(..., style = 1, strip.names = c(T, T)),
  main = "Bioequivalence study: AUC", layout = c(
  1, 2), ylab = "sequence")
beqT <- Bioequi[Bioequi$treatment == "T",  ]
beqR <- Bioequi[Bioequi$treatment == "R",  ]
lgqAUC <- log10(beqT$AUC/beqR$AUC)
beq1 <- Bioequi[Bioequi$period == "1",  ]
beq2 <- Bioequi[Bioequi$period == "2",  ]
lgrAUC <- log10(beq1$AUC/beq2$AUC)
sequence <- paste(as.character(beq1$treatment),
  as.character(beq2$treatment), sep = "")
eqplt[[2]] <- dotplot(sequence ~ lgqAUC, main =
  "Bioequivalence study: Treatment differences",
  xlab = "tdiff=log10(AUC.T)-log10(AUC.R)", ylab
  = "sequence")
eqplt[[1]] <- dotplot(sequence ~ lgrAUC, main =
  "Bioequivalence study: Period differences\nperiod1 - period2",
  xlab = "pdiff=log10(AUC.1)-log10(AUC.2)", ylab
  = "sequence")
trellis.par.set("strip.background", list(col = 0))
trellis.par.set("dot.symbol", list(cex = 1.2, col = 1,
  font = 1, pch = 16))
arrange.plots(eqplt, c(1, 3))
```

6.A.2 Panel Functions to Plot Sputum Eosinophils

```
panel = function(x, y, subscripts, ...)
{
  points(x, y, cex = 2, pch = 1)
  abline(c(0, 1), lty = 2, lwd = 2)
  panel.loess(x, y, lwd = 2)
  abline(lm(y ~ x, na.action = na.omit), lwd = 4)
  text(x, y, Eos$VIOLATOR[subscripts], cex =
    0.6, adj = 0.5)
}

panel = function(x, y, subscripts, ...)
{
  points(x, y, cex = 2, pch = 1)
  abline(c(0, 1), lty = 2, lwd = 2)
  panel.loess(x, y, lwd = 2)
  abline(lm(y ~ x, na.action = na.omit), lwd = 4)
  text(x, y, Eos$PT[subscripts], cex =
    0.6, adj = 0.5)
}
```

6.A.3 Panel Function to Plot the Early and
Late Asthmatic Response

```
panel = function(x, y, subscripts, ...)
{
  lines(x, y, type = "o")
  abline(h = Fev1$FEV1BL[subscripts])
  ivlar <- x > 3.9 & x <
    10.2
  x.1 <- x[ivlar]
  y.1 <- y[ivlar]
  bl <- Fev1$FEV1BL[subscripts]
  bl.1 <- bl[ivlar]
  y.1 <- ifelse(y.1 > bl.1, bl.1, y.1)
  n.1 <- length(x.1)
  abline(v = c(min(x.1), max(x.1)))
  lar <- round((max(x.1) - min(x.1)) * mean(Fev1$
    FEV1BL[subscripts]) - sum(((x.1[2:n.1] -
    x.1[1:(n.1 - 1)]) * (y.1[1:(n.1 - 1)] +
    y.1[2:n.1]))/2), 2)
  text(6, 1, paste("LARabs=", lar), cex =
    0.7, col = 3)
```

```
polygon(c(x.1, max(x.1), min(x.1)), c(y.1, rep(
  mean(Fev1$FEV1BL[subscripts]), 2)),
  density = 20, angle = 90, col = 3)
ivear <- x <= 2.1
x.e <- x[ivear]
x.e <- ifelse(x.e < 0, 0, x.e)
y.e <- y[ivear]
bl.e <- bl[ivear]
y.e <- ifelse(y.e > bl.e, bl.e, y.e)
n.e <- length(x.e)
abline(v = c(min(x.e), max(x.e)))
ear <- round((max(x.e) - min(x.e)) * mean(Fev1$
  FEV1BL[subscripts]) - sum(((x.e[2:n.e] -
  x.e[1:(n.e - 1)]) * (y.e[1:(n.e - 1)] +
  y.e[2:n.e]))/2), 2)
text(1, 1, paste("EARabs=", ear), cex =
  0.7, col = 2)
polygon(c(x.e, max(x.e), min(x.e)), c(y.e, rep(
  mean(Fev1$FEV1BL[subscripts]), 2)),
  density = 20, angle = 90, col = 2)
}
```

7

Graphical Insight and Data Analysis for the 2,2,2 Crossover Design

Bill Pikounis
Merck Research Laboratories, Rahway, NJ, USA

Thomas E. Bradstreet
Merck Research Laboratories, Blue Bell, PA, USA

Steven P. Millard
Probability, Statistics & Information, Seattle, WA, USA

7.1 Introduction

S-PLUS code is presented for the graphical insight into, and the statistical analysis of, a two-treatment, two-period, two-treatment-sequence, or 2,2,2 crossover design. In this introductory section, we describe the 2,2,2 crossover design and its uses in the pharmaceutical industry with emphasis on food interaction studies. We also introduce a specific example and a dataset which will be used pedagogically throughout the chapter. In Section 7.2, we provide a brief introduction to data management in S-PLUS demonstrating just enough manipulations to facilitate the graphical methods and data analyses which follow. Section 7.3 presents a series of graphs for the initial exploration and discovery stage of the analysis of the 2,2,2 crossover design. In Section 7.4, we perform the usual normal theory ANOVA and provide a clear and decision-oriented summary and inference plot. Section 7.5 presents several graphical tools for the "visualization of the ANOVA" and a subsequent model fit assessment, and we end with a summary in Section 7.6.

7.1.1 The 2,2,2 Crossover Design

The 2,2,2 crossover design has been a standard tool of medical researchers for decades. Although it has been used frequently in the pharmaceutical industry for initial studies investigating the safety and efficacy of new drugs, this design is probably best known for its use in evaluating the pharmacokinetics of a drug; most notably average bioequivalence, relative bioavailability, drug interaction, alcohol interaction, and food interaction.

In the simplest form of the 2,2,2 crossover design, half of the subjects are randomized to receive one of two treatments in the first treatment period, and the second of two treatments in the second treatment period. The other half of the subjects receive the two treatments in the reverse sequence. Between the two treatments in each treatment sequence, there is a "washout" period of adequate length to, in concept, prevent the effect of whichever treatment is given in the first treatment period from carrying over to affect the treatment given in the second treatment period.

The 2,2,2 crossover design allows for the evaluation of three effects: treatments, periods, and carryover, although carryover is confounded with both sequence (sometimes referred to as subject group) effects and treatment-by-period interaction. Whether or not to, and how to, evaluate carryover effects, and how to proceed if these are suspected, has been a controversial issue. For further details, see Jones and Kenward (1989) and Senn (1993).

7.1.2   Food Interaction Studies

In some cases, taking a drug with food increases the amount of the drug in the blood stream. This increase may be considered medically safe and provide additional therapeutic benefits to the patient. Or, the increase may be viewed as a potential toxicity problem, regardless of the potential benefit. In other cases, taking the drug with food decreases the amount of drug in the blood stream which may decrease the drug's effectiveness. In still other cases, taking the drug with food may not substantially affect the amount of drug which is delivered to the blood stream, and for all medically meaningful purposes, the effect of the drug is therapeutically equivalent when it is taken either with or without food. Therefore, it is important to evaluate the degree, if any, that a drug interacts with food.

The primary medical objective of a food interaction study is to "prove" the absence of a clinically meaningful food interaction. The clinical investigators may assume initially that there exists some degree of interaction, but it is their intention to show that its magnitude is not of clinical importance, possibly being nonexistent. To accomplish this, food interaction studies investigate the pharmacokinetic properties (ADME: absorption, distribution, metabolism, excretion) of a drug. These properties help to clarify when, where, and how the body processes and uses the drug. By measuring these properties, the bioavailability of the drug both in the presence and the absence of food is compared.

The bioavailability of a drug is often characterized by summarizing its plasma concentration versus time course in the blood with three measurements:

1. Area under the plasma concentration versus time curve (AUC), a measure of total absorption.
2. Maximum plasma concentration (C_{max}), a measure of the extent of absorption.
3. Time to maximum plasma concentration (T_{max}), a measure of the rate of absorption.

If the bioavailability of the drug is clinically similar when taken either with (+) or without (−) food as evaluated through all or a subset of AUC, C_{max}, and T_{max} as is appropriate for a particular drug, then by clinical extrapolation from the pharmacokinetic measurements to expected clinical responses, it should not matter therapeutically whether or not the drug is administered with food. Thus, in a food interaction study, it is important to estimate the relative (Fed versus Fasted) bioavailability of a drug administered both with and without food.

Relative bioavailability can be estimated following a paradigm similar to that for evaluating average bioequivalence. Confidence intervals for the true proportional differences in the mean Fed (+) and mean Fasted (−) values of targeted pharmacokinetic variables (e.g., AUC and C_{max}) can be calculated. Further details on average bioequivalence can be found in Chow and Liu (2000).

7.1.3 Our Example

In our example, eight healthy male subjects participated in a food interaction study to evaluate the magnitude of the food interaction of a new hypertensive therapy, Drug P, and its metabolite, Drug M. Four subjects (numbers 1, 2, 5, 7) were randomized to take Drug P with food in the first treatment period but without food in the second treatment period. The other four subjects (numbers 3, 4, 6, 8) took Drug P without food in the first treatment period but with food in the second treatment period. Four plasma concentration versus time curves were constructed for each of the eight subjects; one for Drug P and one for Drug M in each of the two treatment periods. Drug concentration values were assayed from plasma samples taken at 0, 10, 20, 30, 40, and 50 min, and 1, 1.25, 1.5, 1.75, 2, 3, 4, 5, 6, 8, 10, 12, 18, 24, 30, and 36 h after dosing. AUC was estimated from zero hours to 36 h using the trapezoidal rule, and then using extrapolation from 36 h to infinity incorporating the elimination rate constant. C_{max} and T_{max} were simply observed from the plasma concentration versus time curve. Four AUC, C_{max}, and T_{max} values were obtained for each of the eight subjects in the study, corresponding to the Fed and Fasted states for each of the parent drug and the metabolite. Due to space considerations, in Section 7.2 we present only a subset of these data to analyze, namely the AUC (ng × h/mL) data for the parent drug. The full data set can be retrieved from Part 2, Table 5, at the web site http://www.villanova.edu/~tshort/Bradstreet/ and it is printed in Bradstreet (1992).

7.2 Data Management

In this section, we provide a brief introduction to data management in S-PLUS for the graphical methods and data analysis which follow in Sections 7.3, 7.4, and 7.5. Specifically, we read the data into S-PLUS using a comma-separated text file, and create a data frame named food.df. This is a convenient way to begin the data management process for this type of data as a tabular form with cases as rows and variables as columns. The first row of the food.csv file printed out below will be a header of character strings that S-PLUS will use to label the variables in the resulting data frame:

```
subj,seq,AUC.Fed,AUC.Fasted
1,+/-,809.44,967.82
2,+/-,428,746.45
3,-/+,757.71,901.11
4,-/+,906.83,1146.96
5,+/-,712.24,678.16
6,-/+,561.77,745.51
7,+/-,511.84,568.98
8,-/+,756.6,852.86
```

The command

```
> food.df.orig <- read.table("food.csv", header = T,
    sep = ",", as.is = T)
```

produces an S-PLUS data frame named food.df.orig. The as.is=T argument ensures that character data variables do not get converted to factors here. We'll do this later. To help us later in the chapter with producing graphs, formal statistical analyses, and subsequent model checking, some further manipulations are performed:

```
> food.df <- data.frame(
    subj = rep(food.df.orig$subj, each = 2),
    seq = rep(food.df.orig$seq, each = 2),
    trt = factor(rep(c("Fed", "Fasted"), times = 8),
        levels = c("Fed", "Fasted")),
    per=c(1, 2, 1, 2, 2, 1, 2, 1, 1, 2, 2, 1, 1, 2, 2, 1),
    AUC = as.vector(
        t(food.df.orig[ , c("AUC.Fed", "AUC.Fasted")]))))
```

The new data frame food.df is created by binding columns of data vectors with the help of functions to replicate values (rep), and to transpose (t) and unwind (as.vector) a rectangular block of data in column order.

We also create some factor and numeric variables to impose a desired comparative order when plotting and tabulating the data. The first three statements below create nominal factors so that S-PLUS analysis functions do not interpret values like the subject numbers 1, 2, ..., 8 as having any ordered meaning. The last two statements create numeric codings for sequence (1 = "Fed/Fasted", 2 = "Fasted/Fed") and treatment (1 = "Fed", 2 = "Fasted").

```
> food.df$subjf <- factor(food.df$subj,
    levels = as.character(1:8))
> food.df$perf <- factor(food.df$per,
    levels = as.character(1:2))
```

```
> food.df$seqf <- factor(food.df$seq,
    levels=c("+/-", "-/+"))
> food.df$seqn <- as.numeric(food.df$seqf)
> food.df$trtn <- as.numeric(food.df$trt)
```

Lastly we sort the data frame by sequence number and get a listing by typing its name

```
> food.df <- food.df[order(food.df$seqn), ]
>food.df
```

	subj	seq	trt	per	AUC	subjf	perf	seqf	seqn	trtn
1	1	+/-	Fed	1	809.44	1	1	+/-	1	1
2	1	+/-	Fasted	2	967.82	1	2	+/-	1	2
3	2	+/-	Fed	1	428.00	2	1	+/-	1	1
4	2	+/-	Fasted	2	746.45	2	2	+/-	1	2
9	5	+/-	Fed	1	712.24	5	1	+/-	1	1
10	5	+/-	Fasted	2	678.16	5	2	+/-	1	2
13	7	+/-	Fed	1	511.84	7	1	+/-	1	1
14	7	+/-	Fasted	2	568.98	7	2	+/-	1	2
5	3	-/+	Fed	2	757.71	3	2	-/+	2	1
6	3	-/+	Fasted	1	901.11	3	1	-/+	2	2
7	4	-/+	Fed	2	906.83	4	2	-/+	2	1
8	4	-/+	Fasted	1	1146.96	4	1	-/+	2	2
11	6	-/+	Fed	2	561.77	6	2	-/+	2	1
12	6	-/+	Fasted	1	745.51	6	1	-/+	2	2
15	8	-/+	Fed	2	756.60	8	2	-/+	2	1
16	8	-/+	Fasted	1	852.86	8	1	-/+	2	2

7.3   Initial Exploration and Discovery

Section 7.3 presents graphics for the initial exploration and discovery stage of the data analysis of the 2,2,2 crossover design. Section 7.3.1 presents individual subject (sometimes called "spaghetti") plots ordered both by treatment (Fed, Fasted) within treatment sequence, and also by study period within treatment sequence. Section 7.3.2 presents a graphic useful for evaluating marginal treatment and variance effects. Section 7.3.3 presents a series of three graphs which provide an initial look at the sample bivariate relationships. Section 7.3.4 presents a graph which provides a preliminary look at the 2,2,2 crossover ANOVA.

7.3.1   Individual Subject Plots

Since there are only eight subjects, plotting individual subject profiles over the two study periods should be informative. When there are many subjects, the individual subject profiles may be less informative.

We first set up some parameters for the graphical displays. As opposed to the default style of enclosed box and tick-labels, we request that axis labels be more extreme than any data values for both the x-axis and y-axis, that the box layout in the plot area is open L-shaped, and that the plotting region is square:

```
> par(xaxs = "e", yaxs = "e", bty = "l", pty = "s")
```

The endpoint AUC (ng × h/mL) is analyzed in the log scale, so let us add the log-transformed endpoint to the data frame.

```
> food.df$logAUC <- log(food.df$AUC)
```

To call variables in the food.df data frame more easily, we attach food.df to the S-PLUS search path:

```
> attach(food.df)
```

We want axes in the log scale, but the numeric labels to be in the original scale. The axislog function we create (see Appendix 7.A.1) provides a full range of tick-marks that the usual log="y" or "x" option in the plot function typically misses.

The series of calls starts with setting up the graph without plotting points or axes labels, so that we may later add customized symbols and text. The result is Figure 7.1, which displays the subjects' results ordered by treatment (Fed, Fasted) within each treatment sequence:

```
> plot(trtn, logAUC, xlim = c(0, 5), type = "n",
     axes = F, xlab = "", ylab = "AUC (ng x hr/ml)")
> axislog(AUC, line = 1, srt = 90, cex = 0.9)
> axis(1, at = c(1:2, 4:5), rep(levels(trt), 2), ticks = F)
> subjseq1 <- levels(subjf[seqn == 1, drop = T])
> for(i in seq(along = subjseq1)) {
     points(trtn[subj == subjseq1[i]],
        logAUC[subj == subjseq1[i]],
        pch = subjseq1[i], type = "b")
   }
> subjseq2 <- levels(subjf[seqn == 2, drop = T])
> for(i in seq(along = subjseq2)) {
     points(3 + trtn[subj == subjseq2[i]],
        logAUC[subj == subjseq2[i]],
        pch = subjseq2[i], type = "b")
   }
> mtext(side = 1, at = c(1.5, 4.5),
     text = c("Sequence 1", "Sequence 2"), line = 2)
```

In Figure 7.1, each line represents an individual subject. We see that AUC was less when the subjects received the Fed regimen, most notably for Subject number 2. The only exception was Subject number 5. With such a small sample size, we suspect that the results of Subjects number 2 and 5 may be quite

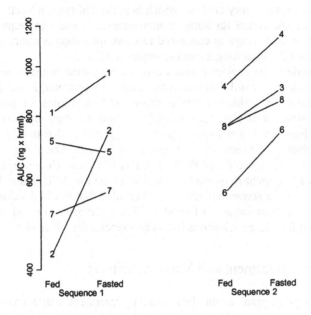

Figure 7.1. Subject plots ordered by treatment sequence.

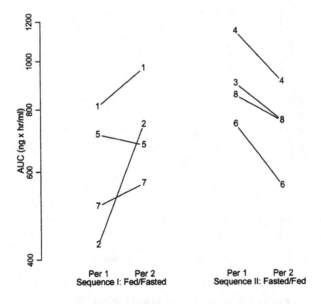

Figure 7.2. Subject plots ordered by period within treatment sequence.

influential on the normal-theory analysis which is presented later in Section 7.4. Also note that the AUC values for Subjects number 1, 2, 5, and 7 in Sequence 1 were somewhat less on average as compared to those for Subjects number 3, 4, 6, and 8 in Sequence 2, suggesting a modest sequence effect.

Figure 7.2 displays the subject profiles ordered by period within treatment sequence. This reflects the order in which the data were collected in the study, and can be insightful for evaluating period effects; and sometimes for discriminating between carryover effects, sequence effects, and treatment-by-period interaction. As in Figure 7.1, a sequence effect is suggested in that the log AUC values for the subjects in Sequence 2 are generally greater than those for the subjects in Sequence 1, regardless of the Fed or Fasted states. However, Figure 7.2 also suggests a treatment-by-period interaction as the log AUC values in the Fed and Fasted state are somewhat similar in Period 2, but the Fed values are much less than the Fasted values in Period 1. The code to accomplish this is very similar to that for Figure 7.1 and is left as an exercise for the reader.

7.3.2   Marginal Treatment and Variance Effects

We next present a point graph of the data sorted by treatment, which enables us to get a first glimpse of marginal location and variance.

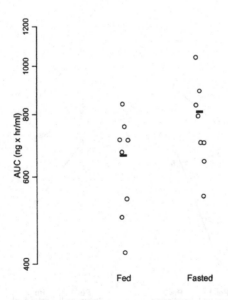

Figure 7.3. Marginal spread and location plot.

The code to produce this plot is given by:

```
> plot(jitter(trtn, factor = 3), logAUC, xlim = c(0, 3),
    axes = F, xlab = "", ylab = "AUC (ng x hr/ml)",
    pch = 1)
> axislog(AUC, line = 1, srt = 90, cex = 0.9)
> axis(1, at = 1:2, labels = levels(trt),ticks = F)
> points(x=1:2, y = c(mean(logAUC[trt == "Fed"]),
    mean(logAUC[trt == "Fasted"])), pch = "-", cex = 3)
```

We applied the jitter function (Chambers et al., 1983) to the x-axis values to
alleviate overlap of points by adding enough random noise without distorting the
structure of the data. The horizontal bars in Figure 7.3 display the geometric
means of the treatment groups. Based on a marginal evaluation, the sample
variability appears similar between the two treatments, and the sample location
differs as we conjectured from examining Figures 7.1 and 7.2. So we have no
data-driven reason to suspect unequal variances in the bivariate treatment population.

7.3.3   Bivariate Treatment and Other Relationships

Figure 7.4 is a scatter plot of the data pairs from the eight subjects. The symbols
indicate which treatment sequence each subject received. In addition to assessing the location of the bivariate point cloud versus the diagonal line, it is important to identify both concordant and discordant outliers. In the food interaction
framework, concordant outliers are those bivariate points that are distant from
the center of the point cloud but vertically close to the diagonal line (and thus
situated near either end of the diagonal line). These points represent those subjects whose paired responses (Fed versus Fasted AUC) are somewhat similar to
each other for that subject, but are notably different (smaller–lower end of the
diagonal line; larger–upper end of the diagonal line) than the magnitudes of the
responses of the other subjects. Discordant outliers are those bivariate points
which stray from the point cloud and stray from the diagonal line. These represent subjects whose paired responses are not similar to each other for that subject, and are also notably different from the responses of the other subjects. The
last statement call is defined in Appendix 7.A.1, and is a wrapper that we will
use repeatedly to place a key in the upper left or other corner of a graph:

```
> plot(logAUC[trt == "Fasted"], logAUC[trt == "Fed"],
    type = "n", axes = F,
    xlab = "Fasted AUC (ng x hr/ml)",
    ylab = "Fed AUC (ng x hr/ml)")
> axislog(AUC, line = 1, srt = 90, cex = 0.9)
> axislog(AUC, side = 1, line = 1)
> points(logAUC[trt == "Fasted" & seqn == 1],
    logAUC[trt == "Fed" & seqn == 1], pch = 1, cex = 1.2)
> points(logAUC[trt == "Fasted" & seqn == 2],
    logAUC[trt == "Fed" & seqn == 2], pch = 2, cex =1.2)
```

```
> # make symbols 20% larger than default
> abline(0, 1) # adds the bivariate identity line
> place.keyseq(0, 1)
```

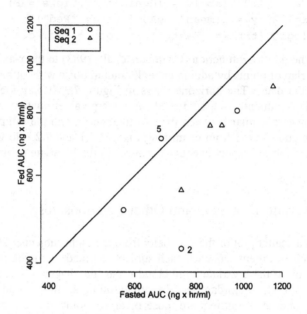

Figure 7.4. Scatter plot.

Again, we see that the Fed AUCs tend to be less than the Fasted AUCs, and the AUCs for subjects in Sequence 2 tend to be larger than those for subjects in Sequence 1, regardless of the Fed or Fasted state. For Subject number 2, who is a discordant outlier with Fed AUC = 428.00 and Fasted AUC = 746.45 ng × h/mL (see Section 7.2 data listing), the difference is relatively large. We identify the bivariate response for Subject number 2 on the plot with

```
> text(x = logAUC[subj == 2 & trt == "Fasted"] + .05,
    y = logAUC[subj == 2 & trt == "Fed"], "2", adj = 1)
```

and with analogous code we mark Subject number 5 as the lone point which is located above the bivariate identity line.

Another informative look at these data is a Tukey sum–difference plot (Cleveland, 1993; Tukey, 1977), which allows us to study the Fed versus Fasted relationship through differences (log Fed – log Fasted) and sums (log Fed + log Fasted) of the bivariate data points. In general, this plot clarifies magnitudes of treatment differences (ratios in our case) and permits discovery of trends across the observed ranges of the data. (Note how it is easier to judge differences from the flat zero line in Figure 7.5 as compared to the bivariate identity line in Figure 7.4.) But we will see in Section 7.5 that the Tukey sum–difference plot has added value for the initial exploration of the data collected from a 2,2,2 cross-over trial as it simultaneously displays information on the carryover and period effects, as well as suggests treatment effects. The code to generate Figure 7.5 is in Appendix 7.A.2.

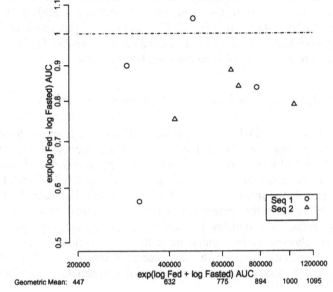

Figure 7.5. Tukey sum–difference plot.

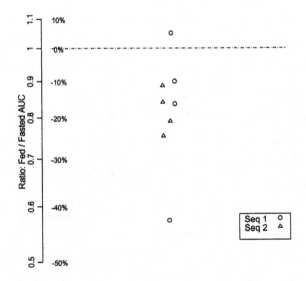

Figure 7.6. Plot of AUC ratios by sequence.

Figure 7.6 is a graph of the Fed/Fasted AUC ratios where the ratio points are labeled by sequence. The code is in Appendix 7.A.2.

It is of interest to observe what shape the sample distribution of ratios has, and how the sample distribution relates to a reference ratio of one. To aid study of the graph, labels in the percentage change scale were placed next to the y-axis ratio labels. We see that six of the eight subjects exhibited roughly a 10–25% decrease in Fed AUC relative to Fasted AUC, and one subject (number 2) exhibited a little more than a 40% decrease. In one subject (number 5), the Fed response increased relative to the Fasted response by about 5%.

7.3.4   A Preliminary Look at the 2,2,2 Crossover ANOVA

Figure 7.7 plots the individual Fed and Fasted AUC values by period along with the corresponding period geometric means. The lines connect the geometric means across periods by sequence. The code is in Appendix 7.A.2. Figure 7.7 indicates an inconsistency from Period 1 to Period 2 in the Fed and Fasted AUCs. In Period 1, the individual Fed AUCs are notably less than the Fasted AUCs; this is reflected in the corresponding geometric means. However, in Period 2, the two groups of individual Fed and Fasted AUC values are quite similar, and this is reflected by the similar geometric mean values which we also note are located centrally between the two geometric mean values from Period 1. Thus, Figure 7.7 alone suggests the presence of a treatment-by-period interaction.

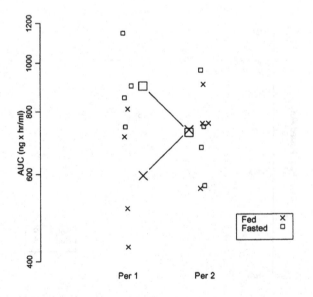

Figure 7.7. Geometric mean and individual responses by period.

7.4   Formal Data Analysis

In this section, we perform the usual normal-theory ANOVA for the 2,2,2 cross-over design. For the full modeling and algebraic details, see Jones and Kenward (1989; pp. 9–10, 22–28, 30–34). The model representation is

$$Y_{ijk} = \mu + s_{ik} + \pi_j + \tau_{d[i,j]} + \lambda_{d[i,j-1]} + e_{ijk}, \tag{7.1}$$

where

- μ = the overall mean effect.
- s_{ik} = the effect of subject k in group i; $i = 1, 2$.
- π_j = the effect of period j; $j = 1, 2$.
- $\tau_{d[i,j]}$ = the effect of treatment administered in period j to group i.
- $\lambda_{d[i,j-1]}$ = the carryover effect of the treatment administered in period j to group i, where $\tau_{d[i,0]} = 0$ (no carryover in the first period).
- e_{ijk} = the random error term for subject k in period j in group i, and is assumed to be normally distributed with mean 0 and variance σ^2.

Given our exploratory graphical work in Section 7.3, we can anticipate what the normal-theory ANOVA may confirm. Specifically, Figures 7.1 to 7.7 indicate a treatment effect in that, on average, Fed AUCs appear to be less than Fasted AUCs. Figures 7.1, 7.2, and 7.4 suggest a modest sequence effect in that on average, AUCs are greater in the subjects in Sequence 2 as compared to Sequence 1. Figures 7.2 and 7.7 suggest a treatment-by-period interaction where in Period 1, on average and individually, Fed AUCs are less than Fasted AUCs but in Period 2 they are quite similar. Figures 7.2 and 7.7 also suggest that there will be no period effect in that, on average, AUCs are similar from Period 1 to Period 2 when disregarding treatment. So, in a relative sense, we might anticipate a smaller p-value for the test of treatment (Fed versus Fasted) effects, a smaller p-value for the test of carryover effects which is confounded with both sequence effects and treatment-by-period interaction, and a larger p-value for period effects.

The initial model fit for log AUC specifies that subjects within sequence constitute a partition term in the error structure:

```
> fit.food <- aov(logAUC ~ seqf + Error(subjf) + trt +
    perf, data=food.df)
```

Application of the summary command demonstrates the two-strata error structure (between and within subjects) of the 2,2,2 crossover design:

```
> summary(fit.food)
Error: subjf
          Df Sum of Sq    Mean Sq  F Value      Pr(F)
    seqf   1 0.1797439  0.1797439 2.096941 0.1977559
Residuals  6 0.5143031  0.0857172
```

```
Error: Within
            Df Sum of Sq    Mean Sq  F Value        Pr(F)
      trt    1 0.1605383  0.1605383 9.026330  0.0238727
     perf    1 0.0000232  0.0000232 0.001305  0.9723585
Residuals  6 0.1067133  0.0177856
```

According to the ANOVA results, the sequence term, seqf, is not statistically significant ($p = 0.198$). Based upon the p-value alone and the intrinsic aliasing of sequence, carryover, and treatment-by-period interaction in a single degree-of-freedom contrast, it might be concluded that none of these effects were an issue with these data. However, Figures 7.1 to 7.7 in Section 7.3 suggest the presence of a modest sequence effect or a treatment-by-period interaction. To choose between these two requires an examination of response-related covariates (e.g., age, gender, weight) as they are distributed among the subjects in the two treatment sequences. Biological carryover is unlikely as the washout period was planned to be of sufficient length, and plasma levels in all eight subjects at the beginning of Period 2 were observed to be zero.

Despite suspicion about a potentially meaningful sequence effect or a treatment-by-period interaction, we continue with the usual analyses of period and treatment effects to continue illustrating S-PLUS. Indeed, the ANOVA results support our initial graphical investigation: there is a statistically significant ($p = 0.024$) treatment effect which we know from our graphical investigations is due to the lower Fed than Fasted AUCs. And, there is no statistically significant ($p = 0.972$) period effect as was seen graphically, as on average, AUCs are not strikingly different from Period 1 to Period 2 regardless of treatment.

In Section 7.5 we present some 2,2,2 crossover ANOVA model checking and diagnostics, but for now let us presume that the underlying assumptions of normality and equal variances are tenable and proceed with formal estimation inference. Our parameter of interest is the proportional difference in AUC for the Fed regimen relative to the Fasted regimen. We need to construct a sample point estimate and a 90% confidence interval for the true value.

Although there are several approaches in S-PLUS to compute the needed quantities, we will use model.tables to compute the means and standard error of the difference in the means on the log scale, and we also use the fit above to calculate the critical value of the t-distribution:

```
> tblmeans <- model.tables(fit.food, type = "means", se=T)
> meandiff.se <- tblmeans$se$trt
> t.crit <- qt(0.95, df = fit.food$Within$df.residual)
```

This code is sufficient to compute the sample point estimate of the difference in log AUC and the 90% confidence interval for the true difference, Fed–Fasted, in the log scale. We exponentiate the log scale results to arrive at the corresponding results for the ratio, Fed/Fasted , in the original scale:

```
> meandiff <- diff(rev(tblmeans$tables$trt))
> orig.ratio <- exp(meandiff)
> # Use Delta method to estimate SE of orig.ratio
```

```
> orig.ratio.se <- orig.ratio * meandiff.se
> lcl.ratio <- exp(meandiff - t.crit * meandiff.se)
> ucl.ratio <- exp(meandiff + t.crit * meandiff.se)
```

See Appendix 7.A.3 for a few more statements that build on these to alternatively express the results as percent change in AUCs. We summarize these results and display them in a table:

```
> tbldiff <- matrix(c(orig.ratio, lcl.ratio, ucl.ratio,
    orig.pctchg, lcl.pctchg, ucl.pctchg), byrow = T,
    nrow = 2)
> # pctchg refers to percent change analogues to ratio.
> dimnames(tbldiff) <- list(c("Ratio", "Pct Chg"),
    c("Point Est", "Lower", "Upper"))

> # Needed to display title with table
> disply <- function()
  {
    cat("Differences with 95% Confidence Bounds", "\n")
    round(tbldiff,2)
  }

> disply()
Differences with 95% Confidence Bounds
        Point Est  Lower  Upper
  Ratio      0.82   0.72   0.93
Pct Chg    -18.15 -28.10  -6.83
```

We use similar code to obtain the individual treatment geometric means and approximated standard errors based on the "Delta" method (Agresti, 1990, Ch. 12).

```
> indv.sem <- meandiff.se / sqrt(2)
> orig.means <- exp(tblmeans$tables$trt)
> orig.sems <-  orig.means * indv.sem
> tblindv <- matrix(c(orig.means, orig.sems), ncol = 2)
> dimnames(tblindv) <- list(c("Fed", "Fasted"),
    c("Geo Mean", "SEM"))

> round(tblindv, 2)
        Geo Mean    SEM
   Fed    662.00  31.21
Fasted    808.84  38.14
```

To summarize the results of our food interaction study, we construct Figure 7.8 which presents the results averaging across periods but also for each period separately, due to the suspected treatment-by-period interaction. The code for all this is in Appendix 7.A.3.

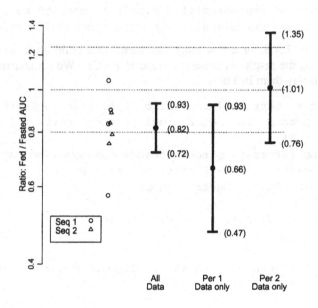

Figure 7.8. Treatment effect, Fed versus Fasted.

The usual average bioequivalence criteria based on ratios of data are shown by the dashed lines located at the *y*-axis ratio values of 0.80 and 1.25. The Fed/Fasted AUC ratios for the individual subjects are represented by the open circles (Sequence 1) and the open triangles (Sequence 2). The geometric mean ratio estimated from all of the data is represented by the solid dot with value 0.82, and the corresponding 90% confidence interval (0.72, 0.93) for the true proportioned difference in mean Fed AUC and mean Fasted AUC is represented by the bold error bars. These results indicate that, on average, the Fed and Fasted states are not bioequivalent; Fed AUC is, on average, less than Fasted AUC.

Also shown are the ratios (Fed/Fasted) of geometric means (solid dots) and corresponding 90% confidence intervals (bold error bars) which were computed separately using the between-subject variability from each study period. This additional part of the display is of interest when there is reasonable evidence of unequal carryover effects (usually only the first period is plotted), or in the case of a suspected treatment-by-period interaction. The results are not surprising given our previous graphical and statistical analyses. The Period 1 data suggest inequivalence of the Fed and Fasted states; the Period 2 data are consistent with equivalence. Thus we have summarized the major individual, mean, and inferential results in a single graphic display which is easily understood by the clinical, pharmacokinetic, and statistical professionals involved in bioequivalence evaluations.

7.5 Visualizing the 2,2,2 Crossover ANOVA, Model Checking, and Diagnostics

In this section we present several valuable graphical tools for visualizing the data which are used in the three single degree-of-freedom ANOVA contrasts testing for carryover effects (confounded with sequence effects and treatment-by-period interaction), period effects, and treatment effects. We also present a series of graphics for assessing model fit. Some of the graphical techniques are reasonably common ones with special adaptations to the 2,2,2 crossover design; others are specific to the 2,2,2 crossover design.

7.5.1 Visualizing the ANOVA

We illustrate two sets of graphical presentations that visualize the data used in the three single degree-of-freedom ANOVA contrasts. The first is based on displaying treatment means by period; the second displays various linear combinations of the two data points collected for each subject.

Plots of Treatment Means by Period

To evaluate period effects, treatment effects, and to help diagnose carryover effects versus sequence effects versus treatment-by-period interaction, we plot the four treatment means, two for Fed and two for Fasted, by study period connecting the treatment means either by treatment sequence (Figure 7.9) or by treatment group (Figure 7.10). The pattern of the connected means illustrates three characteristics:

1. The relative ordering of the two treatments within each study period.
2. The magnitude of the difference between the two treatments within each period.
3. The magnitude of any difference between periods in each treatment group.

The range of the y-axis is determined by the range of the individual data points excluding any outliers which might distort the message in the graph.

We create the needed means with

```
> attach(food.df) # not needed if already attached
> tbpmeans.auc <- aggregate(logAUC,
    list(Treatment = trt, Period = per), mean)
> names(tbpmeans.auc)[3] <- "meanlogAUC"
> mean.fed.per1 <- tbpmeans.auc[1, "meanlogAUC"]
> mean.fasted.per1 <- tbpmeans.auc[2, "meanlogAUC"]
> mean.fed.per2 <- tbpmeans.auc[3, "meanlogAUC"]
> mean.fasted.per2 <- tbpmeans.auc[4, "meanlogAUC"]
```

We construct Figure 7.9 connecting the means by treatment sequence:

```
> plot(rep(1:2, 2), tbpmeans.auc$meanlogAUC,
    type = "n", axes = F, xlab="",
    ylab = "Geometric Mean AUC (ng x hr/ml)",
    xlim = c(0, 3))
> axislog(exp(tbpmeans.auc$meanlogAUC), line = 1, srt = 90,
    cex = 0.9)
> points(c(1, 2), c(mean.fed.per1, mean.fed.per2), pch = 4,
    cex = 1.2)
> points(c(1, 2), c(mean.fasted.per1, mean.fasted.per2),
    pch = 0, cex = 1.2)
> lines(c(1, 2),c(mean.fed.per1, mean.fasted.per2),
    pch=" ", type = "b", cex = 2)
> lines(c(1, 2), c(mean.fasted.per1, mean.fed.per2),
    pch = " ", type = "b", cex = 2, lty = 3)
> axis(1, at = 1:2, c("Per 1", "Per 2"), ticks = F)
> place.keytrt(0,1)
```

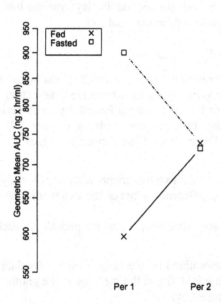

Figure 7.9. Treatment geometric means connected by sequence.

Figure 7.10, which connects the means by treatment, is created by the same sequence of steps except the two `lines` calls are slightly modified to read:

```
> lines(c(1, 2), c(mean.fed.per1, mean.fed.per2),
    pch = " ", type = "b", cex = 2)
> lines(c(1, 2), c(mean.fasted.per1, mean.fasted.per2),
    pch = " ", type = "b", cex = 2, lty = 3)
```

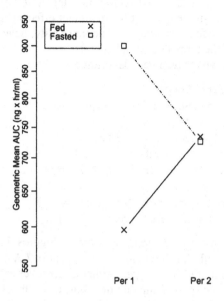

Figure 7.10. Treatment geometric means connected by treatment.

For our example, it is difficult to see much difference in the two display options since the Period 2 means are so close together. We see that in Period 1, the geometric mean AUC for the Fed state is substantially less than that for the Fasted state while in Period 2 they are quite similar. And, the geometric mean for the Fed state is less in Period 1 than in Period 2, and the reverse holds for the Fasted state.

Plotting Linear Combinations of the Data

Recall from Section 7.3 that each of the three single degree-of-freedom ANOVA contrasts can be evaluated using either t-tests or F-tests. More specifically, the contrasts are constructed by comparing between the two treatment sequences various linear combinations of the two observations for each subject. A series of three plots displays the data which are evaluated by each of the three single degree-of-freedom contrasts which test for carryover effects, period effects, or treatment effects. For example, to visualize the data in the contrast testing carryover effects (confounded with sequence effects and treatment-by-period interaction), plot the individual within subject sums (Period 1 log AUC + Period 2 log AUC) for each sequence (Figure 7.11). From the two sample distributions of data and the respective mean sums, the corresponding t-test can be evaluated visually. A difference in location between the two distributions suggests a carryover effect (or a sequence effect or a treatment-by-period interaction). Similarly, plot the individual and mean within-subject treatment differences (log Fed AUC–log Fasted AUC) by treatment sequence to visualize the data in the contrast for period effects (Figure 7.12). And, plot the individual and mean within-subject period differences (Period 2 log AUC–Period 1 log AUC) by treatment sequence to visualize the data in the contrast for treatment effects (Figure 7.13). These plots also allow for a visual assessment of nonnormality, heteroscedasticity, and outliers, for each of the three two-sample *t*-tests.

To construct these plots, we create a new data frame.

```
> food.comb <- aggregate(logAUC,
    list(Subject = subj, Sequence = seqn), sum)
> names(food.comb)[3] <- "sumPer"
> food.comb <- cbind(food.comb, diffTrt =
    aggregate(log(AUC), list(Subject = subj,
      Sequence = seqn), function(x) {-diff(x)})[,3])
> food.comb$diffPer <- c(
    -food.comb$diffTrt[food.comb$Sequence == 1],
    food.comb$diffTrt[food.comb$Sequence==2])
```

We attach our new data frame food.comb and create Figures 7.11 through 7.13 with its variables. The code for these plots is in Appendix 7.A.4.

The shift in location in Figure 7.11 suggests either a carryover effect or a sequence effect or a treatment-by-period interaction. Earlier in the chapter, Figures 7.2, 7.7, 7.9, and 7.10 suggested a treatment-by-period interaction; Figures 7.1, 7.2, and 7.4 suggested a modest sequence effect.

The lack of a shift in location in Figure 7.12 indicates no evidence of a period effect, although the single lower point (Subject number 2) in the Fed then Fasted sequence influences the analysis quite a bit.

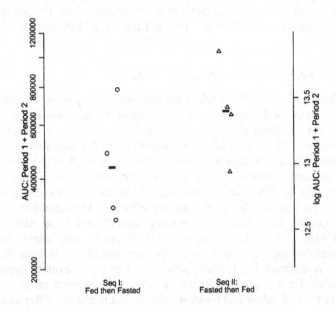

Figure 7.11. Evaluation of carryover effect.

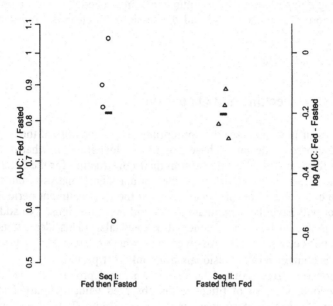

Figure 7.12. Evaluation of period effect.

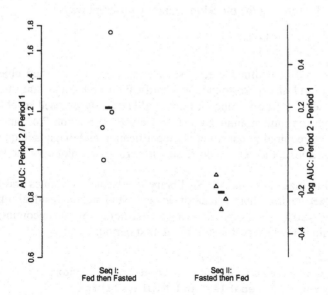

Figure 7.13. Evaluation of treatment effect.

The notable shift in location in Figure 7.13 indicates a treatment effect and again, the single upper point (Subject number 2) in the Fed then Fasted sequence makes this more pronounced. We end the session and clean up with the command

```
> detach("food.comb")
```

7.5.2   Model Checking and Diagnostics

Another series of plots assesses the appropriateness of the normal theory model for the 2,2,2 crossover design. These diagnostic plots are somewhat routine in concept, but there is an interesting twist in their construction for the 2,2,2 crossover design. The residual values from the normal theory analysis sum to zero within a subject. Therefore, the residuals from the two treatment periods for a given subject will have the same magnitudes and opposite signs. No additional information is gained from the second set of residuals. (This idea extends to t treatment, p period, s sequence crossover trials where $t = p$, $p > 2$. All relevant information is contained in the residuals from only $p-1$ periods.)

We demonstrate five plots. The first is a normal probability plot (Figure 7.14); the second is a plot of fitted versus observed values (Figure 7.15); the third is a plot of raw residuals versus fitted values (Figure 7.16); the fourth motivated by John W. Tukey plots the square root of the absolute value of the residuals against fitted values (Figure 7.17); and the fifth illustrates Cook's distance (Figure 7.18).

We begin by refitting the model in a slightly different way:

```
> fit2.food <- lm(logAUC ~ subjf + trt + perf,
    data = food.df)
```

We are interested in within-subject residuals, so calls to the fitted object will provide these and also corresponding identification and diagnostic quantities. Projections (see the proj function) could alternatively be used to obtain the residuals from the multistratum fit.food object in Section 7.4, but manual work would be required to construct the identification and diagnostic quantities. As mentioned above, the two residuals sum to zero, so we plot only the Period 1 residuals.

We need to make sure the MASS library is attached (Venables and Ripley, 1999) in order to use their standardized and Studentized residual functions stdres and studres. Next we create a data frame for our upcoming analysis that contains only observations from the first period:

```
> half.index <- food.df$per == 1
> half.resid <- residuals(fit2.food)[half.index]
> library(MASS) # Venables and Ripley library
> half.studres <- studres(fit2.food)[half.index]
> half.stdres <- stdres(fit2.food)[half.index]
> half.pred <- predict(fit2.food)[half.index]
```

```
> half.cooks <- Cooks.lm(fit2.food)[half.index]
> food.half <- data.frame(food.df[half.index, ],
    resid = half.resid, studres = half.studres,
    stdres = half.stdres, pred = half.pred,
    cooks = half.cooks)
> # synchronize row names and subject ID
> row.names(food.half) <- food.half$subj
```

A listing of this data frame is given in Appendix 7.A.4. The Cooks.lm function is taken from the plot.lm function definition in S-PLUS and is also listed in Appendix 7.A.4. With some modification, we also make use of another function defined in plot.lm called id.n to help us identify values of large magnitude. See Appendix 7.A.1 for its definition.

We do not need to be concerned with transforming back axes from the log scale to the original scale, so our call sequences are simplified. We first look at a normal probability plot:

```
> attach(food.half)
> qqxy <- qqnorm(studres, plot = F)
> qqxy.x <- qqxy$x; qqxy.y <- qqxy$y
> names(qqxy.x) <- names(qqxy.y) <- row.names(food.half)
> plot(qqxy.x, qqxy.y, type = "n",
    xlab = "Standard Normal Quantiles",
    ylab = "Studentized Residuals, Period 1")
> points(qqxy.x[seqn == 1], qqxy.y[seqn == 1], pch = 1)
> points(qqxy.x[seqn == 2], qqxy.y[seqn == 2], pch = 2)
> qqline(studres)
> id.n(qqxy.x, qqxy.y, how.many = 2, offset = 0.2)
> place.keyseq(0, 1)
```

Figure 7.14 plots the eight Studentized residuals from the first period against expected normal order statistics. We observe two notable deviations from linearity: Subjects number 2 and 5, both of whom were randomized to the first treatment sequence, Fed then Fasted. The fact that these two residuals deviate from the other six is not surprising given our initial graphical exploration of this dataset.

Figures 7.15 through 7.17 are generated by similar codes and the full displays of the commands are in Appendix 7.A.4. Figure 7.15 plots the fitted log AUC values from Period 1 against the observed log AUC values for Period 1. The fit is not unreasonable and only one point (Subject number 2) deviates notably from the 45° identity line. Note the location of the open triangles (upper right) for Sequence 2 versus the open circles (lower left) for Sequence 1.

Figure 7.16 plots the raw residuals from Period 1 against the fitted values from Period 1. Subjects number 2 and 5 demonstrate residuals which are the most different from the others.

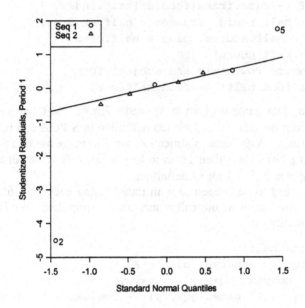

Figure 7.14. Normal probability plot.

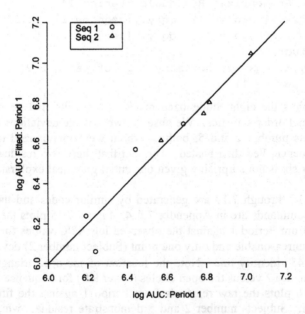

Figure 7.15. Evaluation of model fit.

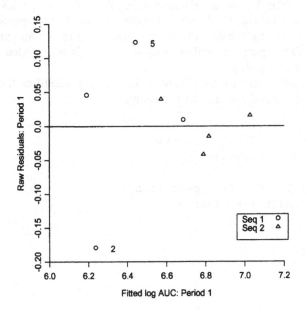

Figure 7.16. Raw residuals versus fitted values.

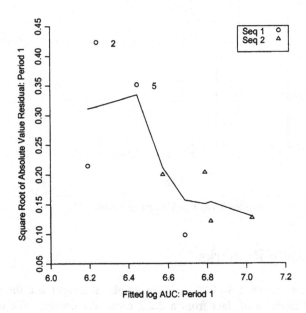

Figure 7.17. Equal variance plot.

Figure 7.17 evaluates the equal variance assumption. The plot method was first suggested by John Tukey; a reference example is shown in Cleveland (1993). A lowess (Cleveland, 1979) robust smoothing curve is superposed as a guide to help discover any trends. Ideally we would like to see an approximately flat line. The apparent volatility of these data is not unexpected given the small number of data points.

And finally, Cook's distance as plotted in Figure 7.18 identifies the data from Subjects number 2 and 5 as the most influential:

```
> plot(1:8, cooks, type = "h", axes = F,
    ylab = "Cook\'s Distance: Period 1",
    xlab="Subject",xlim=c(0,9))
> axis(2)
> mtext(side = 1, at = 1:8, paste(subj))
> detach("food.half") # cleanup
```

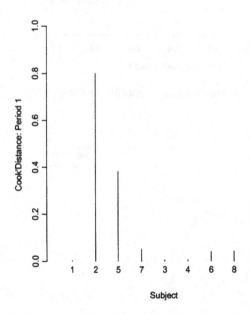

Figure 7.18. Cook's distance plot.

7.6 Concluding Remarks

In this chapter, we provided S-PLUS tools for graphical insight and the normal theory statistical analysis of data from a 2,2,2 crossover design. We demonstrated the value of graphical analyses from initial data exploration and discovery, into understanding the ANOVA with particular emphasis on the single degree of freedom contrast which confounds carryover effects with sequence effects and treatment-by-period interaction, through to model checking and diagnostic work following the ANOVA. We illustrated the use of S-PLUS on a dataset from a food interaction study where there was some evidence of a food effect in the presence of either a modest sequence effect or a treatment-by-period interaction, and where two subjects had notably more influence on the results of the trial than the other six enrolled. And, we provided a useful summary graph for presenting the individual, average, and inferential results from a 2,2,2 crossover average bioequivalence or interaction trial.

7.7   References

Agresti (1990). *Analysis of Categorical Data*. Wiley, New York.

Bradstreet, T.E. (1992). Favorite data sets from early phases of drug research–Part 2. *Proceedings of the Section on Statistical Education*. American Statistical Association, pp. 219–223.

Chambers, J.M., Cleveland, W.S., Kleiner, B., and Tukey, P.A. (1983). *Graphical Methods for Data Analysis*. Wadsworth, Pacific Grove, CA.

Chow, S.–C. and Liu, J.P. (2000). *Design and Analysis of Bioavailability Studies*. Second Edition. Marcel Dekker, New York.

Cleveland, W.S. (1979). Robust locally weighted regression and smoothing scatterplots. *Journal of the American Statistical Association*, **74**, 829–836.

Cleveland, W.S. (1993). *Visualizing Data*. Hobart Press, Summit, NJ.

Jones, B. and Kenward, M.G. (1989). *Design and Analysis of Crossover Trials*. Chapman & Hall, London.

Senn, S. (1993). *Cross–over Trials in Clinical Research*. Wiley, New York.

Tukey, J.W. (1977). *Exploratory Data Analysis*. Addison–Wesley, Reading, MA.

Venables, W.N. and Ripley, B.D. (1999). *Modern Data Analysis*. Third Edition. Springer–Verlag, New York.

7.A. Appendix

7.A.1 Functions for Graphical Annotations

Utility Function for Generating Log Axes Ticks

```
axislog <- function(data, nint = 5, side = 2, opp.side = F,
  ...) {
  orig <- pretty(data, nint)
  orig.px <- orig[orig > 0]
  lograngge <- log(orig.px)
  usrnow <- par("usr")  # graphical layout settings
  if(side == 2) {
    par(usr = c(usrnow[1:2], min(lograngge), max(lograngge)))
    axis(2, at = lograngge, labels = paste(orig.px), ...)
  }
  else if(side == 1) {
    par(usr = c(min(lograngge), max(lograngge), usrnow[3:4]))
    axis(1, at = lograngge, labels = paste(orig.px), ...)
  }
  if(opp.side == T) {
    axis(side + 2, at = pretty(lograngge),
      labels = paste(pretty(lograngge)), ...)
  }
  invisible()
}
```

Utility Function for Placing Sequence Identifier Key on Graphs

```
# x = 0 and y = 1 refers to upper left corner,
# x = 1 and y = 0 refers to lower right corner

place.keyseq <- function(x = 0, y = 1, ...)
{
  oldpar <- par(gr.state)
  par(usr=c(0, 1, 0, 1))
  on.exit(par(oldpar))
  on.exit(par(new = F), add = T)
  key(x, y, text = list(c("Seq 1", "Seq 2", " ")),
    points = list(pch = c(1, 2, 32)), border = T)
# Adds space between text and bottom border
  invisible()
}
```

Similar Function for Treatment Identifier Key

```
place.keytrt <- function(x = 0, y = 1, ...) {
  oldpar <- par(gr.state)
  par(usr = c(0, 1, 0, 1))
  on.exit(par(oldpar))
  on.exit(par(new = F), add = T)
  key(x, y, text = list(c("Fed", "Fasted", " ")),
    points = list(pch = c(4, 0, 32)), border = T)
  invisible()
}
```

Utility Function to Help Identify Values of Large Magnitude on Figures 7.14 – 7.16, Section 7.5.2

```
id.n <- function(x, y, how.many = F, offset = 0)
  if(how.many) {
# Identify "how.many" greatest y-values (in absolute value)
# based on sub-function definition in
# plot.lm function method
    n <- length(y)
    oy <- order(abs(y))
    names(oy) <- names(abs(y))[oy]
    which <- oy[(n - how.many + 1):n]
    text(x[which] + offset, y[which], names(which),
      adj = 1)
  }
```

7.A.2   Additional Code for Section 7.3

Prior to running the code in this section, the data frame food.df should be attached to the search list and graphics parameters should be changed from their default values using the following commands:

```
attach(food.df)
par(xaxs = "e", yaxs = "e", bty = "l", pty = "s")
```

Figure 7.5

```
plot(x = logAUC[trt == "Fed"] + logAUC[trt == "Fasted"],
  y = logAUC[trt == "Fed"] - logAUC[trt == "Fasted"],
  type = "n", axes = F,
  xlab = "exp(log Fed + log Fasted) AUC",
  ylab = "exp(log Fed - log Fasted) AUC")
```

```
axislog(exp(logAUC[trt == "Fed"] -
   logAUC[trt == "Fasted"]), line = 1, srt = 90,
  cex = 0.9)
axislog(exp(logAUC[trt == "Fed"] +
   logAUC[trt == "Fasted"]), side = 1, line = 1,
  cex = 0.8)
points(x = logAUC[trt == "Fed" & seqn == 1] +
   logAUC[trt == "Fasted" & seqn == 1],
 y = logAUC[trt == "Fed" & seqn == 1] -
   logAUC[trt == "Fasted" & seqn == 1], pch = 1,
  cex = 1.2)
points(x = logAUC[trt == "Fed" & seqn == 2] +
   logAUC[trt == "Fasted" & seqn == 2],
 y = logAUC[trt == "Fed" & seqn == 2] -
   logAUC[trt == "Fasted" & seqn == 2], pch = 2,
  cex = 1.2)
abline(0, 0, lty = 3)
mtext(side = 1, at = log(c(150000, 200000, 400000, 600000,
  800000, 1000000, 1200000)),
  text = c("Geometric Mean:",
    paste(round(sqrt(c(200000, 400000, 600000, 800000,
      1000000, 1200000)), 0))), line = 4, cex = 0.8)
place.keyseq(0.8, 0.2)
# a little off lower right-hand corner due to
# x-axis placement
```

Figure 7.6

```
plot(x = rep(1, 8),
  y = log(AUC[trt == "Fed"]/AUC[trt == "Fasted"]),
  type = "n", axes = F, xlab = "",
  ylab = "Ratio: Fed / Fasted AUC")
axislog(AUC[trt == "Fed"]/AUC[trt == "Fasted"], line = 1,
  srt = 90, cex = 0.9)
points(x = jitter(rep(1, 4)),
  y = log(AUC[trt == "Fed" & seqn == 1]/
    AUC[trt == "Fasted" & seqn == 1]), pch = 1)
points(x = jitter(rep(1, 4)),
  y = log(AUC[trt == "Fed" & seqn == 2]/
    AUC[trt == "Fasted" & seqn == 2]), pch = 2)
abline(h = log(1), lty = 3)
axis(2, at = log((seq(0.5, 1.1, by = 0.1))),
  labels = paste(seq(-50, 10, by = 10), "%", sep = ""),
  line = -2, srt = 0, cex = 0.8, ticks = F)
place.keyseq(0.8, 0.2)
```

Figure 7.7

```
plot(per, logAUC, type = "n", axes = F, xlab = "",
  ylab = "AUC (ng x hr/ml)", xlim = c(0, 3))
axislog(AUC, line = 1, srt = 90, cex = 0.9)
points(x = jitter(per[trt == "Fasted"], factor = 3),
  y = logAUC[trt == "Fasted"], pch = 0)
points(x = jitter(per[trt == "Fed"], factor = 3),
  y = logAUC[trt == "Fed"], pch = 4)
lines(c(1.2, 1.8),
  c(mean(logAUC[trt == "Fasted" & per == 1]),
    mean(logAUC[trt == "Fasted" & per == 2])),
  type = "b", pch = 0, cex = 2)
lines(c(1.2, 1.8), c(mean(logAUC[trt == "Fed" & per == 1]),
    mean(logAUC[trt == "Fed" & per == 2])), type = "b",
  pch = 4, cex = 2)
axis(1, at = 1:2, c("Per 1", "Per 2"), ticks = F)
place.keytrt(0.8, 0.2)
```

7.A.3   Additional Code for Section 7.4

Percent Change Expressions

Percent change expressions of Fed relative to Fasted; these build on the ratios computed.

```
orig.pctchg <- (orig.ratio - 1)*100
orig.pctchg.se <- 100*orig.ratio.se
lcl.pctchg <- (lcl.ratio - 1) * 100
ucl.pctchg <- (ucl.ratio - 1) * 100
```

Figure 7.8

```
attach(food.df) # not needed if already attached
plot(x = rep(1, 8),
  y = log(AUC[trt == "Fed"]/AUC[trt == "Fasted"]),
  type = "n", axes = F, xlab = "", xlim = c(0.5, 2.5),
  ylim = log(c(0.5, 1.25)),
  ylab = "Ratio: Fed / Fasted AUC")
axislog(c(0.5, 1.25), line = 1, srt = 90, cex = 0.9)
abline(h = log(c(0.8, 1, 1.25)), lty = 7)
points(x = jitter(rep(1, 4)),
  y = log(AUC[trt == "Fed" & seqn == 1]/
    AUC[trt == "Fasted" & seqn == 1]), pch = 1)
```

```
points(x = jitter(rep(1, 4)),
  y = log(AUC[trt == "Fed" & seqn == 2]/
    AUC[trt == "Fasted" & seqn == 2]), pch = 2)
points(1.4, meandiff, cex = 1.25, pch = 16)
par(lwd = 5)
error.bar(1.4, meandiff, lower = t.crit * meandiff.se,
  add = T, gap = F)
par(lwd = 1)
text(1.5, meandiff - t.crit * meandiff.se,
  paste("(", round(lcl.ratio, 2), ")", sep = ""), adj = 0)
text(1.5, meandiff, paste("(", round(orig.ratio, 2), ")",
    sep = ""), adj = 0)
text(1.5, meandiff + t.crit * meandiff.se,
  paste("(", round(ucl.ratio, 2), ")", sep = ""), adj = 0)

# For Period 1 only
per1data <- t.test(log(AUC[trt == "Fed" & per == 1]),
  log(AUC[trt == "Fasted" & per == 1]), conf.level = 0.9)
meandiff.per1 <-  - diff(per1data$estimate)
lcl.per1 <- per1data$conf.int[1]
ucl.per1 <- per1data$conf.int[2]
points(1.9, meandiff.per1, cex = 1.25, pch = 16)
par(lwd = 5)
error.bar(1.9, meandiff.per1, lower = lcl.per1,
  upper = ucl.per1, incr = F, add = T, gap = F)
par(lwd = 1)
text(2, lcl.per1, paste("(", round(exp(lcl.per1), 2), ")",
    sep = ""), adj = 0)
text(2, meandiff.per1, paste("(", round(exp(meandiff.per1),
    2), ")", sep = ""), adj = 0)
text(2, ucl.per1, paste("(", round(exp(ucl.per1), 2), ")",
    sep = ""), adj = 0)

# For Period 2 only
per2data <- t.test(log(AUC[trt == "Fed" & per == 2]),
  log(AUC[trt == "Fasted" & per == 2]), conf.level = 0.9)
meandiff.per2 <-  - diff(per2data$estimate)
lcl.per2 <- per2data$conf.int[1]
ucl.per2 <- per2data$conf.int[2]
points(2.4, meandiff.per2, cex = 1.25, pch = 16)
par(lwd = 5)
error.bar(2.4, meandiff.per2, lower = lcl.per2,
  upper = ucl.per2, incr = F, add = T, gap = F)
par(lwd = 1)
```

```
text(2.5, lcl.per2, paste("(",
   round(exp(lcl.per2), 2), ")", sep = ""), adj = 0)
text(2.5, meandiff.per2, paste("(",
   round(exp(meandiff.per2), 2), ")", sep = ""), adj = 0)
text(2.5, ucl.per2, paste("(",
   round(exp(ucl.per2), 2), ")", sep = ""), adj = 0)
par(lwd = 1)
axis(1, at = c(1.4, 1.9, 2.4), c("All\nData",
     "Per 1\nData only", "Per 2\nData only"), ticks = F)
place.keyseq(0, 0.2)
detach("food.df") # cleanup
```

7.A.4   Additional Code for Section 7.5

Prior to running the code to create Figures 7.11–7.13, the data frame food.comb should be attached to the search list. Prior to running the code to create Figures 7.14–7.18, the data frame food.half should be attached to the search list.

```
attach(food.comb)
```

Figure 7.11: Carryover Effect

```
plot(as.numeric(Sequence), sumPer,
   type = "n", axes = F, xlab = "",
   ylab = "AUC: Period 1 + Period 2", xlim = c(0.5, 2.5))
axislog(exp(sumPer),opp.side = T, line = 1, srt = 90,
   cex = 0.9)
points(jitter(rep(1, 4), factor = 3),
   sumPer[Sequence == 1], pch = 1)
points(jitter(rep(2, 4), factor = 3),
   sumPer[Sequence == 2], pch = 2)
points(x = c(1, 2), c(mean(sumPer[Sequence == 1]),
     mean(sumPer[Sequence == 2])), pch = "-", cex = 3)
axis(1, at = c(1, 2), labels = c("Seq I:\nFed then Fasted",
     "Seq II:\nFasted then Fed"), ticks = F, cex = 0.9)
mtext("log AUC: Period 1 + Period 2", side = 4, line = 3)
```

Figure 7.12: Period Effect

```
plot(as.numeric(Sequence), diffTrt, type = "n", axes = F,
  xlab = "", ylab = "AUC: Fed / Fasted",
  xlim = c(0.5, 2.5))
axislog(exp(diffTrt), opp.side = T, line = 1, srt = 90,
  cex = 0.9)
points(jitter(rep(1, 4), factor = 3),
  diffTrt[Sequence == 1], pch = 1)
points(jitter(rep(2, 4), factor = 3),
  diffTrt[Sequence == 2], pch = 2)
points(x = c(1, 2), c(mean(diffTrt[Sequence == 1]),
    mean(diffTrt[Sequence == 2])), pch = "-", cex = 3)
axis(1, at = c(1, 2), labels = c("Seq I:\nFed then Fasted",
    "Seq II:\nFasted then Fed"), ticks = F, cex = 0.9)
mtext("log AUC: Fed - Fasted", side = 4, line = 3)
```

Figure 7.13: Treatment Effect

```
plot(as.numeric(Sequence), diffPer, type = "n", axes = F,
  xlab = "", ylab = "AUC: Period 2 / Period 1",
  xlim = c(0.5, 2.5))
axislog(exp(diffPer), opp.side = T, line = 1, srt = 90,
  cex = 0.9)
points(jitter(rep(1, 4), factor = 3),
  diffPer[Sequence == 1], pch = 1)
points(jitter(rep(2, 4), factor = 3),
  diffPer[Sequence == 2], pch = 2)
points(x = c(1, 2), c(mean(diffPer[Sequence == 1]),
    mean(diffPer[Sequence == 2])), pch = "-", cex = 3)
axis(1, at = c(1, 2), labels = c("Seq I:\nFed then Fasted",
    "Seq II:\nFasted then Fed"), ticks = F, cex = 0.9)
mtext("log AUC: Period 2 - Period 1", side = 4, line = 3)
```

Listing of Food Half Data Used to Study Period 1 Residuals in Figures 7.14–7.18

```
> food.half
  subj seq    trt per      AUC subjf perf seqf seqn trtn  logAUC
1   1 +/-    Fed  1   809.44    1   1  +/-    1    1 6.696343
2   2 +/-    Fed  1   428.00    2   1  +/-    1    1 6.059123
5   5 +/-    Fed  1   712.24    5   1  +/-    1    1 6.568415
7   7 +/-    Fed  1   511.84    7   1  +/-    1    1 6.238012
3   3 -/+ Fasted  1   901.11    3   1  -/+    2    2 6.803627
4   4 -/+ Fasted  1  1146.96    4   1  -/+    2    2 7.044870
6   6 -/+ Fasted  1   745.51    6   1  -/+    2    2 6.614069
8   8 -/+ Fasted  1   852.86    8   1  -/+    2    2 6.748595

           resid      studres       stdres     pred        cooks
1   0.009612069   0.1075670    0.1176976 6.686731 0.002308787
2  -0.179138916  -4.4991205   -2.1935151 6.238262 0.801918105
5   0.123479646   1.7543451    1.5119801 6.444935 0.381013961
7   0.046047200   0.5289139    0.5638375 6.191965 0.052985445
3  -0.014709018  -0.1648621   -0.1801086 6.818336 0.005406516
4   0.016085298   0.1803839    0.1969608 7.028785 0.006465593
6   0.040115695   0.4577066    0.4912075 6.573953 0.040214138
8  -0.041491975  -0.4741032   -0.5080598 6.790087 0.043020787
```

S-PLUS Function to Compute Cook's Distance

This function is defined within the S-PLUS function plot.lm.

```
Cooks.lm <- function(fit)
{
  lmi <- lm.influence(fit)
  fit.s <- summary.lm(fit)
  s <- fit.s$sigma
  h <- lmi$hat
  p <- fit$rank
  stdres <- fit$residuals/(s * (1 - h)^0.5)   #standardized
  cooks <- (1/p * stdres^2 * h)/(1 - h)
  if(!is.null(fit$na.action))
    cooks <- nafitted(fit$na.action, cooks)
  return(cooks)
}
```

Figure 7.15: Response versus Fitted

```
plot(pred, logAUC, type = "n", xlab = "log AUC: Period 1",
    ylab = "log AUC Fitted: Period 1")
abline(0, 1)
points(pred[seqn == 1], logAUC[seqn == 1], pch = 1)
points(pred[seqn == 2], logAUC[seqn == 2], pch = 2)
place.keyseq(0.1, 1)
```

Figure 7.16: Raw Residuals versus Fitted

```
plot(pred, resid, type = "n",
  xlab = "Fitted log AUC: Period 1",
  ylab = "Raw Residuals: Period 1")
points(pred[seqn == 1], resid[seqn == 1], pch = 1)
points(pred[seqn == 2], resid[seqn == 2], pch = 2)
abline(0, 0)
id.n(pred, resid, how.many = 2,offset = 0.1)
place.keyseq(0.8, 0.2)
```

Figure 7.17: Equal Variance

```
plot(pred, sqrt(abs(resid)), type = "n",
  xlab = "Fitted log AUC: Period 1",
 ylab = "Square Root of Absolute Value Residual: Period 1")
points(pred[seqn == 1], sqrt(abs(resid))[seqn == 1],
  pch = 1)
points(pred[seqn == 2], sqrt(abs(resid))[seqn == 2],
  pch = 2)
lines(lowess(pred,sqrt(abs(resid))))
id.n(pred, sqrt(abs(resid)), how.many = 2, offset =0.1)
place.keyseq(0.8, 1.0)
```

8

Design and Analysis of Phase I Trials in Clinical Oncology

Axel Benner and Lutz Edler
Biostatistics Unit, German Cancer Research Center, Heidelberg, Germany

Gernot Hartung
Oncological Center, Mannheim Medical School, Heidelberg, Germany

8.1 Introduction

8.1.1 Role and Goal of a Phase I Trial

In oncology the Phase I trial is the first occasion to treat cancer patients experimentally with a new drug with the aim of determining the drug treatment's toxic properties, characterizing its dose-limiting toxicity (DLT), and estimating a maximum tolerable dose (MTD) as a benchmark dose for further clinical trials. Phase I trials are usually accompanied by an elaborate clinical pharmacology and the determination of individual pharmacokinetic parameters. Additionally, there is high interest in screening the drug for early signs of antitumor activity (Von Hoff et al., 1984). At this stage of drug development, a safe and efficacious dose in humans is unknown and information is available only from preclinical in vitro and in vivo studies. Therefore, patients are treated successively and in small groups, beginning at a low dose very likely to be safe (starting dose), and escalating progressively to higher doses until drug-related toxicity manifests and reaches DLT. Doses which induce a predetermined frequency of DLT among the patients treated, define the MTD (Bodey and Legha, 1987).

The statistical evaluation of a Phase I trial has to comprise a complete toxicity analysis considering dose escalation. The toxicity information of all patients should be used to determine the MTD of this trial's mode of administration. Rules for planning and conducting Phase I trials are specifically addressed by the tripartite harmonized ICH guideline, General Considerations for Clinical Trials (ICH, 1997), and the European Agency for the Evaluation of Medicinal Products (EMEA) and its Note for Guidance on Evaluation of Anticancer Medicinal Products in Man (EMEA, 1996).

Statistical methods for the practical conduct of Phase I trials are equally important, both for design and data analysis. We will present both of these below. Although the sample size of a Phase I trial compared to other clinical trials is small in general, and ranges between 10–40 cases per study, the amount of data collected per patient may be considerably larger than for the standard (Phases II–III) clinical trial. This is due to the intrinsic interest of not missing any effect, adverse or beneficial, of the drug while administered to humans for the first time. This implicates a large number of variables recorded for each patient and poses a specific challenge for an adequate, evident, and correct descriptive statistical evaluation.

We will concentrate at first in the presentation of analysis software for a moderate-sized Phase I trial. On the other hand, there is a great need for Phase I design software in order to meet the ethical, medical, and statistical requirements. Therefore, we will present in the second part of this chapter methods for designing and conducting Phase I trials.

8.1.2   Dose Levels

The Phase I clinical trial is characterized by its design parameters. Let us denote the set of dose levels at which patients are treated by $D = \{x_h, h = 1, 2, ...\}$ assuming $x_h < x_{h+1}$. Denote by x_{ih} the dose level of patient i if patients are numbered $i = 1, 2, ...$ Often patients enter the trial in groups of size n_h being treated at the same dose level x_h. A completed Phase I trial will have treated n_h patients at dose level x_h ($h = 1, ..., H$) at a total of H different dose levels. In practice, H varies between a few to about 10–15 levels.

For the planning of a Phase I trial, a small candidate set of M toxicity endpoints is defined which are relevant for the drug being tested. The degree of acceptable and nonacceptable toxicity for each of the M endpoints is determined in advance.

8.1.3   Dose–Toxicity Relationship

One defines dose-limiting toxicity (DLT) for a patient if at least one toxicity of the candidate subset of the M toxicities occurred and was of grade 3 or 4 using the Common Toxicity Criteria of the National Cancer Institute (CTC, 1999). In applications such a toxicity has to be judged as at least "possibly" treatment related before it is considered as approved DLT. For the statistical analysis, each patient should be assessable as having experienced a DLT or not, at his/her dose level.

Given a dichotomous dose toxicity indicator of DLT, the dose–toxicity function $\psi(x, a)$ is a continuous monotone nondecreasing function of the dose x on the real line $0 \leq x < \infty$ defined by

$$\psi(x, a) = \Pr(\text{DLT} \mid \text{Dose} = x) \tag{8.1}$$

with parameter a. Boundary conditions are $\psi(0, a) \geq 0$ and $\psi(\infty, a) \leq 1$. The maximum tolerable dose (MTD) is a dose from D at which an acceptable proportion θ, $0 < \theta < 1$, of toxicity is observed in a set of patients treated at that dose level. Such a MTD is defined as θ-percentile ψ(MTD, a) = θ. Common examples of dose–toxicity models are the probit or the logistic regression model.

8.1.4   Statistical Analysis Strategy

The statistical analysis of the trial data is performed in several steps. At first the study population is described as a whole in terms of patient characteristics. Then each subpopulation, at one of the dose levels, is described separately for treatment, safety, and to a limited extent for efficacy. Finally, the pharmacokinetic (PK) analysis is performed for each patient's blood serum drug concentrations and the PK parameter estimates are summarized for each dose level. This evaluation strategy determines the data organization and the evaluation steps.

8.2   Source of Data

A single-center Phase I study was performed for the dose amount of a new drug MTX-HSA of promising antitumor activity on 17 patients entering the trial between July 1995 and November 1996 (Hartung et al., 1999). The study, denoted as the MTX-HSA-I Trial, was performed under the rules of Good Clinical Practice (GCP) and the Standard Operating Procedures (SOP) of the Phases I/II Study Group of the German Association of Medical Oncology (AIO) in the German Cancer Society (Kreuser et al., 1998).   Data documentation and biometric data analysis were performed according to the SOPs of the study group and the SOPs of the Biometric Center (Edler and Friedrich, 1995).

All 17 patients entering the MTX-HSA-I Trial were eligible and assessed for dose-limiting toxicity. The trial was designed to answer the following questions:

1.   What is the maximum tolerable dose of MTX-HSA in weekly injections administered for 8 weeks?
2.   How is the treatment tolerated and what is the safety/toxicity profile? Can any response be achieved with 20 mg/m^2 or more?

Patient recruitment, eligibility, and treatment modalities were described by Hartung et al. (1999). On the basis of preclinical data and information on the mechanism of the drug, an approximate modified Fibonacci dose scale was chosen starting with 20 mg/m^2, increasing by steps of 20 mg/m^2 to 40 mg/m^2 and 60 mg/m^2 and reducing to 50 mg/m^2. The deterministic 3 + 3 rule was used for dose escalation (van Hoff et al., 1984; Edler, 2001). Three patients were treated at each dose level. If no dose-limiting toxicity (DLT) was observed during the first three administrations of MTX-HSA the next dose level was started. If DLT was observed, up to six patients were treated at the actual dose level. If DLT occurred in two or more of these patients no more patients were treated and the maximum tolerable dose was defined as one dose level below the actual. For each patient, the drug concentration in blood serum was determined for up to

5–15 days. Subsequently a population pharmacokinetic analysis was performed. A second Phase I trial was started in 1998 in order to establish the MTD, if MTX-HSA was administered in a biweekly schedule. It was terminated in 1999. We will use here the complete data of the MTX-HSA-I Trial to illustrate a statistical analysis for Phase I trials using S-PLUS.

8.2.1   Patient Characteristics

The demographic and clinical characteristics of the 17 patients at study entry are based on a total of B quantitative, categorical or free text baseline variables b_r, $r = 1, ... , B$ (> 100), which are documented using the patient initial assessment case report forms.

8.2.2   Phase I Treatment Data

Evaluation of the treatment of oncological patients proceeds in treatment cycles of mostly 2–4 weeks. In this study the treatment schedule was weekly. Each patient i receives at scheduled days t_{is}, $s = 1, ..., S_i$, his/her individual dose x_{is}, calculated on the basis of that patient's body surface in units of mg/m^2, $i = 1, ...,$ 17. A graphical description of these data is given in Figure 8.1. Deviations from the scheduled days occur, in practice, such that the observed days of treatment may show a slightly irregular pattern. The individual dose x_{is} may also change from one day to the next. Treatment of patient i stops after a number of weeks or cycles. The reason for termination, recorded in the case report forms, is typically progression of the disease or occurrence of a dose-limiting toxicity. The MTX-HSA-I trial had a fixed duration of a maximum of eight cycles. However, patients who tolerated the assigned dose well and who were in a stable condition or improving toward a clinical response were treated beyond that limit. The treatment history of those patients was described separately from the day of continuation until termination of the trial medication. These additional data will not be considered here.

8.2.3   Phase I Toxicity Data (Categorical Scale)

The data available after the completion of a Phase I trial are multidimensional longitudinal toxicity data y_{ijm} of the toxicity variable T_m observed for the patient i during the treatment cycle j, $i = 1, ..., n$, $j = 1, ..., J_i$, $m = 1, ..., M$. The data are grouped by a small number of patients, treated at the same scheduled dose levels. Each toxicity variable is categorical with outcomes according to the Common Toxicity Criteria (CTC) of the Phases I/II Study Group of the AIO defined by grade 0–grade 4 (Kreuser et al., 1998). As an example take the classification of loss of appetite and of the platelet count from the hematologic investigation shown in Table 8.1.

Table 8.1. Classification of loss of appetite and platelet count.

	Grade				
	0	1	2	3	4
Appetite	Normal	Slight reduction	Reduced < 1 week	Reduced > 1 week	Complete loss of appetite
Platelets (x 109/l)	≥ 100.0	75.0–99.9	50.0–74.9	25.0–49.9	< 25.0

The CTC list especially includes all relevant laboratory findings related to toxicity. Usually, dose-limiting toxicity is defined by the CTC grades 3–4.

The occurrence of each toxicity is judged by a clinician or by a study nurse for a causal relation to the investigational treatment on a five–point scale: 1 = certain, 2 = probable, 3 = possible, 4 = not probable, 5 = insufficient evidence.

Hence, for each toxicity variable T_m judgment variables u_{ijm} for patient i in cycle j complement the toxicity grading y_{ijm}. The following overview shows the nonhematological toxicity data for the patient with id 106: the treatment week (week), the toxicity endpoint (the number from the CTC catalogue, ctcnr, and the type of toxicity, ctc), the CTC grade (grade), and the judgment of the relation to the treatment (relation):

```
     Patient  week ctcnr       ctc  grade  relation
128    106      2  2.04 Stomatitis     3        1
129    106      2 10.01   Appetite     2        2
130    106      4  2.04 Stomatitis     3        1
131    106      4 10.01   Appetite     2        1
```

The toxicity judgment is often dichotomized as drug-related toxicity (adverse drug reaction, ADR) if $u_{ijm} \leq 3$. Dose-limiting toxicity is defined as a dichotomous function Y_i of (y_{ijm}, u_{ijm}), $j = 1, ..., J_i$, $m = 1, ..., M$, by

$$Y_i = \begin{cases} 0 & \text{no DLT} \\ 1 & \text{DLT} \end{cases} \quad i = 1,...,n. \quad (8.2)$$

8.2.4   Laboratory Data on a Quantitative Scale

Standard clinical laboratory measurements are available as numerical values. A small subset of the laboratory data obtained at weeks 2, 4, 6, and 8 (labweek) for patient 104 of the 50 mg/m^2 group (dose) is given as

```
     Patient  dose labweek  labdate   hb hk platelet leucos
 99    104     50        2 05/29/96 16.0 47      245    7.6
101    104     50        4 06/12/96 14.9 44      148    7.6
103    104     50        6 06/26/96 14.3 40      137    6.7
```

```
105      104     50        8 07/10/96  9.9 26              52     6.4
```

	albumin	bilirub	cholest	ap	ggt	sgpt	ldh	ck
99	49.1	0.5	243	129	28	73	202	30
101	45.4	0.7	200	143	55	150	268	23
103	45.4	0.9	186	185	83	95	382	NA
105	37.0	2.6	128	318	146	60	594	17

8.2.5   Response Data

Although response is not the primary goal of the Phase I trial, all patients are evaluated for possible clinical response.   Tumor staging and response data documentation follows the same procedure as in a Phase II trial. Because of an expected small number of responding patients, a case-by-case description is in most cases sufficient.   This has to include dosing information, the time to response, and the time to later relapse.   Occasionally, survival data may be reported.   However, it is important to report the type of tumor of the patients who showed response, and information about improvement or at least a stabilization of the cancer disease.   It is usually sufficient to summarize this information in listings.

8.2.6   Pharmacokinetic Data

Measurements of blood serum concentrations of the drug $c_{is} = C(t_{is})$ are assumed to be available at predetermined times t_{is}, $s = 1, \ldots , S_i$, for patient i. The time scale of the data is usually in hours. The pharmacokinetic data, describing time of measurement since application (Time), the plasma concentration at that time (conc), and the individual dose of MTX-HSA (dose.i) for patient 104 is given as

```
           Day   time conc Patient dose.i
 1 11/12/97      0.0  3.7     104    91.5
 2 11/12/97      0.5 65.8     104    91.5
 3 11/12/97      1.0 65.3     104    91.5
 4 11/12/97      2.5 57.5     104    91.5
 5 11/12/97      4.0 59.5     104    91.5
 6 11/12/97      8.0 46.5     104    91.5
 7 11/13/97     24.0 37.5     104    91.5
 8 11/14/97     48.0 30.7     104    91.5
 9 11/17/97    120.0 22.6     104    91.5
10 11/19/97    168.0 18.2     104    91.5
```

8.3 Data Processing

The original data were available from a SAS database which has been constructed in the Biometric Center of the Phases I/II Study Group responsible for the documentation and the biometric analysis of the data. The structure of the database and the description has been documented in the SOPs of the Biometric Center (Edler and Friedrich, 1995).

The complete dataset of the MTX-HSA-I Trial was available for this investigation as an SAS data file created from this database. Modules from the biometric analysis system were used to preprocess the data and to create the files for the specific analyses. The function sas.get was used to import the datasets into S-PLUS, e.g.

```
> toxicity <- sas.get("./datasets", "toxicity")
```

loads the SAS dataset of toxicity information into S-PLUS.

Processing of these data is determined by the type and aim of the statistical analysis steps. S-PLUS, Version 3.4 for Sun Solaris was used for all statistical computations. Data management and preprocessing was performed using SAS, Version 6.12.

8.4 Descriptive Statistical Analysis

The baseline variables are summarized for the description of the study population distinguishing quantitative, categorical, and textual data types and applying standard functions as boxplots or summary statistics, using S-PLUS functions like bwplot, summary, etc.

All results obtained in a Phase I trial have to be reported in a descriptive statistical analysis which accounts for the dose levels. This is somehow cumbersome since each dose level has to be described as a separate stratum even if the number of patients treated at that dose is very small. A comprehensive and transparent report of all toxicities observed in a Phase I trial is an absolute must for both the producer's (drug developer) and the consumer's (patient) risks and benefits. Therefore graphical displays for descriptive analysis are developed using the Trellis library of S-PLUS.

For example, Figure 8.3 is produced using function dotplot to display the grade of toxicity, the labels of the toxicity variables, and the relation to treatment for each individual patient. To tailor the dotplot according to our needs we only have to alter the panel function which describes what the display method should do. Here we especially use gray-shading to support the toxicity grading (see Appendix 8.A.6 for code to produce Figures 8.1–8.4).

8.4.1   Trellis Displays

The Trellis library of S-PLUS implements Cleveland's ideas of data visualization (Cleveland, 1993). One of the main features of Trellis displays is multipanel conditioning where panels are laid out in a regular trellis-like structure into columns, rows, and pages. All panels of a Trellis display show the same graphical method for a defined subset of the data. One-, two-, and three-dimensional plots are drawn depending on one or more conditioning variables by categorizing the data points into groups corresponding to the realization of the conditioning variables. If a partitioning variable is categorical then for each level of this variable a subplot of the corresponding subset is drawn. If the partitioning variable is continuous then subsets will be defined on overlapping intervals ("shingles") where endpoints of the intervals are computed to make the counts of points in the subsets as nearly equal as possible.

8.4.2   Evaluation of Treatment

Course of the trial, individual patient's study duration, and individual dosing are summarized for each dose level by an application of Trellis displays. For each patient the individual administered doses and the date of the start of all treatment courses/cycles is shown in panels corresponding to the designed dose levels (Figure 8.1).

8.4.3   Evaluation of Categorical Toxicity Data

Among the methods describing toxicity of a Phase I trial are those for categorical data most important because of the common basis of the Common Toxicity Criteria (CTC) classification–formerly the WHO criteria–for all oncological studies. Notice that this outline applies both to nonlaboratory and laboratory toxicity data which then have been transformed into the categorical scale according to Common Toxicity Criteria.

Contingency tables are generated to present absolute and relative frequencies of the occurrences of grades 1–4 toxicity separately for each of the four categories, for the two high categories 3–4 combined and for all four categories 1–4 combined. In principle, toxicity can be evaluated for each cycle separately. In most analyses, however, the toxicity is summarized per patient over all observed cycles of each patient and the worst category is recorded for the analysis. From these data, we calculate the frequencies mentioned above by use of common S-PLUS functions. We use functions like `table` to build multiway contingency arrays and `apply` for handling specified functions to sections of the arrays. In addition a very useful presentation for communicating the toxicity of the complete trial to clinicians is obtained by the use of multipanel displays (Figure 8.2).

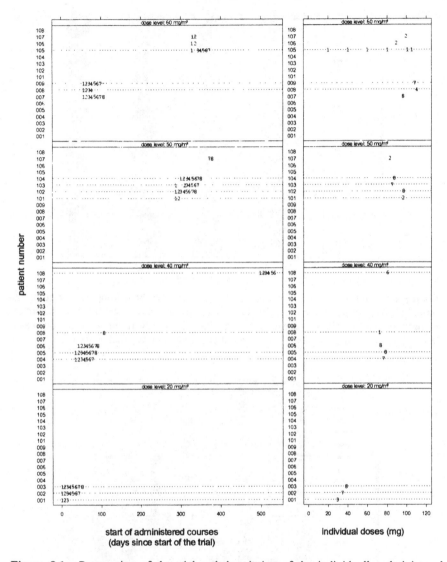

Figure 8.1. Progression of the trial and description of the individually administered doses of MTX-HSA shown separately in panels corresponding to the four designed dose groups. The numbers on the lines count the corresponding course since the start of the trial (left column) and denote the total number of administered courses at the patient's individual dose levels (right column).

The frequencies of the occurrences of the toxicity categories and their combinations are supplemented by information about the relation to treatment together with each toxicity outcome. Furthermore, this judgment is dichotomized into the occurrence or nonoccurrence of an adverse drug reaction. This evaluation becomes rather complicated if the judgment of the relation to treatment of a

198   A. Benner, L. Edler, and G. Hartung

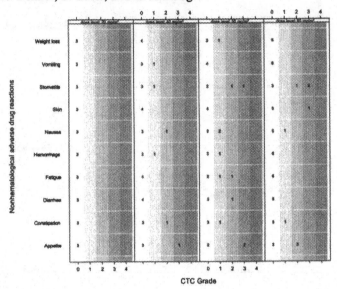

Figure 8.2. Dotplot of grade of toxicity for the main nonhematological toxicities observed displaying the absolute number of patients for each dose level (mg/m^2), where the highest grade of toxicity is recorded for each patient if the adverse effect was judged as at least possibly treatment related. For each initial dose level one panel is drawn.

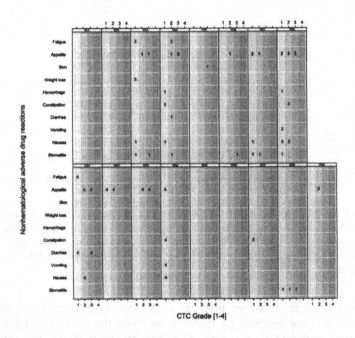

Figure 8.3. Dotplot of grade of toxicity for the main nonhematological toxicities observed, labeled by an indicator for the strongest relation to treatment observed, drawn separately for each of the 17 patients.

given toxicity changes over the cycles within the same patient. In such a case an adequate assessment of adverse drug reaction has to be discussed with the clinician and a decision has to be found as to which cycle is taken as representative for the patients adverse drug reaction.

There are two methods for presenting this toxicity information per dose level. One approach uses the grading of the toxicity as main descriptor and another uses the relation to treatment as main descriptor. In the first case the panels are organized separately for each dose group with the toxicity endpoints as rows, provided at least one toxicity of that type of grades 1–4 occurred at all, and the grading as columns. The judgment information will then be inserted into the cells of this array (Figure 8.2).

A description of the individual load of toxicity of each patient is required for indicating multiplicity of dose-limiting toxicities which may go almost as far as a case-wise individual description, again supported by the use of multipanel displays (Figure 8.3). This kind of graphical display can be complemented by a second one of the same structure with relation to treatment as main descriptor.

8.4.4   Evaluation of Hematological and Laboratory Data

Hematological and laboratory reactions can be presented using Trellis displays, either showing the quantitative data directly (Figure 8.4) or by categorizing using the Common Toxicity Criteria grading resulting in figures like Figure 8.3.

8.5   Exploratory Data Analysis

8.5.1   Pharmacokinetic Data Analysis

An important secondary objective of a Phase I trial is the assessment of the distribution and elimination of the drug in the body. Specific parameters that describe the pharmacokinetics of the drug are the absorption and the elimination rate, the drug half-life, the peak concentration, and the area under the time–drug concentration curve (AUC). Drug concentration measurements $C_{is} = C(t_{is})$ of patient i at time t_{is}, $s = 1, \dots , S_i$, are usually obtained from blood samples taken regularly. They are analyzed using pharmacokinetic (PK) models (Gibaldi and Perrier, 1982). Mostly, one- and two-compartment models have been used to estimate the pharmacokinetic characteristics. In practice, statistical methodology for pharmacokinetic data analysis is primarily based on nonlinear curve fitting using least squares methods or their extensions (Edler, 1998). The goal is to estimate the drug time course $C(t)$ and the key pharmacokinetic parameters. One distinguishes between an individual pharmacokinetic analysis of each patient's pharmacokinetic and a population pharmacokinetic analysis which combine the information obtained for all patients treated with the aim to predict future patients pharmacokinetic reaction.

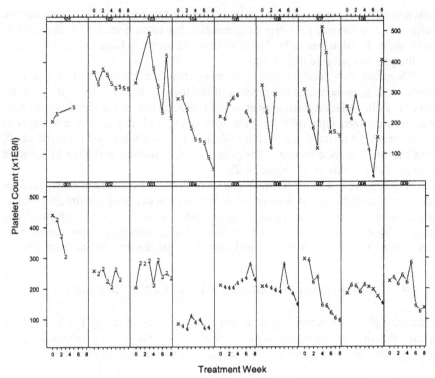

Figure 8.4. Platelet counts ($\times 10^9$/l) during the 8-week study period, determined on a weekly basis. The plotting character describes the corresponding dose group by the first digit of the dose level (e.g., 4 denotes 40 mg/m^2). An "x" marks the platelet count at treatment start or during treatment interruption. In patient 008, therapy began with 60 mg/m^2 MTX-HSA. After four administrations, stomatitis (CTC grade 3) occurred. The treatment was interrupted until recovery 3 weeks later, deescalated, and continued with weekly injections of 40 mg/m^2.

For individual pharmacokinetic analyses standard functions for nonlinear modeling like `nls` or `nlminb` can be used. For population pharmacokinetic analyses, software for nonlinear mixed effects models can be used (Bates and Pinheiro, 1997). Bates and Pinheiro use the Trellis approach and functions `nlsList` and `nlme` for a class of object, called `groupedData`, which is used to define a group of variables and their relationship for modeling and graphical presentation.

The model used here is the one-compartment model with first-order absorption and elimination. The plasma concentration $C(t)$ is modeled by

$$C(t) = \frac{x_0 \, K_e \, K_a}{Cl \, (K_a - K_e)} \left(e^{-K_e t} - e^{-K_a t} \right) \tag{8.3}$$

where x_0 is the initial dose, K_a is the absorption rate constant, K_e is the elimination rate constant, and Cl is the drug clearance.

As an example, the dataset pkpd, describing the pharmacokinetic data of a group of six patients (ids h1, h2, h3, vor, 104, and 107; the last two are from the MTX-HSA-I Trial), is converted to a groupedData object PKPD:

```
> PKPD <- groupedData(conc ~ time | Patient, data = pkpd,
    labels = list(x = "Time (hours since administration)",
      y = "MTX-HSA (\265mol/l)"))
```

Using Version 3.3 of the nlme library that comes with S-PLUS 6.0, a preliminary fit can be obtained using nlsList together with the self-starting nonlinear regression model function SSfol:

```
> PKPD.list <- nlsList(conc ~ SSfol(Dose = dose.i,
    input = time + 0.01, lKe, lKa, lCl), data = PKPD)
```

where SSfol evaluates the first-order compartment function and its gradient. Arguments are Dose (initial dose), input (times $\neq 0$ at which to evaluate the model), and the model parameters lKe, lKa, and lCl (the natural logarithms of the elimination rate constant, the absorption rate constant, and the clearance). Individual parameter estimates are

```
> coef(PKPD.list)
          lKe       lKa        lCl
  h1 -4.850819 3.722281 -3.343159
  h2 -4.400866 1.487227 -3.482554
  h3 -3.915599 2.189390 -3.285989
 107 -4.681108 1.999659 -4.092716
 104 -4.609915 2.335856 -4.173909
 vor -4.920996 2.630922 -4.679261
```

The individual fits can be analyzed using a Trellis plot of the 95% confidence intervals on the parameter estimates:

```
> plot(intervals(PKPD.list))
```

All parameters (especially lCl) seem to vary between the individual fits, and a full nonlinear mixed effects model seems adequate. The individual regression fits can now be used to describe the full nonlinear mixed effects model:

```
> PKPD.nlme <- nlme(PKPD.list, control = list(gradHess=F))
```

A plot of the standardized residuals against the fitted values indicates that a heteroscedastic model may provide a better fit, which is finally supported by likelihood comparison:

```
> plot(PKPD.nlme)
> PKPD.nlme2 <- update(PKPD.nlme, weights=varPower(),
    control = list(gradHess=F))
```

```
> anova(PKPD.nlme, PKPD.nlme2)
           Model df      AIC      BIC      logLik
PKPD.nlme      1 10 440.8272 462.7238 -210.4136
PKPD.nlme2     2 11 427.9314 452.0176 -202.9657


           Test  L.Ratio p-value
PKPD.nlme
PKPD.nlme2 1 vs 2  14.8958  0.0001
```

The `compareFits` method can be used to compare the individual estimated coefficients of the `nlme` fit `PKPD.nlme2` to those of the individual nonlinear fits of `PKPD.list`:

```
> plot(compareFits(coef(PKPD.list), coef(PKPD.nlme2)))
```

A Trellis scatterplot of the predictions versus time using the `augPred` (augmented prediction) method is shown in Figure 8.5. Original observations are presented by circles, predicted values are joined by lines. The panels are drawn for each patient individually:

```
> plot(augPred(PKPD.nlme2))
```

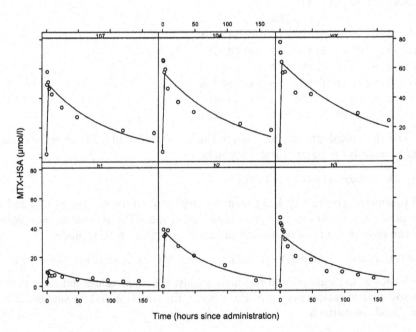

Figure 8.5. MTX-HSA plasma concentrations (μmol/L) during the first days of six patients (ids : h1, h2, h3, 104, 107, and vor) treated with 20 (h1), 40 (h2, h3), or 50 (104, 107, vor) mg/m² in weekly courses. The plots show the fit of the full nonlinear mixed effects model (solid lines) together with the original observations (circles).

8.6   Design Analysis

8.6.1   Dose Escalation Schemes

Having defined the set of dose levels, a rule has to be found according to which the patients, who, consecutively entering the Phase I trial, are assigned to those doses. Traditionally, deterministic assignment rules have been used. Most prominent is the traditional escalation rule (TER), which escalates step by step. Patients are treated in groups of three, each receiving the same dose. If none of the three patients shows a dose-limiting toxicity (DLT), the next group of three receives the next higher dose. Otherwise, a second group of three is treated at the same dose level again. If only one out of the six patients exhibits a DLT, the trial continues at the next higher level. If two or more patients exhibit DLT, the increase stops at that level. After the increase has stopped, some alternatives of treating a few more patients at that level or at the next lower level (if that level did not already have six patients) have been proposed and used.

8.6.2   Estimation of the Maximum Tolerable Dose

In standard Phase I designs the estimation of the maximum tolerable dose (MTD) is part of the study designs and an estimate of the MTD results directly from there. In the traditional escalation design, the MTD is taken as the dose level next lower to the dose level where a predefined proportion of patients experienced dose-limiting toxicity (DLT). A dichotomous summary variable Y_i is derived which characterizes each patient as having experienced DLT or not (8.2). From this, one can in principle calculate a toxicity rate for each dose level. An estimate of a standard error of the rate is impaired by the small number of cases and will be useless in most practical situations. However, the DLT information of all patients can be used to derive an estimate of the maximum tolerable dose using a dose–toxicity model as defined in (8.1).

A general method for analyzing dose–toxicity data is the logistic regression of the dichotomized toxicity variable Y_i on the actually applied doses x_i of all n patients treated in the Phase I trial. This would disregard any dependency of the dose–toxicity data on the design that had created the data and contains therefore some bias. Assuming independent sampling of pairs (Y_i, x_i), $i = 1, ..., n$, the logistic regression model is given by

$$\Pr\left(Y_i = 1 \mid x_i\right) = \frac{1}{1 + \exp\left(a_0 + a_1 x_i\right)} \tag{8.4}$$

Logistic (or probit) regression via the S-PLUS function glm provides the maximum likelihood estimate (MLE) of $a = (a_0, a_1)$ which is standard in biostatistical toxicology (Finney, 1978; Morgan, 1992). The maximum tolerable dose is given as the dose x_θ for which the probability of dose-limiting toxicity is θ:

$$x_\theta = \frac{\log\left[\theta/(1-\theta)\right] - a_0}{a_1} \tag{8.5}$$

Estimates and (approximate) confidence limits can be obtained by using `glm` together with the `dose.p` function of the MASS library (Venables and Ripley, 1999, pp. 221–222).

With $\theta = 1/3$ and Y_i as defined in (8.2), the S-PLUS code to compute the estimate of MTD_θ, together with an approximate 95% confidence interval, is given in Appendix 8.A.7. The essence of the code is as follows

```
> fit <- glm(y ~ dose, family = binomial)
> xp <- dose.p(fit, p = 0.33)
> xp.ci <- xp + c(-qnorm(0.975),
    qnorm(0.975)) * attr(xp, "SE")
> mtd <- cbind(xp.ci[1], xp, xp.ci[2])
> dimnames(mtd)[[2]] <- c("LCL", "xp", "UCL")
> round(mtd, 1)
          LCL   xp  UCL
p = 0.33: 32.1 47.1 62.1
```

8.6.3 Dose Titration Designs

Inclusion of grade 2 toxicity was formalized in the dose titration design proposed by Simon et al. (1997), who modified the traditional escalation rule (TER). Let Y_i now be a categorical endpoint variable with four levels: acceptable toxicity (grade ≤ 1), conditionally acceptable toxicity (grade $= 2$), dose-limiting toxicity (DLT; grade $= 3$), unacceptable toxicity (grade $= 4$). A new within-patient dose escalation strategy was considered: dose increase as long as only acceptable toxicity occurs (grade ≤ 1), decrease in case of DLT or unacceptable toxicity (grade ≥ 3). In the case of moderate toxicity (grade $= 2$), the dose remains unchanged.

Simon et al. (1997) formulated three accelerated titration designs (design 2–design 4) by combining an accelerated group escalation with this intrapatient dose escalation strategy. (Their design 1 is the standard TER design described above.)

- **Design 2.** Escalate the dose, after each patient, by one level as long as no grade 3 or worse toxicity occurs in the first course, and at most one patient shows grade 2 toxicity in the first course. If a DLT occurs in the first course or if grade 2 has occurred twice in the first course, switch to the TER.

- **Design 3.** Same as Design 2, except that a two-dose-step increase is used in the first stage before switching to TER.

- **Design 4.** Same as Design 3, but no restrictions on the increase with respect to the treatment course. If one DLT occurred in any course or if grade 2 toxicity occurred twice in any course, switch to the TER.

To implement the Accelerated Titration Design (ATD), toxicity data for patient i in course (or cycle) j must be described by y_{ij} on a four-grade scale. Simon et al. related it to the Common Toxicity Criteria of the National Cancer Institute as nonmild (grades 0–1), moderate (grade 2), dose-limiting (grade 3), and unacceptable (grades 4–5). The individual doses per treatment cycle are denoted as x_{ij} and the cumulative dose z_{ij}.

A regression model for the data analysis is given by a mixed effects model

$$y_{ij} = \log\left(x_{ij} + \alpha\, z_{ij}\right) + \beta_i + \varepsilon_{ij} \tag{8.6}$$

where α represents the influence of cumulative toxicity and β_i describes a random effect caused by interpatient variability. A normal distribution $N(\mu_\beta,\ \sigma_\beta^2)$ is assumed for β_i and $N(0,\ \sigma_\varepsilon^2)$ for e_{ij}.

The S-PLUS function phlatd is used to fit this model. It was written by B. Freidlin, in collaboration with R. Simon and L. Rubinstein, and is available on the Internet at http://linus.nci.nih.gov/~brb/Methodologic.htm. The function computes the maximum likelihood estimates (together with 95% confidence limits) of the model parameters and draws graphs of the probabilities of moderate and dose-limiting toxicities. An application of this procedure (design 4) to simulated data according to the data of the MTX-HSA-I Trial resulted in maximum likelihood estimates and their 95% confidence limits:

	MLE	Lower Bound	Upper bound
K1	3.782	3.779	3.785
K2	5.149	NA	NA
K3	6.237	6.234	6.240
Alpha	0.008	NA	NA
Sigma Beta	1.262	1.256	1.268

Now y_{ij} translates to the toxicity levels by means of $K_1 < K_2 < K_3$, which divide the range of toxicities into four regions representing the grades of toxicity. If y_{ij} was less than a prespecified constant K_1, then toxicity less than grade 2 was considered for patient i during course j, given dose x_{ij}. Toxicity of grade 2 was considered, if $K_1 < y_{ij} < K_2$. If $K_2 < y_{ij} < K_3$, dose-limiting toxicity, and if $K_3 < y_{ij}$ unacceptable toxicity was considered.

In the model, K_1 was prespecified by $[K_1 - \log(\textit{starting dose})]/\log(1.4)$ to represent 40% dose steps between the starting dose and the dose at which an average patient has a 50% chance of experiencing at least grade 2 toxicity. The distance between the other K-values is defined in a similar way. The 40% increment is close to the 33% increment of the modified Fibonacci approach, which was used for the MTX-HSA-I Trial.

The probability of toxicity grade 2 or worse is given by

$$\Phi\left[\frac{\log(x+\alpha z)+\mu_\beta - K_1}{\sqrt{\sigma_\beta^2 + \sigma_\varepsilon^2}}\right] \tag{8.7}$$

Probabilities for grade 3+ and grade 4+ toxicity are computed by replacing K_1 by K_2 and K_3. Resulting estimates of the probabilities of grade 2+, 3+, and 4+ toxicities at the various dose levels of the simulation are given in Figure 8.6, which was created with the following commands:

```
> phlatd("demo.pat.txt", "demo.dose.txt")
```

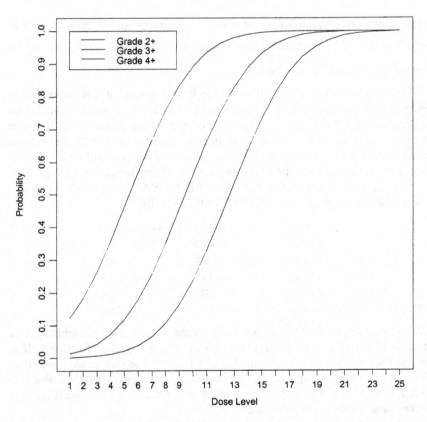

Figure 8.6. Graph of the estimated probabilities of moderate toxicity and dose-limiting toxicity at the various dose levels of a simulation of an accelerated titration design using toxicity data from the MTX-HSA-I Trial.

Continual Reassessment Method (CRM)

A Bayesian-based dose escalation rule was introduced by O'Quigley et al. (1990), which selects doses from a predetermined set of dose levels using a prior distribution of the maximum tolerable dose (MTD) and updating that with each new patient's toxicity response. Each patient is treated at the dose level closest to the currently estimated MTD.

Formally, let Y_j denote the dichotomous response variable of the jth patient, as defined in (8.2). Summarizing sequentially collected dose–toxicity information $\{x_1, ..., x_{j-1}; Y_1, ..., Y_{j-1}\}$ by Ω_j, with Ω_1 as a priori dose–toxicity information, one describes the information upon the parameter a of the dose–toxicity function $\psi(x, a)$ by the posterior density function $f(a, \Omega_j)$. This is a conditional density normalized by

$$\int_0^\infty f(a, \Omega_j)\, da = 1, \qquad a > 0, \ j \geq 1. \tag{8.8}$$

Denote by MTD_θ the MTD corresponding to the target toxicity probability θ, $0 < \theta < 1$. The CRM then works as follows: Given the previous information Ω_j the next dose level is determined such that it is closest to the current estimate of MTD_θ. For this, the probability of a toxic response is calculated for each dose $x_h \in D$ given Ω_j as

$$\theta_{hj} = \int_0^\infty \psi(x_h, a) f(a, \Omega_j)\, da \tag{8.9}$$

for $h = 1, ..., H$. The dose level for the jth patient is then selected from D such that the distance of θ_{hj} to the target toxicity rate θ is minimal.

After observing at dose level x_j the toxicity Y_j, the posterior density of the parameter a is obtained from the prior density $f(a, \Omega_j)$ and the likelihood of the jth observation

$$L(Y_j, x_j, a) = \psi(x_j, a)^{Y_j} (1 - \psi(x_j, a))^{1 - Y_j}, \tag{8.10}$$

using Bayes Theorem as

$$f(a, \Omega_{j+1}) = \frac{L(Y_j, x_j, a) f(a, \Omega_j)}{\int_0^\infty L(Y_j, x_j, u) f(u, \Omega_j)\, du}. \tag{8.11}$$

The iterative procedure starts with an a priori density $f(a)$. The maximum tolerable dose is estimated by the last dose level of the prefixed number of n patients. Modifications of the CRM have been suggested restricting the choice of a starting dose, not skipping consecutive dose levels and allowing groups of patients at one dose level (see references in Hanauske and Edler, 1996, or Edler, 2001).

FORTRAN code for the CRM design is available on the Internet at http://odin.mdacc.tmc.edu/anonftp. Version 1.0 of this code is written in FORTRAN90 and FORTRAN77 by J.J. Venier, B.W. Brown, and P.F. Thall. It is denoted **crm** and is described as the first regular release of a menu-driven program for the continual reassessment method. A second program, denoted **phase1**, written by P.F. Thall and B.W. Brown in FORTRAN77 runs the modified continual reassessment method as described in Goodman et al. (1995), and is available at the same Internet address.

8.7   References

Bates, D.M. and Pinheiro, J.C. (1997). Software design for population pharmacokinetic analysis. Presented at the *International Association for Statistical Computing Conference*, Los Angeles, CA.

Bodey, G.P. and Legha, S.S. (1987). The Phase I study: General objectives, methods and evaluation. In: Muggia, F.M. and Rozencweig, M., eds. *Clinical Evaluation of Antitumor Therapy*. Dordrecht Nijhoff, pp. 153–174.

Cleveland, W.S. (1993). *Visualizing Data*. Hobart Press, Summit, NJ.

CTC, *Common Toxicity Criteria, Version 2.0* (Cancer Therapy Evaluation Program of the National Cancer Institute, 1999; http://ctep.info.nih.gov/CTC3).

Edler, L. (1998). Computational statistics for pharmacokinetic data analysis. In: Payne, R. and Green, P., eds. *COMPSTAT. Proceedings in Computational Statistics*. Physica, Heidelberg, pp. 281–286.

Edler, L. (2001). Phase I Trials. In: Crowley, J., ed., *Handbook of Statistics in Clinical Oncology*. Marcel Dekker, New York, pp. 1–34.

Edler, L. and Friedrich G. (1995). Standard-Arbeitsanweisungen - SOPs: Standard Operating Procedures für das Biometrische Zentrum (BZ) der Phase I/II Studiengruppe der Arbeitsgemeinschaft Internistische Onkologie (AIO) in der Deutschen Krebsgesellschaft (DKG). Technical Report 2/94. DKFZ, Heidelberg.

EMEA. (1996). European Agency for the Evaluation of Medicinal Products; Note for Guidance on Evaluation of Anticancer Medicinal in Man (EMEA, London; http://www.eudra.org/PDFs/EWP/020595en.pdf).

Finney, D.J. (1978). *Statistical Methods in Biological Assay*. C. Griffin, London.

Gibaldi, M. and Perrier, D. (1982). *Pharmacokinetics*. Marcel Dekker, New York.

Goodman, S.N., Zahurak, M.L., and Piantadosi, S. (1995). Some practical improvements in the continual reassessment method for Phase I studies. *Statistics in Medicine* 14, 1149–1161.

Hanauske, A.R. and Edler, L. (1996). New clinical trial designs for Phase I studies in hematology and oncology: Principles and practice of the continual reassessment model. *Onkologie* **19**, 404–409.

Hartung, G., Stehle, G., Sinn, H., Wunder, A., Schrenk, H.H., Heeger, S., Kränzle, M., Edler, L., Frei, E., Fiebig, H.H., Heene, D.L., Maier-Borst, W., and Queisser, W. (1999). Phase I trial of methotrexate–albumin in a weekly intravenous bolus regimen in cancer patients. *Clinical Cancer Research* **5**, 753–759.

ICH (1997). International Conference on Harmonization; General Considerations for Clinical Trials (ICH Secretariat, c/o IFPMA, Geneva, Switzerland, http://www.ifpma.org/ich5e.html#Design).

Kreuser, E.D., Fiebig, H.H., Scheulen, M.E., Hanauske, A., Keppler, B.K., Mross, K., Schalhorn, A., Eisenbrand, G., Edler, L., Höffken, K., and Berdel, W.E. (1998). Standard operating procedures and organization of German Phase I, II and III study groups, new drug development group (AWO) and study group of pharmacology in oncology and hematology (APOH) of the German Cancer Society. *Onkologie* **21** (Suppl. 3), VI+70.

Morgan, B.J.T. (1992). *Analysis of Quantal Response Data*. Chapman & Hall, London.

O'Quigley, J., Pepe, M., and Fisher, L. (1990). Continual reassessment method: a practical design for Phase I clinical trials in cancer. *Biometrics* **46**, 33–48.

Simon, R.M., Freidlin, B., Rubinstein, L.V., Arbuck, S., Collins, J., and Christian, M. (1997). Accelerated titration designs for Phase I clinical trials in oncology. *Journal of the National Cancer Institute* **89**, 1138–1147.

Venables, W.N. and Ripley, B.D. (1999). *Modern Applied Statistics with S-PLUS*. Third Edition. Springer-Verlag, New York.

Von Hoff, D.D., Kuhn, J., and Clark, G.M. (1984). Design and conduct of Phase I trials. In: Buyse, M.E., Staquet, M.J., and Sylvester, R.J., eds. *Cancer Clinical Trials: Methods and Practice*. Oxford University Press, Oxford, UK, pp. 210–220.

8.A   Appendix

8.A.1   Common S-PLUS Functions

```
apply(X, MARGIN, FUN, ...)
```
apply a function to sections of an array
```
bwplot(formula, ...)
```
box and whisker plot

```
dotplot(formula, ...)
```
   multiway dot plot

```
glm(formula, family, data, ...)
```
   fits a generalized linear model

```
nlminb(start, objective, ...)
```
   minimization for smooth nonlinear functions

```
nls(formula, data, ...)
```
   fits a nonlinear regression model via least squares

```
sas.get(library, member, ...)
```
   converts an SAS dataset into an S-PLUS data frame

```
summary(object, ...)
```
   summarize an object

```
table(...)
```
   create contingency table from categories

```
xyplot(formula, ...)
```
   conditioning scatter plots

8.A.2   Software from the MASS Library
        (Venables and Ripley, 1999)

```
dose.p(object, ...)
```
   function for LD50-like fits

Software for the third edition of *Modern Applied Statistics with S-PLUS* by W.N. Venables and B.D. Ripley, Springer, 1999, is available from http://www.stats.ox.ac.uk/pub/MASS3/Software.html

8.A.3   S-PLUS Functions for Nonlinear Mixed Effects Models
        (Version 3.3 of NLME)

```
augPred(object, ...)
```
   computes augmented predictions

```
compareFits(object1, object2, ...)
```
   compares fitted objects

```
groupedData(formula, data, ...)
```
   constructs a grouped Data object

```
intervals(object, ...)
```
   computes confidence intervals on coefficients

```
nlsList(formula, data, cluster, control)
```
   creates a list of nls objects with common regression model

```
nlme(object, fixed, random, cluster, data, ...)
```
   fits a nonlinear mixed effects model

8.A.4   Program for the Analysis of Accelerated Titration Designs

```
phlatd("inputfile", "doselist")
```
   provides the analysis of Accelerated Titration Designs

Available from http://linus.nci.nih.gov/~brb/Methodologic.htm.

8.A.5   Programs for Continual Reassessment Designs

FORTRAN code is available from http://odin.mdacc.tmc.edu/anonftp.

8.A.6   S-PLUS Code to Create Figures

Figure 8.1: Progression of Trial and Individually Administered Doses

```
# Start trellis device
trellis.device(device = "postscript",
  file = "Figure.8.1.eps", hor = F, width = 9, height = 11,
  pointsize = 12)
trellis.par.set("strip.background", list(col = 0))
attach(therapy)
lev.ad <- levels(as.factor(admin))
n.ad <- length(lev.ad)
trellis.par.set("superpose.symbol",
  list(pch = lev.ad, col = rep(1, n.ad),
  cex = rep(0.65, n.ad), font = rep(2, n.ad) )) #
# Plot for each patient the dates of administration
# by dose level and draw the actual administration
# number in a 4 column layout
z1 <- dotplot(Patient ~ thdate |
  paste("dose level:", dose, "mg/m\262"),
  groups = admin, layout=c(1, 4),
  strip = function(...) strip.default(..., style = 1),
  xlab = paste("start of administered courses",
    "(days since start of the trial)", sep = "\n"),
  ylab="patient number",
  panel=function(x, y, ...) {
    panel.abline(h = unique(y), lwd = 0.4, lty = 1,
      col = 2)
    panel.superpose(x, y, ...) }) #
```

```
# Draw the left-hand display
print(z1, position = c(0, 0, 0.67, 1), more = T) #
# Right display: z2
# Use the number of courses (cycles) ncycles as
# plot character
ncycles <- tapply(admin, list(Patient, dose, dose.i),
  FUN = function(x) sum(!is.na(x)))
positions <- tapply(admin, list(Patient, dose, dose.i))
l.nc <- levels(as.factor(ncycles))
n.nc <- length(l.nc)
trellis.par.set("superpose.symbol",
  list(pch = l.nc, col = rep(1, n.nc),
    cex = rep(0.65, n.nc), font = rep(2, n.nc))) #
# Plot for each patient the individual dose by dose level
# of design and draw the total number of the
# administrations of this dose
z2 <- dotplot(Patient ~ dose.i |
  paste("dose level:", dose, "mg/m\262"),
  groups = ncycles[positions], layout = c(1, 4),
  strip = function(...) strip.default(..., style = 1),
  xlab = "individual doses (mg/m\262)",
  ylab = "", xlim = c(0, 120),
  panel = function(x, y, ...) {
    panel.abline(h = unique(y), lwd = 0.4, lty = 1,
      col = 2)
    panel.superpose(x, y, ...)}) #
# Draw the right-hand display
print(z2, position= c(0.58, 0, 1, 1))
detach("therapy") #
dev.off()
rm(z1, z2, l.nc, n.nc, n.ad, lev.ad, ncycles, positions)
```

Figure 8.2: Number of Patients at Each Toxicity Grade by Dose Level

For Figures 8.2–8.4, if you are using S-PLUS on a UNIX or Linux platform, use the following commands to define colors:

```
colors <- ps.colors.rgb[c("black", "grey", "gray72",
  "gray78", "gray84","gray90","gray96"), ]
colors <- ps.rgb2hsb(colors)
```

If you are using S-PLUS on a Windows platform, use the following command to define colors:

```
colors <- c(0, seq(0.6, 1, len = 10))
```

Now create Figure 8.2:

```
thertox <- merge(therapy, toxicity, by.x = "Patient",
  by.y = "Patient")
thertox <- thertox[!(thertox$Patient == "008" &
  thertox$dose == 40), ]
thertox <- thertox[!(thertox$Patient == "107" &
  thertox$dose == 50), ]
tox.table <- double()
for(ic in levels(thertox$ctc)) {
  for(ip in unique(thertox$Patient)) {
    tmp.tg <- max(thertox$grade[thertox$Patient == ip &
      thertox$ctc == ic & thertox$relation < 4], na.rm = T)
    tmp.tg <- ifelse(is.na(tmp.tg), 0, tmp.tg)
    tmp.td <- thertox$dose[thertox$Patient == ip][1]
    tox.table$td <- c(tox.table$td, tmp.td)
    tox.table$tg <- c(tox.table$tg, tmp.tg)
    tox.table$tp <- c(tox.table$tp, ip)
    tox.table$tc <- c(tox.table$tc, ic)
  }
}
tox.table <- as.data.frame(tox.table)
attach(tox.table)
judgement <- tapply(tg, list(tc, td, tg),
  function(x) sum(!is.na(x)))
position <- tapply(tg, list(tc, td, tg))
n.ctc <- length(levels(tc)) #
# Start trellis device
trellis.device(device = "postscript",
  file = "Figure.8.2.eps", hor = F, width = 9, height = 11,
  pointsize = 12, colors = colors)
trellis.par.set("strip.background", list(col = 0))
dotplot(tc ~ tg | paste("dose level:", td, "mg/m\262"),
  data = tox.table, layout = c(4, 1), cex = 0.9,
  xlim=c(-0.5, 4.5), xlab = "CTC Grade",
  ylab = "Nonhematological adverse drug reactions",
  strip = function(...) strip.default(..., style = 1),
  scales = list(x = list(ticks = T, at = 0:4, cex = 0.7),
  y = list(cex = 0.7)), groups = judgement[position],
  panel =
    function(x, y, subscripts, groups, cex = cex, ...) {
      xy <- par("usr")
      polygon(c(-0.5, 0.5, 0.5, -0.5), xy[c(3, 3, 4, 4)],
        border = F, col = 7)
      polygon(c(0.5, 1.5, 1.5, 0.5), xy[c(3, 3, 4, 4)],
        border = F, col = 6)
      polygon(c(1.5, 2.5, 2.5, 1.5), xy[c(3, 3, 4, 4)],
```

```
       border = F, col = 5)
    polygon(c(2.5, 3.5, 3.5, 2.5), xy[c(3, 3, 4, 4)],
       border = F, col = 4)
    polygon(c(3.5, 4.5, 4.5, 3.5), xy[c(3, 3, 4, 4)],
       border = F, col = 3)
    abline(h = 0.5:11.5, col = 0, lwd = 1)
    text(x, y, groups[subscripts], cex = 0.6,
       font = 5, ...)
  }
)
dev.off()
detach("tox.table")
rm(tmp.td, tmp.tg, tox.table, judgement, position, n.ctc,
  thertox)
```

Figure 8.3: *Toxicity Grade for Each Patient*

```
trellis.device(device = "postscript",
  file = "Figure.8.3.eps", hor = F, width = 11,
  height = 10, pointsize = 12, colors = colors)
trellis.par.set("strip.background", list(col = 0))
attach(toxicity)
maxrelated <- tapply(relation, list(ctc, Patient, grade),
  min, na.rm = T)
positions <- tapply(relation, list(ctc, Patient, grade))
dotplot(ctc ~ grade | Patient, layout = c(9, 2), cex = 0.9,
  xlim = c(0.5, 4.5), xlab = "CTC Grade [1-4]",
  ylab = "Nonhematological adverse drug reactions",
  strip = function(...) strip.default(..., style = 1),
  scales=list(x = list(ticks = T, at = 1:4, cex = 0.6),
  y = list(cex = 0.6)), groups = maxrelated[positions],
  panel =
    function(x, y, subscripts, groups, cex = cex, ...) {
    xy <- par("usr")
    polygon(c(0.5, 1.5, 1.5, 0.5), xy[c(3, 3, 4, 4)],
       border = F, col = 6)
    polygon(c(1.5, 2.5, 2.5, 1.5), xy[c(3, 3, 4, 4)],
       border = F, col = 5)
    polygon(c(2.5, 3.5, 3.5, 2.5), xy[c(3, 3, 4, 4)],
       border = F, col = 4)
    polygon(c(3.5, 4.5, 4.5, 3.5), xy[c(3, 3, 4, 4)],
       border = F, col = 3)
    abline(h = 0.5:11.5, col = 0, lwd = 1)
    text(x, y, groups[subscripts], cex = 0.5,
       font = 5, ...)
```

```
        }
    )
dev.off()
detach("toxicity")
rm(maxrelated, positions)
```

Figure 8.4: Platelet Counts by Week and Patient

```
attach(laboratory)
trellis.device(device = "postscript",
    file = "Figure.8.4.eps", hor = F, width = 11, height = 9,
    pointsize = 12, colors = colors)
trellis.par.set("strip.background", list(col = 0))
realdose <- ifelse(!is.na(dose.i), dose, 0)
trellis.par.set("superpose.symbol",
    list(pch = c("x", "2", "4", "5", "6"),
    col = rep(1, 5), cex = rep(1, 5), font = rep(2, 5)))
xyplot(platelet ~ labweek | Patient, groups = realdose,
    layout=c(9, 2), cex = 0.7, xlab = "Treatment Week",
    ylab = "Platelet Count (x1E9/l)",
    strip = function(...) strip.default(..., style = 1),
    scales = list(x = list(ticks = T, at = 0:8, cex = 0.5),
    y = list(cex=0.6)),
    panel = function(x, y, subscripts, ...) {
        lines(x, y, type = "b", cex = 0.5, pch = " ")
        panel.superpose(x, y, type = "p", subscripts,
            cex=0.5, ...)
    }
)
dev.off()
detach("laboratory")
rm(realdose, colors)
```

8.A.7 S-PLUS Code to Estimate Maximum Tolerable Dose (Section 8.6.2)

```
thertox <- merge(therapy, toxicity, by.x = "Patient",
    by.y = "Patient")
thertox <- thertox[!(thertox$Patient == "008" &
    thertox$dose == 40), ]
thertox <- thertox[!(thertox$Patient == "107" &
    thertox$dose == 50), ]
tox.table <- double()
```

```
for(ip in unique(thertox$Patient)) {
  tmp.tg <- max(thertox$grade[thertox$Patient == ip &
    thertox$relation < 4], na.rm = T)
  tmp.tg <- ifelse(is.na(tmp.tg), 0, tmp.tg)
  tmp.td <- thertox$dose[thertox$Patient == ip][1]
  tox.table$td <- c(tox.table$td, tmp.td)
  tox.table$tg <- c(tox.table$tg, tmp.tg)
  tox.table$tp <- c(tox.table$tp, ip)
}
tox.table <- as.data.frame(tox.table)
y <- as.numeric(tox.table$tg > 2)
fit <- glm(y ~ td, data = tox.table,
  family = binomial(link="logit"))
library(MASS)
xp <- dose.p(fit, p = 0.33)
xp.ci <- xp + c(-qnorm(0.975),
  qnorm(0.975)) * attr(xp, "SE")
mtd <- cbind(xp.ci[1], xp, xp.ci[2])
dimnames(mtd)[[2]] <- c("LCL", "xp", "UCL")
mtd
```

9

Patient Compliance and its Impact on Steady State Pharmacokinetics

Wenping Wang

Pharsight Corporation, Cary, NC, USA

9.1 Introduction

Physicians commonly prescribe drug products in a multiple dosage regimen (e.g., one daily or q.d., twice daily or b.i.d, etc.) for prolonged therapeutic activity. The purpose of a multiple dose regimen is to maintain the drug plasma concentration within its *therapeutic window* (i.e., the concentration of drug in the serum or plasma should yield optimal benefit at a minimal risk of toxicity). To study the effect of multiple dosing on drug concentration in blood, researchers often employ a deterministic model with the assumptions that drugs are administered at a fixed dosage, with fixed (usually constant) dosing intervals. In practice, as is well known in the medical community, patients may not follow such a rigid schedule. Hence, two possible scenarios might occur: patients might not take the prescribed dosage, resulting in irregular dosing amounts; or they might not adhere to the dosing schedule, resulting in irregular dosing times. This chapter intends to lay out a probability framework to model these two types of noncompliance and consequently study their impact on the steady state pharmacokinetics in a rigorous setting.

To study the steady state drug concentration in multiple dose pharmacokinetics, the principle of superposition is the key tool. In this chapter, the principle of superposition in the presence of noncompliance is formulated generally as a recursive formula. With this formula, we are able to generalize the notion of steady state in multiple dose pharmacokinetics given noncompliance. Using the compliance models and the principle of superposition, important pharmacokinetic parameters are rigorously studied. Factors affecting the steady state trough concentration are characterized through a simulation study. The relationship between the compliance index and the average concentration at steady state is established. This result generalizes the classic result about the equality between the single dose area under the curve (AUC) and the multiple dose AUC at steady state. Using theophylline (an antiasthma agent) as an example, we demonstrate that noncompliance causes the drug concentration time curve to exhibit an increased fluctuation. The increase in fluctuation due to noncompliance cannot be explained with the use of the classical deterministic multiple dose model.

9.2   Probability Foundation for Patient Compliance

9.2.1   Three Components in Multiple Dose Pharmacokinetic Studies

Medication errors is formally termed as compliance in the literature. As discussed by Wang et.al. (1996), compliance composes two sub-categories: "dosage compliance" - taking the drug at the prescribed dose, and "dosing time compliance" - taking the drug at the scheduled times.

The difference of these medication errors prompts separate modeling for them. Let X_n denote the relative dosage taken at the nth nominal dose. Thus $X_n = 0$ means that the patient skips the nth dose, $X_n = 1$ means that the patient takes the prescribed dosage, and $X_n = 2$ indicates that the patient takes double doses at the nth dose, etc. X_n is referred to as the compliance variable at the nth nominal dose.

The compliance variable series $\{X_1, X_2, \ldots\}$ is a stochastic process. Girard et.al. (1996) proposed a Markov chain model for the compliance variable series

$$p_{ij} = P(X_{n+1} = j \mid X_n = i)$$

We shall assume an irreducible ergodic (positive recurrent and aperiodic) Markov chain model for $\{X_n\}$ with stationary distribution

$$\pi_j = \lim_{n \to \infty} p_{ij}^n$$

Note that π_j equals to the long-term proportion of time that the process will be in state j.

The compliance variable series is further assumed to be stationary over time. The stationary assumption is rather stringent mathematically. In practice, however, this seemingly strong assumption is usually adequate for describing a patient's dosing pattern.

To model time compliance, Wang et al. (1996) used an additive model to describe the relationship between the actual dosing times (T_n) and the prescribed dosing times (t_n):

$$T_n = t_n + \epsilon_n, \ n = 1, 2, 3, \ldots$$

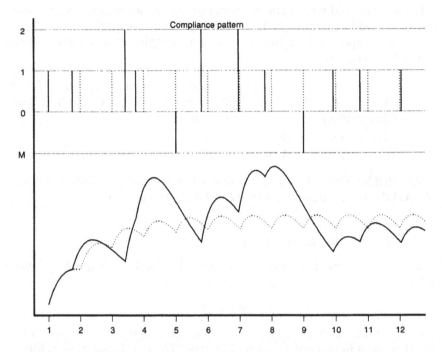

Figure 9.1. A simulated noncompliant patient's concentration profile after 12 nominal doses. The upper half of the plot describes the dosing pattern for the patient. Each dotted vertical bar represents one scheduled dosing time; and each solid vertical bar shows one actual dosing time. The height of the solid bar can be: -1 = no dose taken at this time; 1 = 1 dose taken; and 2 = 2 doses taken.

where $\{\epsilon_n\}$ are independently and identically distributed (iid) random variables with mean 0 and variance σ^2. Based on the knowledge of a patient's dosing time pattern, a parametric model may be used to describe the distribution of the error term. Examples are normal distribution and truncated normal distribution. Note that with the assumption of normality, there is always a small, however positive, possibility that $T_n < T_{n-1}$ for some combination of ϵ_{n-1} and ϵ_n. The use of the truncated normal distribution can avoid this technical difficulty. However, with moderate σ ($\sigma/\tau \leq 0.5$, where $\tau = t_{i+1} - t_i$ is the length of dosing interval), there are no differences between normal and truncated normal distributions in our numerical analyses.

It is noted that Girard et al. (1996) also mentioned the dosing time error (the discussion after equation (3) on page 268), although they did not give a stochastic model for the error term. Figure 9.1 illustrates the two processes related to compliance.

For a compound with linear kinetics (i.e., the concentration of the compound in blood is proportional to the dosage), the concentration–time profile of the compound following a multiple dose regimen can be determined by the following three components:

1. $f(t)$: the single dose concentration curve with unit dose (e.g., $f(t) = 1/V \exp(-k_e t)$ for a one-compartment i.v. bolus model with volume of distribution V and elimination rate k_e);
2. $\{T_i\}$: actual dosing times, $i = 1, 2, \ldots$
3. $\{X_i\}$: dosage taken at each dosing time, $i = 1, 2, \ldots$

In the triplets $(f(t); \{T_i\}, \{X_i\})$, the two-dimensional stochastic processes $\{T_i, X_i\}$ describe a patient's practice of taking medications.

9.2.2   Metrics for Compliance

The dosage compliance may be measured by the following compliance index:

$$p = E(X_\infty)$$

where $E(.)$ is the mathematical expectation, and X_∞ denotes the stationary distribution of the compliance variable series $\{X_n\}$. The compliance index can be regarded as a mathematical reformulation of pill-counting, a practice commonly employed by researchers to measure patients' compliance. For example, if a patient misses 20% of his prescribed dose, his compliance index is 80%. The time compliance can be measured by the *adherence index* (Wang et al. (1996)):

$$\lambda = (1 - 2\sigma/\tau) \cdot 100\%$$

For example, for a twice daily regimen ($\tau = 12$), if there is no dosing time error ($\sigma = 0.0$ h), then $\lambda = 100\%$ (perfect time compliance). If, with 95% probability, the patient's dosing times are ± 2.4 h within schedule ($\sigma = 1.2$ h), then $\lambda = 80\%$ (fair time compliance). Similarly, if the patient's dosing time intervals are 12 ± 4.8 h with 95% probability ($\sigma = 2.4$ h), then $\lambda = 60\%$ (poor time compliance).

9.3   The Principle of Superposition: A Tool to Link $(f(t); \{T_i\}, \{X_i\})$

The principle of superposition is essential for studying the concentration–time course following a multiple dose regimen. The concentration–time curve of a multiple dose regimen is the sum of those following each single dose. Examples of applying the principle of superposition to establish

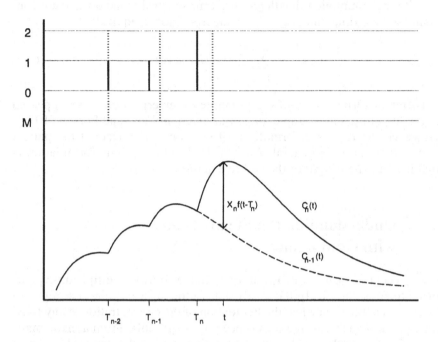

Figure 9.2. Illustration of the principle of superposition. The term $X_n f(t - T_n)$ is the contribution of the nth dose.

fundamental results for multiple dose pharmacokinetics for low-order compartment models can be found in the standard textbooks such as Wagner (1975), Gibaldi and Perrier (1975), and Ritschel (1986). Other examples can be found in Thron (1974) and Weiss and Förster (1979).

The principle of superposition reflects the additive nature of the linear kinetics. The additiveness is captured by the formulation given by Wang et al. (1996):

$$C_n(t) = C_{n-1}(t) + X_n f(t - T_n) \qquad (9.1)$$

which means that the concentration at time t after n doses is equal to the sum of the concentration at t if the patient takes exactly $n - 1$ doses and the single dose concentration at $t - T_n$ (of the last dose), adjusting the actual dosage of the last dose; or the contribution of the last dose can be superimposed/added to that of the first $n - 1$ doses. Figure 9.2 illustrates the principle of superposition with compliance.

With the assumption that the dosing time interval τ and dosage are constant over the time, the principle of superposition is equivalent to

$$C_n(t) = \sum_{i=1}^{n} f(t - (i - 1)\tau) \tag{9.2}$$

Different formulations of the principle of superposition have appeared in the literature. For example, (9.2) was given in Weiss and Förster (1979). However, the recursive formula in (9.1), taking into account the patient compliance, is more general. An added advantage of this formulation is that it keeps the insight of the additive nature.

9.4   Understanding the Steady State with Compliance

The concept of steady state is important in the study of multiple dose pharmacokinetics. Intuitively, the steady state following a multiple dose regimen is reached at the time when the fluctuation of the concentration at any fixed time point of each dosing interval becomes negligible. From a mathematical point of view, the steady state is a limiting state. In the standard books on the multiple dose pharmacokinetics, e.g., the trough concentration series, $\{C_{min}^n = C_n(T_{n+1}-)\}$ are shown to converge to a constant for low-order compartment models, where $C_n(T_{n+1}-)$ is the concentration just prior to the $n + 1^{st}$ dose; i.e., the trough concentration after the nth dose. For a general concentration curve $f(t)$, Wang et al. (1996) provided a set of sufficient conditions to ensure the convergence of the trough concentration series:

(C1) $f(t) = 0$ when $t < 0$ and $f(t)$ is continuous for $t \geq 0$
(C2) eventually non-increasing (when t and s are large); $f(t) \leq f(s)$ if $t \geq s$
(C3) $AUC_{0,\infty} = \int_0^\infty f(t)dt < \infty$

The concentration curve from a multicompartment model satisfies (C1)–(C3). Hence, these conditions do not seem to be too stringent for most drug products in real life situations.

When noncompliance occurs, each C_{min}^n is a random variable, and fluctuates around a constant. As the number of doses increases, the limit of $\{C_{min}^n\}$ is no longer a constant, but a random variable (see Figure 9.3):
The following result captures the intuition.

Result 1. *Suppose that the principle of superposition holds and the nonnegative function $f(\cdot)$ satisfies the conditions (C1), (C2), and (C3). Then $C_{min}^n = C_n(T_{n+1}-)$ converges in distribution as $n \to \infty$.*

We call the limiting random variable "limiting state trough concentration," and denoted by $C_{\min}^{\infty}$. Similarly, we can define the "limiting state peak concentration," $C_{\max}^{\infty}$.

Wang et al. (1996) studied factors that may influence the limiting state distribution of the trough concentration series, including the pharmacokinetics of the drug (the function $f(\cdot)$ and the elimination half-life), the dosing regimen $\{t_n\}$, the compliance index p, and the adherence index λ. Major findings from their Monte Carlo (simulation) study are summarized below:

1. When $p = 100\%$, the distribution has a shape similar to a log-normal distribution with mode $C_{\min}^{SS}$ and a long right tail
2. Poor compliance (as measured by p) makes the distribution bimodal
3. Poor time compliance increases the dispersion of the distribution
4. Shorter half-life leads to larger dispersion of the distribution
5. Shorter half-life results in more pronounced bi-modal shape (two modes are more separated)
6. Longer dosing interval leads to larger dispersion of the distribution

9.5 The Steady State Average Concentration and the Compliance Index

As another application of the probability model in Chapter 9.2, we study the relationship between the average concentration at steady state and patients' compliance index in this section.

When patients' dosing times change over time as modeled by Girard et al. (1996), defining the average concentration at steady state is not straightforward due to the random nature of the dosing intervals. Let us look at the definition of the steady state average concentration with total compliance more closely and try to gain some insight. Note that, with total compliance, the value of C_{av}^{SS} does not depend upon the starting point at which the AUC is calculated; i.e., when t is large enough, the quantity

$$C_{av}^{SS} = \frac{\int_t^{t+\tau} C(u)\,du}{\tau}$$

is independent of t. Therefore,

$$
\begin{aligned}
C_{av}^{SS} &= \frac{\int_t^{t+n\tau} C(u)\,du}{n\tau}, \quad n = 1, 2, 3, \ldots \\
&= \lim_{n \to \infty} \frac{\int_t^{t+n\tau} C(u)\,du}{n\tau}
\end{aligned}
$$

This consideration leads us to the following definition:

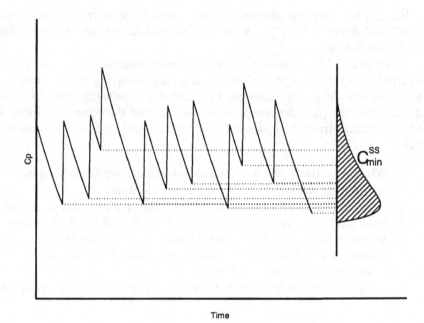

Figure 9.3. The steady state with patient compliance is a probabilistic distribution. The shaded area represents the sampling distribution of the steady state trough concentrations with compliance included.

Definition. The limiting state average concentration in the presence of noncompliance is defined by

$$C_{av} = \lim_{n \to \infty} \frac{\int_{T_1}^{T_n} C_n(t)dt}{T_n - T_1}$$

With this definition, we have

Result 2. *Suppose that the principle of superposition holds and the non-negative function $f(\cdot)$ satisfies conditions (C1) – (C3) and*

(C4) $$\int_0^\infty \left(\int_x^\infty f(t)dt \right) dx < \infty$$

Then

$$\lim_{n \to \infty} \frac{\int_{T_1}^{T_n} C_n(t)dt}{T_n - T_1} = p \cdot \frac{AUC_{0,\infty}}{\tau} a.s. \qquad (9.3)$$

where p is the compliance index.

A proof for Result 2 can be found in Wang and Ouyang (1998).

Remark. Condition (*C4*) requires the drug be eliminated from the body not too slowly. It holds for any drug with linear elimination kinetics, for which $f(t)$ is an exponential function. This means that from a practical point of view, condition (*C4*) does not seems to be a stringent condition.

9.5.1   A Numerical Example

We present a numerical example to illustrate the use of (9.3) and evaluate the precision of this approximation.

The data used in this example were collected by the Medication Event Monitoring System (MEMS, APREX Corp., Fremont, CA) over 3 months from 24 patients infected with the Human Immunodeficiency Virus (HIV), who had been prescribed AZT three times a day ($\tau = 8$ h). These patients are part of the ACTG 175 clinical trial designed to determined the relative efficacy of nucleoside analog mono- (AZT or DDI) versus combination- (AZT+DDI or AZT+DDC) therapy (Kastrisios et.al. (1995). Girard et al. (1996) used this dataset to demonstrate the impact of patients' compliance in population pharmacokinetic analyses.

Following the approach of Girard et al. (1996), we assume a one-compart-ment model with bolus input in our example. In addition, we assume that the half-life of the drug is equal to the dosing interval ($T_{1/2} = \tau = 8$ hr). The exact AUC over 5 weeks is then calculated and the actual average concentration is calculated as the ratio of the exact AUC and the total length of dosing time (in hours). The predicted average concentration is calculated using formula (9.3). Table 9.1 shows the compliance index for the 4 patients reported in Girard et al. (1996) and their predicted and actual average concentrations; Figure 9.4 presents the result graphically.

Table 9.1. Patients' compliance and the actual and predicted concentrations

Patient ID	Compliance index	Actual average concentration	Predicted average concentration
6	0.953	1.361	1.362
3	0.972	1.386	1.389
1	0.953	1.357	1.362
2	0.377	0.536	0.539

From Table 9.1 and Figure 9.4 we see that in general the prediction using formula (9.3) agrees well with the actual values, although Table 9.1 seems to suggest that formula (9.3) tends to slightly overpredict the actual average concentration.

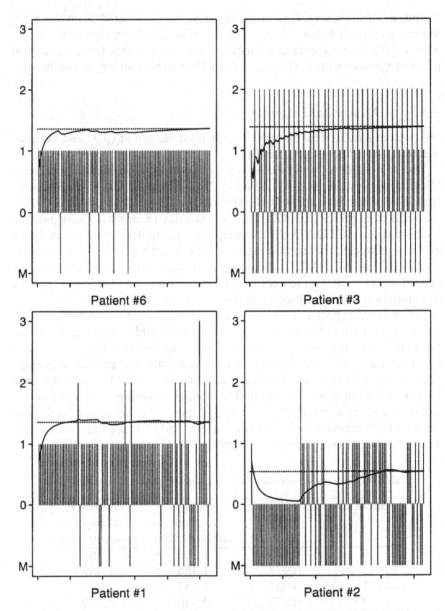

Figure 9.4. Dosage pattern, predicted and actual accumulated average concentrations for patients #1, #2, #3, and #6 over 5 weeks (Girard et al. (1996)), assuming that the patient's nominal dosage is one tablet every 8 h. Each vertical bar represents one nominal dosage time. The height of the bar can be: -1 = no dose taken at this time; 1 = 1 dose taken; 2 = 2 doses taken; 3 = more than 2 doses taken. The actual average concentration is plotted as a solid line; while the predicted is a dashed line.

The dosing pattern of Patient #3 in this example prompts some notes on the compliance index defined in Glanz et.al. (1984). Patient #3 represents a typical example of noncompliant patients: they may skip one dose (perhaps because of inconvenience), and later decided to make-up the skipped dose by a double dose. The dosing pattern of these patients has the tendency of regressing toward the prescribed dosage. Consequently, if the method of pill-counting were used, this patient would be considered as compliant because of the compliance index. Only the more advanced system (MEMS) would exhibit the opposite. When the safety and efficacy of an agent depends solely upon the extent of absorption, the dosing pattern of Patient #3 may not cause serious medical problems. However, if an agent has a narrow therapeutic window, the dosing pattern of Patient #3 may induce adverse medical consequences.

9.6 Fluctuation of Blood Concentration due to Noncompliance

The fluctuations of drug concentration in the blood are inevitable for most dosing regimens. Excessive fluctuations may have unacceptable clinical consequences (see Weinberger et.al. (1981)). In this section, we use theophylline as an example to study the impact of noncompliance on the fluctuation of blood concentration.

Theophylline is a bronchodilator commonly used in a multiple dose regimen as a prophylactic agent to relieve and/or prevent symptoms from asthma and reversible bronchospasm associated with chronic bronchitis and emphysema. Its therapeutic window is within the range of 10 to 20 mcg/ml. Persistent adverse effects, including nausea, vomiting, headache, diarrhea, irritability, insomnia, cardiac arrhythmias, and brain damage, would occur when serum concentrations rise above 20 mcg/ml. See Weinberger et al. (1981) and Hendeles et.al. (1986) for details. Within the therapeutic window, a linear two-compartment model adequately describes the concentration–time profile of theophylline (Mitenko and Oglivie (1973)). The elimination half-life of theophylline has a wide range: 3.7 h on average for children, 8.2 h for adults, and 14 h for patients with congestive heart failure, liver dysfunction, alcoholism, and respiratory infections.

The common dosing regimens are four times a day (q.i.d.) per oral for children and twice a day (b.i.d.) for adults. The amounts of drug are computed so that for 100% compliance and 100% time compliance:

$$10 \approx C_{min}^{SS} < C^{SS}(t) < C_{max}^{SS} \approx 20$$

When noncompliance occurs, the fluctuation of blood concentration will increase. The following two probabilities are of interest to quantify the adverse impact due to poor compliance: $p_1 = P(C_{min}^{\infty} < 0.8 C_{min}^{SS})$, $p_2 =$

$P(C_{\max}^{\infty} > 1.2 C_{\max}^{SS})$. Here, we arbitrarily choose 20% as a cutoff for clinically noneffective (in p_1) and toxic (in p_2).

A noncompartmental model approach, proposed by Weinberger et al. (1981) will be used to generate data. Let $AUC_{0,i}$ denote the area under the concentration curve from time 0 to time t_i, and let F_i denote the fraction of a dose absorbed from time 0 to time t_i. Following oral administration of an uncoated theophylline tablet, the mean absorption fractions are 0.0 after 0 h; 0.82 after 1 h; 0.92 after 2 h; 0.99 after 3 h; and 1.0 after 4 h or more (Weinberger et.al. (1978)). We obtain the serum concentration of theophylline at any time using the following recursive formula:

$$f(t_i) = \frac{F_i \cdot AUC_{0,\infty} - f(t_{i-1})(t_i - t_{i-1})/2 - AUC_{0,i-1}}{(t_i - t_{i-1})/2 + 1/\beta}$$

which can be derived from the Nelson–Wagner equation (Wagner (1975)) and the trapezoidal rule,

$$F_i = \frac{AUC_{0,i} + f(t_i)/\beta}{AUC_{0,\infty}}$$

$$AUC_{0,i} = AUC_{0,i-1} + \frac{f(t_i) + f(t_{i-1})}{2}(t_i - t_{i-1})$$

We first examine the dosing regimen of q.i.d. and assume the elimination half-life $T_{1/2}$ to be 3.7 h. For the four combinations of compliance index and adherence indices, we generated 2000 random samples, and from the simulated data we calculated p_1 and p_2. Plotted in Figure 9.5 are 50 of the simulated trough and peak concentrations at steady state (after 20 scheduled doses).

The plots reveal the following results:

1. As the adherence index λ decreases, both p_1 and p_2 increase
2. As the compliance index decreases, p_1 increases and p_2 decreases
3. As Figure 9.5(b) indicates, poor time compliance might result in exposure of many patients (10%) to dangerous drug concentration levels
4. For poor compliance (Figure 9.5(d)), about 42% patients' concentration is 20% below the designed steady state trough concentration. Therefore, the consequence of poor compliance is that patients did not receive the designed therapeutic effect. This finding is consistent with previous studies (see Glanz et al. (1984) and Cramer et.al. (1989))

We also study the fluctuation of serum theophylline concentration for adult patients. The elimination half-life is 8.2 h, and the dosing regimen is twice a day. Results reveal similar patterns.

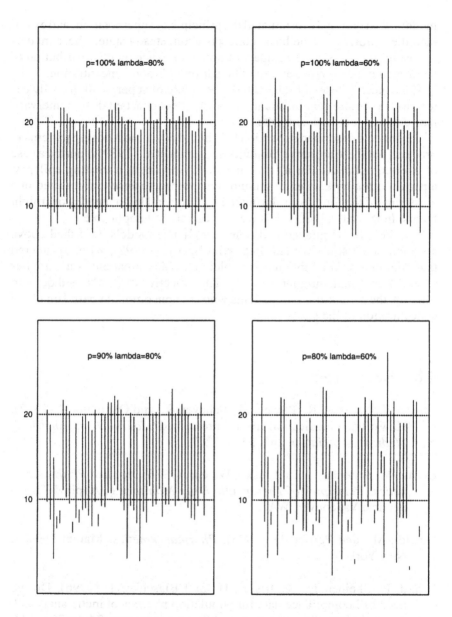

Figure 9.5. The fluctuation of concentration ($T_{1/2} = 3.7$ h, q.i.d.).

9.7   Discussion

In this chapter, we have separated the concepts of dosage compliance and
time compliance in multiple dose regimens, and proposed some statistical

models. We rigorously formulated the principle of superposition, and examined the validity of some basic concepts about steady state. These models are useful for studying the impact of noncompliance on the distribution of the trough concentration and the fluctuation of blood concentration.

The concept of steady state and the principle of superposition easily extend to other more general situations, such as nonconstant dosing intervals, half-dosing, double-dosing, etc.

The models described in Section 9.2 should cover most dosing patterns in practice, but there are some limitations. Cramer et al. (1989) reported a case of over-compliance. For a four-times-a-day regimen in treating epilepsy, a patient often took an extra midmorning valproate dose that resulted in a clustering of the four doses during the daytime and left a lengthy overnight hiatus. In this case, the models we propose are not applicable.

The dosage compliance and time compliance models described above are more sophisticated than that given by Lim (1992), who considered that compliance is a binomial variable; i.e., either a patient is a complier ($Y = 0$), or a noncomplier ($Y = 1$). The objective of Lim's method is to estimate the proportion ρ of patients who are noncompliant based on blood concentration of the study drug.

9.8 References

Cramer, J., Mattson, R., Prevey, M., Scheyer, R., and Ouellette, V. (1989). How often is medication taken as prescribed?. *Journal of the American Medical Association* **261**, 3273–3277.

Evans, W., Schentag, J., and Jusko, W., eds. (1986). *Applied Pharmacokinetics: Principles of Therapeutic Drug Monitoring*. Applied Therapeutics, Vancouver, WA.

Gibaldi, M., and Perrier, D. (1975). *Pharmacokinetics*. Marcel Dekker, New York.

Girard, P., Sheiner, L., Kastrisios, H., and Blaschke, T. (1996). Do we need full compliance data for population pharmacokinetic analysis?. *Journal of Pharmacokinetics and Biopharmaceutics* **24**(3), 265–282.

Glanz, K., Fiel, S., Swartz, M., and Francis, M. (1984). Compliance with an experimental drug regimen for treatment of asthma: its magnitude, importance, and correlates. *Journal of Chronic Disease* **37**, 815–824.

Hendeles, L., Massanari, M., and Weinberger, M. (1986). Theophylline. In: Evans et.al. (1986).

Kastrisios, H., Suares, J., Flowers, B., and Blaschke, T. (1995). Could decreased compliance in an aids clinical trial affect analysis of outcomes?. *Clinical Pharmacology and Therapeutics* **57**, 190–198.

Lim, L. (1992). Estimating compliance to study medication from serum drug levels: Application to an aids clinical trial of zidovudine. *Biometrics* **48**, 619–630.

Mitenko, P., and Oglivie, R. (1973). Pharmacokinetics of intravenous theophylline. *Clinical Pharmacology and Therapeutics* **14**, 509–513.

Ritschel, W. (1986). *Handbook of Basic Pharmacokinetics*. 3rd ed. Drug Intelligence Publications.

Thron, C. (1974). Linearity and superposition in pharmacokinetics. *Pharmacology Review* **26**(1), 3–31.

Wagner, J. (1975). *Fundamentals of Clinical Pharmacokinetics*. Drug Intelligence Publications.

Wang, W., and Ouyang, S. (1998). The formulation of the principle of superposition in the presence of non-compliance and its applications in multiple dose pharmacokinetics. *Journal of Pharmacokinetics and Biopharmaceutics* **26**(4), 457–469.

Wang, W., Hsuan, F., and Chow, S. (1996). The impact of patient compliance on drug concentration profile in multiple dose. *Statistics in Medicine* **15**, 659–669.

Weinberger, M., Hendeles, L., and Bighley, L. (1978). The relation of product formulation to absorption of oral theophylline. *New England Journal of Medicine* **299**, 852–857.

Weinberger, M., Hendeles, L., Wong, L., and Vaughan, L. (1981). Relationship of formulation and dosing interval to fluctuation of serum theophylline concentration in children with chronic asthma. *Journal of Pediatrics* **99**, 145–152.

Weiss, M., and Förster, W. (1979). Pharmacokinetic model based on circulatory transport. *European Journal of Clinical Pharmacology* **16**, 287–293.

9.A Appendix

9.A.1   S-PLUS Code for Figure 9.1

```
# Create random dosage X_n
# n.dose  - # doses
random.dosing <- function(n.dose)
{
  x <- rnorm(n.dose)
  out <- rep(1, length(x))
  out[x < 0] <- 0
  out[x < -1] <- 2
  out
}

# ---
# Concentration curve with partial compliance
#
# n.dose  - number of doses
# tau   - length of dosing interval
# dd    - compliance variable series
# sigma  - s.d. of the dosing time error
# beta1  - parameter for 1-cmpt model
# beta2  - parameter for 1-cmpt model
# ---
calc.conc <- function(n.dose, tau, dd, sigma, beta1 =
  0.3, beta2 = 0.5)
{
  maxt <- n.dose * tau * 2
  min.inc <- tau/24
  x <- seq(0, maxt, min.inc)
  # single dose concentration curve
  y <- exp( - beta1 * x) - exp( - beta2 * x)
  j1 <- cbind(x, y)
  for(i in 1:n.dose) {
    # random dosing time
    a1 <- round((tau * i + sigma * rnorm(
      1))/min.inc) * min.inc
    # contribution of ith dose if exactly one dose taken
    j1 <- cbind(j1, append(rep(0, match(
      a1, x) - 1), y)[1:length(y)])
  }
  y <- 0
  # principle of superposition
  for(i in 1:n.dose)
```

```
    y <- y + dd[i] * j1[, i + 1]
  list(x = x, y = y, all.dat = j1)
}


# plot compliance info in the upper half
calc.comp <- function(dd, n.dose, all.dat)
{
  x <- 0
  for(i in 2:n.dose) {
    a1 <- all.dat[, i + 1]
    a1 <- all.dat[length(a1[a1 == 0]), 1]
    if(dd[i] == 0)
      a1 <- (i - 1) * 3
    x <- append(x, a1)
  }
  y <- dd
  y[y == 0] <- -1
  list(x = x, y = y)
}


conc.plot <- function(n.dose, tau)
{
  # if perfect compliance
  dd <- rep(1, n.dose)
  tmp0 <- calc.conc(n.dose, tau, dd, 0)
  # partial compliance
  dd <- random.dosing(n.dose)
  tmp1 <- calc.conc(n.dose, tau, dd, tau/3)
  hi.y <- max(tmp0$y, tmp1$y)
  #setup the frame
  par(mar = c(2, 2, 1, 1))
  plot(1:2, 1:2, xlim = c(0, (n.dose - 1) * tau +
    1), ylim = c(0, 2 * hi.y), xlab = "",
    ylab = "", type = "n", axes = F)
  box(bty = "l", lwd = 2)
  axis(1, at = (1:n.dose - 1) * tau, labels = 1:
    n.dose, lwd = 2)
  axis(2, at = c(1.1,
    1.4, 1.7, 2) * hi.y,
    labels = c("M", 0, 1, 2), tck = 1, lty
    = 1, lwd = 2)
  lines(tmp0$x, tmp0$y, lty = 2, lwd = 2)
  lines(tmp1$x, tmp1$y, lty = 1, lwd = 2)
  # compliance pattern
  tmp <- calc.comp(dd, n.dose, tmp1$all.dat)
```

```
for(i in 1:n.dose) {
  segments(tmp$x[i], 1.4 *
    hi.y, tmp$x[i],
    1.4 * hi.y +
    0.3 * hi.y *
    tmp$y[i], lwd = 2)
  segments(tau * (i - 1),
    1.4 * hi.y, tau *
    (i - 1), 1.7 * hi.y, lty = 2,
    lwd = 2)
}
text((n.dose/2 - 1) * tau, 2.03 *
  hi.y, "Compliance pattern")
}
```

Figure 9.1 is created by:

```
> set.seed(61)
> conc.plot(12,3)
```

9.A.2   S-PLUS Code for the Simulation in Section 9.6

```
fluc <- function(compliance, adherence, TAU = 6, T.half
  = 3.7, ITER = 2000, N.DOSE = 20,
  MAXT = 300)
{
  BETA <- log(2)/T.half
  # targeted SS avg concentration
  C.avg <- 14.8
  AUC.all <- C.avg * TAU
  # sigma, absolute dosing time error
  sigma <- 0.5 * (1. - adherence) * TAU
  out <- matrix(0, ITER, 3)
  single.dose.curve <- fill.C(BETA, 1., AUC.all,
    MAXT)
  for(iter in 1:ITER) {
    tmp <- get.D.time(N.DOSE, TAU, sigma,
      compliance)
    tmp <- calc.fluc(N.DOSE, TAU, tmp$D.time,
      tmp$D.indx, single.dose.curve,
      AUC.all)
    out[iter, ] <- unlist(tmp)
```

```
  }
  out
}

# calculate the trough and peak concentration
# N.dose - # nominal doses
# D.Indx - compliance variable series D.indx
# D.time - and actual dosing times
calc.fluc <- function(N.dose, tau, D.time, D.indx,
  SD.curve, auc.all)
{
  MD.conc <- rep(0, tau + 1)
  for(time in 1:(tau + 1))
    MD.conc[time] <- sum(D.indx * SD.curve[
      D.time[N.dose] - D.time + time])
  if(MD.conc[2] < MD.conc[1])
    MD.conc[2] <- MD.conc[tau]
  else {
    C.min <- ifelse(MD.conc[tau] < MD.conc[
      1], MD.conc[tau], MD.conc[1])
    MD.conc[1] <- MD.conc[2]
    MD.conc[2] <- C.min
  }
  list(trough = MD.conc[2], peak = MD.conc[1],
    F.idx = ((MD.conc[1] - MD.conc[2]) *
    tau)/auc.all)
}

# calculate concentration curve using (10) - (12)
fill.C <- function(beta, delta.t, auc.all, maxt)
{
  conc <- rep(0, maxt)
  AUC <- 0
  for(time in 2:maxt) {
    conc[time] <- (FA(time) * auc.all - 0.5 *
      delta.t * conc[time - 1] - AUC)/
      (0.5 * delta.t + 1./beta)
    AUC <- AUC + 0.5 * (conc[time] + conc[
      time - 1]) * delta.t
  }
  conc
}

# oral bioavailability for theophylline
# data from Weinberger et al., (1981)
FA <- function(time)
```

```
{
  # from Weinberg et al.
  temp <- c(0., 0.82,
    0.92, 0.99
    )
  if(time < 5)
    return(temp[time])
  else return(1.)
}

# generate the compliance variables and dosing times
get.D.time <- function(N.dose, tau, sigma, P.comp)
{
  D.indx <- rep(0, N.dose)
  D.indx[runif(N.dose) < P.comp] <- 1
  D.time <- (1:N.dose - 1) * tau + round(sigma *
    rnorm(N.dose))
  list(D.indx = D.indx, D.time = D.time)
}
```

The top left panel of Figure 9.5 is created by:

```
out <- fluc(1, 0.8, ITER = 50)
plot(1:2, 1:2, type = "n", xlim = c(0, 51), ylim = c(
  0, 30), axes = F, xlab = "", ylab = "", cex =
  1.1)
axis(2, at = c(10, 20), tck = 1, lty = 2)
box(bty = "o", lwd = 2)
for(i in 1:50) {
  x <- matrix(c(i, out[i, 1], i, out[i, 2]), ncol
    = 2, byrow = T)
  lines(x, lwd = 1.5)
}
x <- matrix(c(0, 10, 51, 10), ncol = 2, byrow = T)
lines(x, lwd = 1.5, lty = 2)
x <- matrix(c(0, 20, 51, 20), ncol = 2, byrow = T)
lines(x, lwd = 1.5, lty = 2)
text(25, 27, "p=100% lambda=80%")
```

10

Analysis of Analgesic Trials

Ene I. Ette
Vertex Pharmaceuticals, Inc., Cambridge, MA, USA

Peter Lockwood and Raymond Miller
Pfizer Global Research and Development, Ann Arbor, MI, USA

Jaap Mandema
Pharsight Corporation, Mountain View, CA, USA

10.1 Introduction

Analgesic clinical trials are usually complex. Because of the complexity of these trials knowledge of pharmacokinetic/pharmacodynamic relationships of analgesics is limited. The consequence of this is that some recommended analgesic doses may not be optimal. The design of an analgesic clinical trial is usually of the following pattern; patients receive a single dose of an analgesic or a placebo after a pain-initiating event such as surgery, and pain intensity or pain relief is measured to assess drug efficacy at specific times after drug administration. On ethical grounds, the patients can demand a rescue medication of a known effective analgesic at any time if their pain relief is inadequate. Pain relief and remedication time—two clinical efficacy endpoints—are compared between placebo and the administered active doses of the analgesic tested.

The analysis of analgesic trial data are complicated by several factors:

- Repeated measurements are obtained per patient.
- The responses measured are not continuous—pain relief is often measured as an ordered categorical variable, while time for remedication is a survival variable.
- Due to remedication, pain relief is nonrandomly censored.

The nonrandom censoring creates a biased sample of patients, especially at the later time points. This is because patients sensitive to drug treatment and who experience pain relief are the ones who will not remedicate. To derive the pharmacodynamic relationships and decide on the appropriate dose required to achieve adequate pain relief, only the unconditional pain relief measurements are relevant to address the question of whether the drug causes pain relief relative to placebo.

Traditional (ANOVA) analysis of analgesic clinical trials (i.e., testing the null hypothesis when comparing treatment and placebo groups) have dealt inadequately with the complexities of pain relief data collected in these studies (Laska et. al., 1991; Sheiner, 1994). When patients have required rescue medication before the end of the study, scores of unobserved subsequent pain and pain relief (PR) scores have historically been imputed according to predetermined rules. Pain scores are set to baseline (or highest possible rating) and zero values are imputed for censored pain relief observations; or pain and pain relief scores recorded at the time of rescue are used for imputation—the so-called last observation carried forward (LOCF) imputation scheme. The imputation schemes are used to artificially create data which are based on no explicit assumption. Evaluation of pain relief data with these imputation schemes have been shown to significantly underestimate the response to treatment, overestimate placebo-corrected drug response, and yield a biased dose–response relationship (Mandema, 1997). In addition to the problem associated with the imputation schemes discussed above, the traditional approach used in the analysis of analgesic clinical trials results in a loss of all information on the individual patient, and fails to render any insight into the intersubject variability associated with the "population average" dose–response relationship. Dosing guidelines are therefore hard to design.

Sheiner (1994) developed a subject-specific random effects model for the analysis of analgesic clinical trials, which accounted for the distribution of pain relief scores, time, drug concentration, and other covariates. This approach has been subsequently used for the development of a model for ketorolac analgesia (Mandema and Stanski, 1996). Liu and Sambol (1995) introduced a slight modification of it which involved the use of a model-independent method (empirical convolution) to generate effect site concentrations. The effect site concentration is the concentration of drug at the site of action or biophase.

In the subsequent sections we present the methodology of Sheiner's approach for analyzing analgesic data, and apply it to data obtained from a study which investigated the efficacy of a newly formulated nonsteroidal anti-inflammatory analgesic agent to relieve pain following a third molar extraction. In the example, we use S-PLUS to create plots and simulate responses of hypothetical patients to compare the dose–response time course after administration of an immediate release (IR) and a modified release (MR) formulation. The overall objective was to derive the dose of a newly formulated modified release product that was comparably efficacious to a single dose of a 500 mg IR product. This so-called Monte Carlo simulation will be discussed. This methodology is not limited to analgesic studies but is applicable wherever the outcome is measured as a categorical variable.

10.2 Methodology

The approach involves a semimechanistic or mechanistic model that describes the joint probability of the time of remediation and the pain relief score (which is related to plasma drug levels). Plasma drug levels are determined by the pharmacokinetics of the drug and the dose(s) administered. This joint probability can be written as the product of the conditional probability of the time of remediation given the level of pain relief and the probability of the pain relief score. First, a population pharmacokinetic model is developed using the nonlinear mixed effects modeling approach (Racine-Poon and Wakefield, 1998; Ette and Ludden, 1995; Mandema et al., 1992). With this approach both population (average) and random (inter- and intraindividual) effects parameters are estimated. When the pharmacokinetic model is linked to an effect (pharmacodynamic model) the effect site concentration (C_e) as defined by Sheiner et al. (1979) can be generated. The effect site concentration is useful in linking dose to pain relief and subsequently to the decision to remediate.

To model the distribution of pain relief scores and remediation times, subject-specific random effects models are developed. Let the vector of pain relief scores for an individual be $Y = (Y_1, Y_2, ..., Y_N)$. At time t the pain relief score is denoted by Y_t and the time at which an individual remediates is denoted by the variable T. The pharmacodynamic model parameter estimates are obtained by maximum likelihood which estimates the model parameters most probable for the observed data. $P(T, Y)$ denotes the likelihood of an individual's data, and it is expressed by the following equation:

$$
\begin{aligned}
P(T, Y) &= \int P(T, Y \mid \eta) \, P(\eta) \, d\eta \\
&= \int P(T \mid Y, \eta) \, P(Y \mid \eta) \, P(\eta) \, d\eta
\end{aligned}
\tag{10.1}
$$

where η is a vector of subject specific random effects, assumed to be (multivariate) normally distributed with a mean of zero and variance Ω. The likelihood can be factored out in two terms: one related to pain relief [$P(Y \mid \eta)$], and the other related to the remediation behavior conditional on pain relief [$P(T \mid Y, \eta)$]. In the subsequent sections, models for subject specific distributions $P(Y \mid \eta)$ and $P(T \mid Y, \eta)$ are discussed.

10.2.1 Pain Relief Model [$P(Y \mid \eta)$]

Pain relief is a categorical variable that can take a value of 0 (no pain relief), 1 (a little pain relief), 2 (some pain relief), 3 (a lot of pain relief), or 4 (complete pain relief). The probability that Y_t is greater than or equal to the score m ($m = 1, ..., 4$) is given by

$$
g\left\{ P(Y_t \geq m \mid \eta) \right\} = f_p(m, t) + f_d(C_e) + \eta_Y
\tag{10.2}
$$

where $g\{x\}$ denotes the logit transform of a probability to ensure probability values between 0 and 1, f_p is the function describing the time course of the placebo effect, the f_d function describes the drug effect, and η_Y is a random individual effect determining individual sensitivity. The η_Y are assumed to be normally distributed with standard deviation ω_Y.

The probability distribution of pain relief scores is given by

$$P(Y_t \geq m) = \frac{\exp\left[f_p(m, t) + f_d(m, C_e, t) + \eta_Y\right]}{1 + \exp\left[f_p(m, t) + f_d(m, C_e, t) + \eta_Y\right]} \qquad (10.3)$$

where

$$f_p(m, t) = \sum_{k=0}^{m} \beta_k + A\left[e^{-\alpha t} - e^{-\lambda t}\right] \qquad (10.4)$$

characterizes the placebo effect, and α and λ are first-order rate constants of the offset and onset of the placebo effect. A is a scaling parameter which determines the size of the placebo effect, and β_k specifies the baseline set of probabilities of the various degrees of pain relief. There is no estimate for β_0 in the model because $P(Y_t \geq 0) = 1$. The probability that $P(Y_t = m)$ is then equal to the difference in probabilities of two subsequent pain relief scores, i.e., $P(Y_t \geq m) - P[Y_t \geq (m-1)]$.

Models of varying complexities can be used to describe the placebo effect. Models for drug effect can be semimechanistic (i.e., link model, Sheiner et al., 1979) or mechanistic (i.e., indirect response model, Dayneka et al., 1993). A semimechanistic drug effect model can be expressed as

$$f_d(m, C_e, t) = \frac{E_{max} \times C_e}{EC_{e50} + C_e} \qquad (10.5)$$

where E_{max} is maximum drug effect, and EC_{e50} is the effect compartment drug concentration at 50% of the maximal drug effect. C_e denotes effect site concentration given by the following:

$$C_e(t) = k_{eo} \times e^{-k_{eo}t} \times C_p(t) \qquad (10.6)$$

where k_{eo} is the first-order rate constant that characterizes the delay between plasma and effect site concentrations. The population average pharmacokinetic parameters derived from the pharmacokinetic analysis are used to calculate $C_p(t)$, the concentration of the drug in the plasma at time t. Assuming that pain relief scores within an individual at distinct times are independent, the vector of pain relief scores $P(Y \mid \eta)$ is given by:

$$P(Y \mid \eta) = \prod_{t \leq T} P(Y_t \mid \eta) \qquad (10.7)$$

10.2.2   Model for Remediation $P(T \mid Y, \eta)$

The time to remediation can be viewed as a survival variable (McCullagh and Nedler, 1989). By definition, a survival function, $S(t)$, is the probability that a person remains in the study (does not remedicate) up to time t, and is given by

$$P(T > t \mid Y, \eta) = S(t) = \exp\left[-\int_0^t \lambda(t)\, dt \right] \qquad (10.8)$$

where $\lambda(t)$ is the hazard function. The hazard function can be interpreted as an instantaneous risk, in that $\lambda(t)\, \delta t$ is the probability that a subject remedicates in the next small interval of time δt, given that he has not remedicated. A constant hazard over a fixed interval of time indicates that a constant proportion of patients that are still in the study are expected to remedicate. A constant hazard function implies an exponential distribution of remedication with mean $1/\lambda$ and hence a Poisson process. Several models can be used to evaluate the hazard function, and the model that best describes the data is used to describe remedication (Sheiner, 1994; Mandema, 1997; Mandema and Stanski, 1996).

A model that allows the baseline hazard to change linearly over time but remain constant over a time interval δt is expressed as

$$\lambda(t, Y_t, \eta) = \lambda_m \left[1 + FH(t-1)_+ \right] \exp(\eta_T) \qquad (10.9)$$

where λ_m is the baseline hazard rate, FH is the fractional change in λ_m with time, $(t-1)_+ = t-1$ for $t > 1$ and zero otherwise, and η_T is a random individual effect. The assumption implicit in this model is that $P(T \mid Y, \eta)$ depends only on η_T and the observable elements of Y. $P(T > t \mid Y, \eta)$ is set to 1 for the time points before remedication is allowed. This time can be either 1, 2, 3, or 4 h, depending upon the study design. However, the hazard is allowed to accumulate according to (10.9) independent of the first time remedication is allowed. It follows from (10.8) that the probability that an individual will remedicate at time t given they are still in a study at the previous observation time $t-1$, is given by

$$P(T = t \mid T \geq t) = 1 - \frac{S(t)}{S(t-1)} = 1 - \exp\left[-\int_{t-1}^t \lambda(t)\, dt \right] \qquad (10.10)$$

10.2.3   Estimation and Inference

The Laplacian estimation method as implemented in the NONMEM program (a program the authors use for nonlinear mixed effects modeling) is used to provide maximum likelihood estimates of model parameters (Beal and Sheiner, 1982, 1992). Assuming the individuals to be independent, the likelihood L for all the data from N subjects can be specified by the product of the probability of each subject's data:

$$L = \prod_{i=1}^{N} \int P(T, Y | \eta) \, P(\eta) \, d\eta$$

$$(10.11)$$

$$= \prod_{i=1}^{N} \int P(T | Y, \eta) \, P(Y | \eta) \, P(\eta) \, d\eta$$

In order to simplify the calculations, it can be assumed that η_Y and η_T are independent, that is, $\mathrm{cov}(\eta_Y, \eta_T) = 0$. The implication of this is that pain relief data can be fitted separately from the remedication data by independent maximization of the following likelihoods:

$$L = \prod_{i=1}^{N} \int P(Y | \eta_Y) \, P(\eta_Y) \, d\eta_Y \qquad (10.12)$$

$$L = \prod_{i=1}^{N} \int P(T | Y, \eta_T) \, P(Y | \eta_T) \, P(\eta_T) \, d\eta_T \qquad (10.13)$$

Model selection is based on the likelihood ratio test with $p < 0.001$ and diagnostic plots. The difference in minus twice the log of the likelihood ($-2LL$) between a full and a reduced model is asymptotically χ^2 distributed with degrees of freedom equal to the difference in the number of parameters between two models. At $p < 0.001$ a decrease of more than 6.6 in $-2LL$ is significant. Asymptotic standard errors are obtained from the asymptotic covariance matrix. Alternatively, confidence intervals on parameters can be computed for this very nonlinear situation from the likelihood profile plot (Bates and Watts, 1988).

10.2.4 Prediction

Once the population model has been developed, interesting population statistics (time to onset of effect, percent of patients at peak effect) can be computed by means of Monte Carlo integration with respect to η. By simulating η-values from the estimated distribution, response profiles for individual subjects are generated. The population mean probability of having a certain pain relief score and the population mean expected pain relief score at a specific time and dose can be computed from these profiles. Virtual patient populations can be simulated for all doses. The goodness-of-fit of the model to the data can be judged by comparing model-generated simulations of the probability that pain relief is greater than or equal to m conditional on the remedication times, $P(Y \geq m | T \geq t)$, and model-generated estimates of the probability that a patient will remedicate at time t given they are still in the study up to that time point, $P(T = t | T \geq t)$, with data-derived estimates of these probabilities.

With Monte Carlo simulations the following questions can be answered:

- Is the analgesia caused by the drug greater than that produced by placebo?

- What dose of the drug should be recommended if the drug is efficacious?

In the example described below, the primary objective was to determine the dose of a newly developed modified release formulation of a nonsteroidal anti-inflammatory drug, which produced the same pain relief as an immediate release dosage form.

10.3 Application

In this section we apply the aforementioned methodology to the analysis of data derived from a study which investigated the analgesic efficacy of two formulations of a nonsteroidal anti-inflammatory drug. The formulations had different in vivo release rates and the study was conducted in individuals who had undergone oral surgery for molar extraction. Initially we describe the study design, study objectives, data characteristics, and population pharmacokinetic and pain relief model development. This is followed by simulation of the joint probability of remediation and pain relief. The application of S-PLUS at each step is noted and the code for each procedure is presented in the Appendix.

10.3.1 Study Design

Data were derived from a double blind, parallel group study to compare an immediate release (IR) formulation of a nonsteroidal analgesic (NSAID), a modified release (MR) formulation, a narcotic analgesic (NAD), and a placebo (control) treatment in patients with at least moderate pain following molar extraction. Patients received a study drug or placebo at intervals of four hours as displayed in Table 10.1. When patients reported pain of at least moderate intensity after molar extraction (pain intensity score ≥ 2), they received the first dose of the study drug in the treatment group to which they were assigned.

Table 10.1. Drug/placebo administration by period for different treatment groups. The number of individuals randomized to each treatment group is indicated in parentheses.

Group	Period 1	Period 2	Period 3
1	NSAID 500 mg; MR ($n = 30$)	Placebo	Placebo
2	NSAID 2×500 mg; MR ($n = 29$)	Placebo	Placebo
3	NSAID 500 mg; IR ($n = 30$)	Placebo	Placebo
4	NAD 30 mg (N/A)	NAD 30 mg	NAD 30 mg
5	Placebo ($n = 20$)	NAD 30 mg	NAD 30 mg

N/A—Data from this group was not used in the analysis.

10.3.2 Objectives

The objectives of the analysis were:

- To derive a population PK–PD model that characterized the distribution of pain relief scores following administration of either the IR or MR NSAID formulations.
- To determine how differing in vivo release rates of the NSAID formulation would affect the pain relief outcome.
- To determine equivalent doses for an IR and MR formulation of the NSAID.
- To determine the time to achieve maximal pain relief and the duration of pain relief for equally efficacious doses of the IR and MR NSAID formulations.

10.3.3 Data

Pain relief was recorded on a five-point scale [0, 1, 2, 3, 4] as described in Section 10.2 at intervals of 0, 15, 30, 45, 60, 90, and 120 min, and then hourly through to 12 h. Pharmacodynamic evaluations were recorded in patient diaries and all patients remained at the clinical research center for at least 2 h. A total of 1292 observations was obtained from 109 subjects. Samples of venous blood for pharmacokinetic evaluation were obtained at 15, 30, 45, 60, 90, and 120 min, and then every 2 h through 12 h with a 24 h sample the next day. Fifteen subjects were assigned to both the 500 mg IR and 2 × 500 mg MR treatment groups, and 14 subjects were assigned to the 500 mg MR treatment group.

10.3.4 Population Pharmacokinetic and Pain Relief Model Development

The development of a population pharmacokinetic model is the first step in the analysis because of the link between pain relief and the decision to remedicate. This involves estimating the population pharmacokinetic parameters using a nonlinear mixed effects model (NONMEM ver 5). Interindividual variability in parameters was modeled according to an exponential variance model as

$$P_{ki} = P_k \exp(\eta_{ki}) \tag{10.14}$$

where P_{ki} is the estimate of the kth pharmacokinetic parameter in the ith individual, P_k is the estimate of the kth pharmacokinetic parameter in the population (i.e., in a typical individual), and η_{ki} is the individual random effect for the kth parameter. It is assumed that the pharmacokinetic parameters are log-normally distributed. The values for η_{ki} are assumed to be independently multivariate normally distributed, with mean zero and diagonal variance–covariance matrix

Ω with diagonal elements (ω_1^2 ,..., ω_m^2). Residual error was modeled using a combination of additive and constant coefficient of variation error models as

$$C_{ij} = C_{mij} + C_{mij} \times \varepsilon_{1ij} + \varepsilon_{2ij} \tag{10.15}$$

where C_{ij} is the jth measured plasma concentration in the ith individual, and C_{mij} is the model predicted j^{th} concentration in the ith individual.

A two-compartment model with first-order absorption and an absorption lagtime best described the pharmacokinetic profiles of the NSAID in the subject population. The model parameters were: apparent clearance (CL/F), apparent volumes of distribution of the central (V_c/F) and peripheral compartments (V_p/F), intercompartmental clearance (Q/F), absorption rate constant (Ka), and absorption lag time (T_{lag}). The body mass index (percent ideal body weight) was a covariate for clearance and the volume of distribution. Two absorption rate constants, Ka_1 and Ka_2, were used to characterize absorption from MR and IR dosage forms, respectively. The parameters were reasonably well estimated, except for variability in V_p (Table 10.2). This parameter was poorly estimated because too few samples were located in the critical portion of the plasma concentration profile. Also, there was no information in the dataset to explain the large variability in Q. The parameter Q is derived in part from the volume terms, and any inadequacy in the estimation of the volume parameters indirectly affects the estimation of this parameter. The clearance and volume parameter estimates were scaled with respect to the oral bioavailability (F) because no intravenous dose was available.

Table 10.2. Parameter estimates for NSAID population pharmacokinetic models.

Parameter	Estimate (Percent RSE)[a]	Interindividual variability (%RSE)
CL/F*BMI (L/h)	0.0048 (41.67%)	20.98% (42.50%)
V_c/F*BMI (L)	0.016 (29.19%)	67.74% (37.69%)
Q/F (L/h)	2.08 (28.56%)	108.16% (48.46%)
V_p/F*BMI[b] (L)	0.044 (69.59%)	21.38% (140.48%)
Ka_1 (h^{-1})	0.16 (18.00%)	20.95% (18.00%)
Ka_2 (h^{-1})	0.49 (27.36%)	21.00% (27.80%)
T_{lag} (h)	0.22 (41.16%)	93.91% (42.30%)

[a] %RSE—Percent relative standard error.
[b] BMI—Body mass index.

The population pharmacokinetic parameters in combination with the individual patient body mass index (as covariate) were used to predict drug plasma levels in all the patients in the study. Predicted individual plasma drug concentrations and individual patient pain relief data were in turn used to generate individual patient effect-site concentrations. Effect site concentrations (10.6) were used as a covariate for estimating the probability of having a specific pain relief score.

For each level of pain relief (m), the probability of obtaining a pain relief score of an equal or greater value was determined at each time point (i.e., $P(Y_t \geq m)$). This was estimated using a logistic function that accounts for the drug effect, the placebo/time effect, and the individual random effect. The individual random effect accounts for the difference between the probability of a pain score predicted for the entire population and the probability of a pain score specific for each individual (see (10.3) to (10.5)). The same relationship was assumed for all levels of pain relief.

For each level of pain relief (m), the probability of remedicating at time t, given that subjects were in the study at the previous observation time $t-1$, was determined based on (10.9) and (10.10). The Laplacian method, as implemented in the NONMEM program, was used to provide estimates of model parameters. Parameter estimates for the model describing the probability of a pain relief score for each level of pain relief and for remedication are displayed in Tables 10.3 and 10.4.

Table 10.3 shows parameters describing the baseline probabilities for each category of pain relief, $(\beta_1 - \beta_4)$ were reasonably well estimated. These values, when substituted in (10.3) to (10.6), give the probability of an outcome in the absence of any placebo or drug effect. The poor precision associated with estimation of EC_{50} was possibly due to the fact that insufficient pharmacokinetic samples were located in the appropriate region of the pharmacokinetic profile necessary for the efficient estimation of this parameter. Similarly, there was insufficient data available for the characterization of k_{eo} and the onset rate (λ) of the placebo effect. The rate of offset (α) of the placebo effect was fixed at an infinitely small value. This is because this parameter approached very small values during the fitting procedure, a consequence of pain subsiding during the course of the study.

Table 10.3. Pain relief model parameter estimates.

Parameter	Value	Std Error
$\beta 1$	−2.00	fixed
$\beta 2$	−2.17	0.179
$\beta 3$	−1.54	0.132
$\beta 4$	−2.41	0.221
k_{eo} (h^{-1})	0.50	0.578
EC_{50} (ng/ml)	45.4	83.3
E_{max}	5.23	3.35
α (h^{-1})	0.00001	fixed
λ (h^{-1})	1.26	1.00
A	3.07	1.00
ω^2	5.24	1.24

Table 10.4. Remedication model parameter estimates.

Parameter	Estimate
λ_0	0.413
λ_1	0.0058
λ_2	0.0043
λ_3	0.0008
λ_4	0.00001
FH	0
ω_T^2	0

The model-derived probabilities for each level of pain relief, compared with the probabilities derived from the raw data after the administration of either placebo, 500 mg or 1000 (2 × 500) mg NSAID to subjects, are shown in Figure 10.1. The pain relief model shown as the solid line adequately describes the data shown by the plotting symbol, for the 500 and 1000 mg dose groups. The placebo data is not adequately described by the model. This is possibly due to too few data points because many subjects in this group remedicated with NAD. The figure (middle and right panels) suggests the appropriateness of the model for simulating outcomes in a patient population unbiased by censoring.

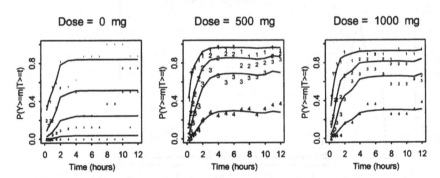

Figure 10.1. Plots of the mean Bayesian-derived probabilities for the pain relief data for the placebo, 500 mg and 1000 mg dose. Each profile corresponds to each level of pain relief. Obtaining a pain relief score ≥ 1 has the highest probability. The size of the plotting symbol is proportional to the square root of the number of data points.

The plots were generated using S-PLUS. The code for the procedure appears in Appendix 10.A.1. S-PLUS loops through each dose and generates a data frame containing all pain relief data at each time point for each dose and the model predicted probability of this observation. At each time point, the ratio of the number of scores greater than or equal to each level of pain relief relative to the number of observations is calculated. This ratio is the conditional probability of the observation and is plotted against the mean probability derived from the model.

The hazard calculated for each level of pain relief at each time point is shown in Figure 10.2. The figure indicates that within each time interval, the probability to remedicate was less than 0.12 (i.e., $1-\exp(-0.25 \times 0.5)$ for all pain relief categories greater than 0. Additionally, the figure indicates that remedication occurred predominantly in the placebo group with the remedication rate being greatest between 1 and 2 h. Few individuals remedicated after 4 h for any level of pain relief. Figure 10.2 suggests that the hazard model used was not the most appropriate. In this instance, a more flexible hazard model, rather than the step function type of hazard model used here, may have been more suitable. However, at the time the analysis was conducted, the routines allowing for such nonlinear mixed effects analysis in the NONMEM program were unavailable. The S-PLUS code for producing Figure 10.2 is given in Appendix 10.A.2. The total number of observations at each time point was determined for each level of pain relief in addition to the number of remedication observations for that interval. This is expressed as the log of (1–the probability of remedicating)/Δt = the hazard at time t.

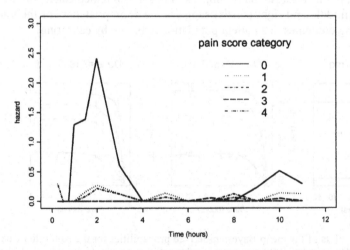

Figure 10.2. Calculated hazard for each pain relief category.

Figure 10.3 displays the model-predicted and observed remedication probability at each time point for each pain relief category. The model predictions suggest that the probability to remedicate was constant during the course of the study for each level of pain relief, and that the model is a reasonable predictor of remedication conditional on pain relief scores of 1, 2, 3, and 4. The S-PLUS code to produce Figure 10.3 is given in Appendix 10.A.3.

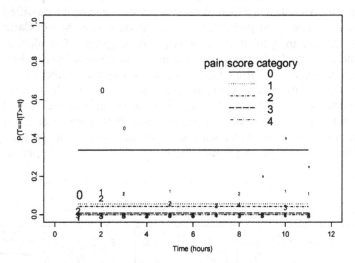

Figure 10.3. Model predictions of remedication probability versus time for all pain relief categories. The size of the plotting symbol is proportional to the square root of the number of observations.

10.3.5   Simulation of Joint Probability of Remedication and Pain Relief

Probabilities were simulated for doses of 0, 50, 150, 300, 500, and 1000 mg for a population of 1000 subjects using model parameter estimates in conjunction with simulated values of the random effect (η) (based on assuming a standard deviation of ω_Y). For the IR and MR dosage forms, the unconditional probability (derived from a completely nonremedicating population) and the conditional probability (derived from a population, biased by individuals remedicating) of a pain relief score being equal to or greater than each pain relief category (i.e., $P(Y_t \geq m)$) was determined at each time point. The probability of individual responses was then averaged. The simulated unconditional probability $P(Y_t \geq m)$ was compared with simulated conditional probabilities $P(Y_t \geq m \mid T > t)$ (that is, pain relief score probabilities derived from a patient population with simulated remedication) to visually gauge the extent of bias or erroneous sensitivity introduced by analyzing data only from individuals who remained in the study.

    To address the fundamental goal of the analysis, the relationship between dose, time, and the probability that the pain relief score was at least 2 ($P(Y_t \geq 2)$) was evaluated for each formulation. This clinical endpoint was selected based on analysis of remedication data from this study and on the work of other investigators (Sheiner, 1994; Mandema 1997; Mandema and Stanski, 1996). To address the duration of drug action, the probability that subjects will have a 75% chance of having a pain relief score of at least 2 at various doses was also computed.

250   E.I. Ette, P. Lockwood, R. Miller, and J. Mandema

Figure 10.4 displays fitted and observed conditional probabilities of pain relief scores for the individuals administered either a placebo dose or the modified
release formulation. The fitted probabilities were obtained after Monte Carlo
integration with respect to η of 1000 subjects per dose group. The plotting
symbols indicate the data-derived estimates for a specific value of pain relief.
The size of the symbols is proportional to the square root of the number of observations on which the probability estimates are based. These results suggest
that the joint probability model adequately predicts the response in the "dental
pain" study population which included some drop outs. Within one dose group
the size of plotting symbols decreases with time because some of the subjects
remedicate. The S-PLUS code to produce Figure 10.4 is given in Appendix
10.A.4.

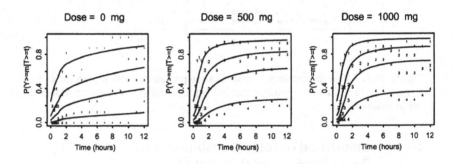

Figure 10.4. Probability of pain relief scores over a study period for subjects receiving
placebo or MR formulation. The solid lines represent model prediction and the numerals
reflect data-derived probabilities. Probabilities are conditional on individuals remaining
in the study.

Figure 10.5 compares the simulated conditional and unconditional probabilities of obtaining a pain relief score greater than or equal to m for each pain relief
category following oral administration of 500 mg of an IR product (left panel)
and an MR product (right panel). The figure demonstrates the bias (i.e., a higher
probability of an outcome) introduced into the analysis when probabilities are
determined only from individuals who remain in the study. Analysis of data
obtained only from these individuals suggests that the probability of obtaining a
pain relief score equal to or greater than m, conditional on remedication, tends to
have plateaued after 4 h for the IR product and to be marginally increasing at
this time for the MR product. In reality, had no individuals remedicated, the interpretation would be different: the probability of an outcome would decline
after 4 h for individuals administered the IR product, but remain constant for
those administered the MR product. The S-PLUS code to produce Figure 10.5 is
given in Appendix 10.A.5.

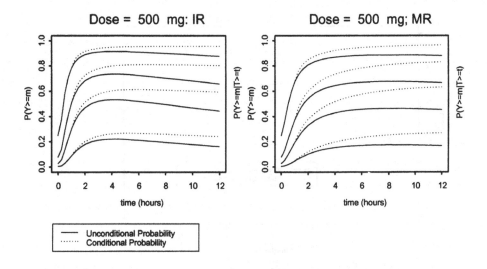

Figure 10.5. A comparison of the simulated conditional and unconditional probability of obtaining a pain relief score $\geq m$ for each pain relief category following oral administration of the 500 mg IR product (left panel) and MR product (right panel).

10.3.6 Dose Selection

The simulated unconditional probabilities can be used to make pharmacodynamic comparisons and guide dosing. A comparison of the unconditional probability of obtaining a pain relief score greater than or equal to m following the administration of the IR and MR formulations at an equivalent dose (500 mg) indicated a delayed onset time for the MR formulation (Figure 10.5). However, the duration of effect is longer following the administration of the modified release product. This prompted a direct comparison of the pain relief following the administration of the 500 mg dose of the IR formulation with the 2 × 500 mg dose of the MR dosage form.

Figure 10.6 demonstrates the results of the comparison of efficacy from the two dosage forms, the primary objective of the analysis. If adequate pain relief is considered as the probability of a pain relief score greater than or equal to 2, it can be seen that patients experience a comparable pain relief up to 3 h with the 2 × 500 mg dose of the MR dosage form and 500 mg of the IR dosage form. However, the pain relief provided by the MR formulation lasts longer than that provided by the IR formulation. Figure 10.6 also suggests that greater than 90% of the maximal effect is obtained within 2 h for the NSAID. This figure confirms the similar performance of the 500 mg IR and 2 × 500 mg MR formulations during the first 4 h following drug administration. The S-PLUS code to produce Figure 10.6 is given in Appendix 10.A.6.

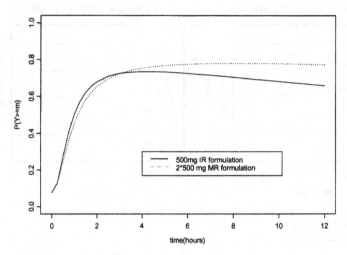

Figure 10.6. Unconditional probability that the pain relief score is greater than or equal to 2. Comparable efficacy of a 2 × 500mg MR dose and a 500 mg IR dose. The duration of effect of the MR formulation is longer.

Immediate Release Dosage Form   Modified Release Dosage Form

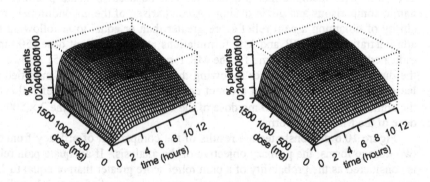

Figure 10.7. Surface response plots for different doses of the immediate release and modified release formulations. The surface represents the percentage of patients that would have a 75% chance of a pain relief score greater than or equal to 2. The plots are obtained after Monte Carlo simulations of 1000 patients per dose group.

Immediate Release Dosage Form   **Modified Release Dosage Form**

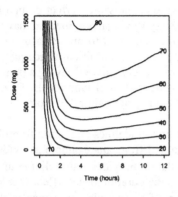 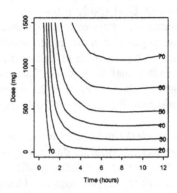

Figure 10.8. Comparitive contour plots for 500 mg IR and 2 × 500 mg MR doses.

A further goal of the analysis was to compare the simulated outcomes of adequate pain relief, dose, and time for the two dosage forms at different doses. Adequate pain relief was defined as a 75% chance that pain relief is equal to or greater than 2 ($P(Y_t \geq 2) > 0.75$). This level was adequate based on the remedication behavior displayed in Figure 10.2, that is, the probability to remedicate was low given a pain relief score of 2 or greater.

The plots in Figure 10.7 indicate that at maximal drug effect, 80% of patients will have adequate pain relief, and that the time to onset of PR is approximately 1 h. Additional insight into the dose response relationship can be obtained with the contour plots displayed in Figure 10.8. The S-PLUS code for plotting these figures is given in Appendix 10.A.7. Contour plots of the surface response figures confirm the similar performance of the 500 mg IR and 2 × 500 mg MR formulations during the first 4 h following drug administration. The performance of the 2 × 500 mg MR formulation is superior to the 500 mg IR beyond the 4 h time point as evidenced by the higher percentage of patients with $P(Y_t \geq 2) > 0.75$.

10.4   Conclusions and Summary

A new methodology for the analysis of analgesic trials has been discussed. The approach involves a model that links pain relief and the decision to remedicate to pharmacokinetics, and dosage, within a semimechanistic pharmacological framework. The advantages of this methodology over the traditional ANOVA approach has been presented. The use of this methodology together with the simulation of pain relief in a "virtual patient" permitted a comparison of the efficacy of two formulations of an NSAID with different in vivo release rates.

The conclusions are based on the uncensored distribution of pain relief scores which can only be determined from simulated data. A new aspect of this method was the use of population pharmacokinetic parameters and individual measures of BMI (body mass index) to predict individual effect compartment concentrations. Previous investigators using this analysis method have only used the population mean pharmacokinetic parameters to predict drug concentrations. Comparison of efficacy was based on the definition of adequate pain relief. In this analysis it was defined as a 75% probability of having a pain relief score greater than or equal to 2. Approximately 20% of individuals did not receive adequate PR and about 10% achieved adequate pain relief from a placebo dose.

The method can be used to develop dosing guidelines and guide formulation development. Dosing guidance was illustrated from this example as evidenced by the fact that the 2 × 500 mg MR formulation provided pain relief equivalent to 500 mg of the IR formulation. However, the duration of the effect was longer for the 2 × 500 mg MR dose. Projection of the simulations suggests that at equivalent doses the MR formulation is more effective after 4 h when compared with the IR formulation. Thus, with this methodology evidence of equivalent efficacy of two formulations with different in vivo absorption rates was obtained.

10.5   Acknowledgements

The authors would like to Dr. Jackie Butler and Dr. John Devane of the Elan Corporation, Athlone, Ireland, for their willingness to share the data used to derive the models described in this chapter.

10.6   References

Bates, D.M. and Watts, D.G. (1988). *Nonlinear Regression Analysis and Its Application*. Wiley, New York.

Beal, S.L. and Sheiner, L.B. (1982). Estimating population pharmacokinetics. *CRC Critical Review in Biomedical Engineering* 8, 195–222.

Beal, S.L. and Sheiner, L.B. (1992). *NONMEM Users Guides*, Parts I–VII. University of California, San Francisco, CA.

Dayneka, N.I., Garg, V., and Jusko, W.J. (1993). Comparison of four basic models of indirect pharmacodynamic responses. *Journal of Pharmacokinetics and Biopharmaceuticals* 21, 457–478.

Ette, E.I. and Ludden, T.M. (1995). Population pharmacokinetic modeling: The importance of informative graphics. *Pharmaceutical Research* 12, 1845–1855.

Laska, E.M. Siegel, C., and Sunshine, A. (1991). Onset and duration: Measurement and analysis. *Clinical Pharmacology and Therapeutics* **49**, 1–5.

Liu, C.Y. and Sambol, N.C. (1995). Pharmacodynamic analysis of analgesic trials using empirical methods. *Pharmaceutical Research* **12**, 438–445.

Mandema, J.W. (1997). Population pharmacokinetics/pharmacodynamics of analgesics: theory and applications. In: Aarons, L., Balant, L.P., Danhof, M., et al., eds. *The Population Approach: Measuring and Managing Variability in Response, Concentration and Dose.* Commission of the European Communities. European Cooperation in the Field of Scientific and Technical Research, Brussels, pp. 74–82.

Mandema, J.W. and Stanski, D.R. (1996). Population pharmacodynamic model for ketorolac analgesia. *Clinical Pharmacology and Therapeutics* **60**, 619–635.

Mandema, J.W., Verotta, D., and Sheiner, L.B. (1992). Building population pharmacodynamic models. I. Models for covariates. *Journal of Pharmacokinetics and Biopharmaceuticals* **20**, 58–528.

Max, M.B., Portenoy, R.K., and Laska, E.M., eds. (1991). *Advances in pain research and therapy.* Volume 18: *The Design of Analgesic Clinical Trials.* Raven Press, New York.

McCullagh, P. and Nelder, J.A. (1989). *Generalized Linear Models.* Second edition. Chapman & Hall, New York.

Racine-Poon, A. and Wakefield, J. (1998). Statistical methods in population pharmacokinetic modeling. *Statistical Methods in Medical Research* **7**, 63–84.

Sheiner, L.B. (1994). A new approach to the analysis of analgesic drug trials, illustrated with bromfenac data. *Clinical Pharmacology and Therapeutics* **56**, 3098–322.

Sheiner, L.B., Stanski, D.R., Vozeh, D.R., and Miller, R.D. (1979). Simultaneous modeling of pharmacokinetics modeling of pharmacokinetic and pharmacodynamics: Application to D-tubocurarine. *Clinical Pharmacology and Therapeutics* **25**, 358–371.

Velagapudi, R., Harter, J.G., Brueckner, R., and Peck, C.C. (1991). Pharmacokinetic/pharmacodynamic models in analgesic study design. In: Max, M.B., Portenoy, R.K., Laska, E.M., eds. *Advances in Pain Research and Therapy,* Vol. 18. Raven Press, New York, pp. 559–562.

Wajdula, J.S., Vavra, I., Sullivan, D., Ermer, J., and Osman, M. (1994). Bromfenac. In: Lewis, A.J., Furst, D.E., eds. *Nonsteroidal Anti-Inflammatory Drugs: Mechanisms and Clinical Uses.* Second edition. Marcel Dekker, New York, pp. 267–284.

10.A Appendix

10.A.1 Code for Figure 10.1

```
# Plot of model-derived probabilities for
# each level of pain relief compared with the probabilities
# derived from the raw data read in Nonmem output table
aa <- matrix(scan("r4.out.tbl"), ncol = 20, byrow = T)
#id time dose pid quit mdvp p0 prl tqt ce eta p1 p2 p3 p4 ty
# 1    2    3   4   5    6   7  8   9  10  11 12 13 14 15 16
# levels of cumalitive pain relief defined
prl <- c(1, 2, 3, 4)
# dose levels defined
dose <- sort(unique(aa[, 3]))
nd <- length(dose)
dev.ask(ask = T)
par(mfrow = c(2, 3), mar = c(6, 6, 5, 1))
for(k in dose) {
# select data for specific dose
  a <- aa[aa[, 3] == k,  ]
# design times
  td <- sort(unique(a[, 2]))
  ntd <- length(td)
  frq <- matrix(nrow = ntd, ncol = 1)
  tot <- matrix(nrow = ntd, ncol = 1)
  frqf <- matrix(nrow = ntd, ncol = 1)
  plot(0, 0, type = "n", xlim = c(0, 12), ylim = c(0, 1),
    xlab = "Time (hours)", ylab = "P(Y>=m|T>=t)",
    cex = 0.9, lwd = 2)
  title(paste("Dose = ", k, " mg"), cex = 0.8)
  for(j in prl) {
    for(i in seq(ntd)) {
# total number of observations at specific time
      tot[i] <- length(a[a[, 2] == td[i], 8])
# total number of PRL>=j at specific time
      totp <- length(a[a[, 2] == td[i] & a[, 8] >= j, 8])
# mean of  model derived  estimate PRL>=j
      frqf[i] <- mean(a[a[, 2] == td[i], j + 11])
      frq[i] <- (totp)/(tot[i])
    }
    lines(td, frqf, lwd = 2)
    text(td, frq, j, cex = sqrt(tot)/10)
  }
}
```

10.A.2   Code for Figure 10.2

```
# This file plots the hazard at each time point
# for each level of pain relief
a <- matrix(scan("prdat2.txt", skip = 1), ncol = 10,
  byrow = T)
# id time dose form prl q  mdvq bmi po tqt
#  1   2    3    4   5 6    7   8  9  10
td <- sort(unique(a[, 2]))
ntd <- length(td)
prl <- c(0, 1, 2, 3, 4)
dose <- sort(unique(a[, 3]))
nd <- length(dose)
frq <- matrix(nrow = ntd, ncol = 1)
tot <- matrix(nrow = ntd, ncol = 1)
frqf <- matrix(nrow = ntd, ncol = 1)
i <- duplicated(a[, 1])
pr <- a[i == F, c(1, 2, 3, 4, 5, 6, 7, 8)]
nid <- matrix(nrow = nd, ncol = 1)
for(i in seq(nd)) {
  nid[i] <- length(pr[pr[, 8] == dose[i], 8])
}
par(mfrow = c(1, 1), mar = c(5, 5, 5, 1))
# generate hazard estimates
plot(0, 0.1, type = "n", xlim = c(0, 12), ylim = c(0, 3),
  xlab = "Time (hours)", ylab = "hazard", lwd = 2)
# title("Hazard vs Pain Relief Category and Time")
for(j in prl) {
  for(i in seq(ntd)) {
    tot[i] <- length(a[a[, 2] == td[i] & a[, 5] == j, 1])
    frq[i] <- length(a[a[, 2] == td[i] & a[, 5] == j &
      a[, 6] == 1, 1])
    frq[i] <- frq[i]/tot[i]
    if(tot[i] == 0) {
      frq[i] <- 0
    }
    if(i == 1) {
      frq[i] <-  - log(1 - frq[i])/td[i]
    }
    if(i > 1) {
      frq[i] <-  - log(1 - frq[i])/(td[i] - td[i - 1])
    }
  }
  lines(td, frq, lty = j + 1, lwd = 3)
```

```
  text(td, frq, j, cex = sqrt(tot)/4)
}
legnam <- c("0", "1", "2", "3", "4")
legend(7, 2.5, legend = legnam, lty = c(1:5), cex = 1.5,
  bty = "n")
text(8.5, 2.7, "pain score category", cex = 1.5)
```

10.A.3   Code for Figure 10.3

```
# This file plots the observed probability of remedication
# and the model predictions as determined by
# the exponent of the -ve hazard for each Dt
a <- matrix(scan("remed.dat"), ncol = 12, byrow = T)
dimnames(a) <- list(NULL, c("id", "time", "dose", "form",
  "prl", "q", "mdvq", "bmi", "po", "tqt"))
# parameters of fit from hazard model
bh <- c(0.413, 0.0579, 0.0435, 0.00773, 1e-005)
th <- 1.68e-008  #design times
td <- c(1, 2, 3, 4, 5, 6, 7, 8, 9, 10, 11)
t1 <- c(0, 1, 2, 3, 4, 5, 6, 7, 8, 9, 10)
t2 <- c(1, 2, 3, 4, 5, 6, 7, 8, 9, 10, 11)
ntd <- length(td)
pty1 <- matrix(1, nrow = 11, ncol = 5)
# calculate p(T!=t|Y=k)
# this calculates the survival ratio in each interval
# for each value of PRL
# for patients first allowed to quit at t==1
for(j in seq(5)) {
  surv <- 1
  for(i in seq(ntd)) {
    if(td[i] < 1) {
      lhaz <- 0
    }
    if(td[i] == 1) {
      lhaz <- bh[j]
    }
    if(td[i] > 1) {
      lh1 <- bh[j] * (td[i] + th * (0.5 * td[i] * td[i] -
        td[i]))
      lh2 <- bh[j] * (td[i - 1] + th * (0.5 * td[i - 1] *
        td[i - 1] - td[i - 1]))
      lhaz <- lh1 - lh2
    }
    pty1[i, j] <- exp( - lhaz)
```

```
      }
}
pty <- 1 * pty1  #levels of pain relief
prl <- c(0, 1, 2, 3, 4)
frq <- matrix(nrow = ntd, ncol = 1)
tot <- matrix(nrow = ntd, ncol = 1)
par(mfrow = c(1, 1), mar = c(5, 5, 3, 1))
plot(0, 0.1, type = "n", xlim = c(0, 12), ylim = c(0, 1),
  xlab = "Time (hours)", ylab = "P(T==t|T>=t)", lwd = 2)
# title(paste("Remedication Probability vs",
#    "Pain Relief Category and Time"))
for(j in prl) {
  for(i in seq(ntd)) {
# number of patients in time interval with PRL==j
    tot[i] <- length(unique(a[a[, 2] > t1[i] & a[, 2] <=
      t2[i] & a[, 5] == j, 1]))
# number of patients quitting in time interval with PRL==j
    frq[i] <- length(a[a[, 2] > t1[i] & a[, 2] <= t2[i] &
      a[, 5] == j & a[, 6] == 1, 6])
    frq[i] <- frq[i]/tot[i]
  }
  text(td, frq, j, cex = sqrt(tot)/5)
  lines(td, (1 - pty[, j + 1]), lty = j + 1, lwd = 2)
}
legnam <- c("0", "1", "2", "3", "4")
legend(7, 2.5, legend = legnam, lty = c(1:5), cex = 1.5,
  bty = "n")
text(8.5, 2.7, "pain score category", cex = 1.5)
```

10.A.4   Code for Figure 10.4

```
# This file plots pain relief and remedication model
# to the pain relief data
# Matrices "pop", "surpop", and "dospop"
# are derived in Appendix 10.A.5
#
# read in nm output table
aa <- matrix(scan("nlog.1.tbl"), ncol = 20, byrow = T)
#id time dose pr tqt mdvp p0 prl tqt ce eta p1 p2 p3 p4 ty
# 1    2   3   4  5   6   7  8   9  10 11 12 13 14 15 16
#
# levels of cumalitive pain relief
prl <- c(1, 2, 3, 4)  # dose levels
dose <- sort(unique(aa[, 3]))
```

```
nd <- length(dose)
options(object.size = 6000000)
# design times
td <- c(0.25, 0.5, 0.75, 1, 1.5, 2, 3, 4, 5, 6, 7, 8, 9,
  10, 11, 12)
ntd <- length(td)
frq <- matrix(nrow = ntd, ncol = 1)
tot <- matrix(nrow = ntd, ncol = 1)
frqf <- matrix(nrow = ntd, ncol = 1)
# time points for simulation
tt <- c(seq(0, 1.75, 0.25), seq(2, 12, 0.5))
nt <- length(tt)  # dev.ask(ask=T)
par(mfrow = c(2, 3), mar = c(6, 6, 5, 1))
for(k in dose) {
# select data for specific dose
  a <- aa[aa[, 3] == k,   ]
  plot(0, 0, type = "n", xlim = c(0, 12), ylim = c(0, 1),
    xlab = "Time(hours)", ylab = "P(Y>=m|T>=t)", cex = 0.9,
    lwd = 2)
  title(paste("Dose = ", k, " mg"), cex = 0.8)
  for(j in prl) {
    for(i in seq(ntd)) {
#total number of observations at specific time
      tot[i] <- length(a[a[, 2] == td[i], 8])
#total number of PRL>=j at specific time
      totp <- length(a[a[, 2] == td[i] & a[, 8] >= j, 8])
      frq[i] <- (totp)/(tot[i])
    }
    yy1 <- pop[dospop == k, j] * surpop[dospop == k]
    yy1 <- matrix(yy1, nrow = ni, ncol = nt, byrow = T)
    yy2 <- matrix(surpop[dospop == k], nrow = ni,
      ncol = nt, byrow = T)
    for(i in seq(nt)) {
      y1[i] <- mean(yy1[, i])/mean(yy2[, i])
    }
    lines(tt, y1, lwd = 2)
    text(td, frq, j, cex = sqrt(tot)/10)
  }
}
```

10.A.5   Code for Figure 10.5

```
# This file simulates and plots the conditional and
# unconditional probability of obtaining a
```

```
# pain relief score m for each pain relief category
# for different doses of the MR formulation.
#
# Probabilities for the IR formulation can be simulated
# following substitution of the rate constant (ka)
# for the IR formulation.
#
# The pharmacokinetics are described by a
# two compartment model
#
#model parameters
alag <- 0.16  # lag time
ka <- 0.49  # form=1 for cr & 0 for ir
cl <- 0.5136  # clearance (0.0048*107.088(mean bmi))
v2 <- 0.1713  # vc (0.0016*bmi;)
q <- 2.08  # intercomp clearance
v3 <- 6.07  # vp (0.044*bmi + 1.36)
k20 <- cl/v2
keo <- 0.5
bl <- c(-2, -2 - 2.17, -2 - 2.17 - 1.54,
  -2 - 2.17 - 1.54 -2.41)
ec5 <- 45.4
em <- 5.23
bl1 <- 1e-005  # placebo offset rate
bl2 <- 1.26  # placebo onset rate
bb1 <- 3.07  # placebo scalar
ome <- 5.24  # mean squared error, typical size of
#    difference between subjects and popln mean
bh <- c(0.413, 0.0579, 0.0435, 0.00773, 1e-005)
#hazard for
# each level of pain relief
th <- 1.68e-008
options(object.size = 50000000)
set.seed(4663)  #time points for simulation
tt <- c(seq(0, 1.75, 0.25), seq(2, 12, 0.5))
te <- tt - alag
te <- ifelse(te < 0, 0, te)
# puts 0 in vector if tt-alag<0
nt <- length(tt)  #doses for simulation
dose <- c(0, 50, 100, 500, 1000, 1500)
nd <- length(dose)  #number of subjects
ni <- 1000  #simulated probability matrix
pop <- matrix(0, nrow = ni * nt * nd, ncol = 4)
temp <- matrix(seq(ni), nrow = nt, ncol = ni, byrow = T)
idpop <- matrix(temp, nrow = ni * nt * nd, ncol = 1)
```

```
temp <- matrix(dose, nrow = ni * nt, ncol = nd, byrow = T)
dospop <- matrix(temp, nrow = ni * nt * nd, ncol = 1)
tpop <- matrix(tt, nrow = ni * nt * nd, ncol = 1)
k23 <- q/v2
k32 <- q/v3
bet1 <- k23 + k32 + k20
bet2 <- sqrt(bet1^2 - 4 * k32 * k20)
beta <- 0.5 * (bet1 - bet2)
alfa <- (k32 * k20)/beta
bsl <- (keo * ka)/v2
# put dose in here if need to (ie keo*ka*dose/v2)
p1 <- alfa - ka
p2 <-  - p1
q1 <- beta - ka
q2 <-  - q1
r1 <- keo - ka
r2 <-  - r1
s1 <- beta - alfa
s2 <-  - s1
t1 <- keo - alfa
t2 <-  - t1
u1 <- keo - beta
u2 <-  - u1
z1 <- k32 - ka
z2 <- k32 - alfa
z3 <- k32 - beta
z4 <- k32 - keo
e1 <- exp( - ka * te)
e2 <- exp( - alfa * te)
e3 <- exp( - beta * te)
e4 <- exp( - keo * te)
ce1 <- (z1 * e1)/(p1 * q1 * r1)
ce2 <- (z2 * e2)/(p2 * s1 * t1)
ce3 <- (z3 * e3)/(q2 * s2 * u1)
ce4 <- (z4 * e4)/(r2 * t2 * u2)
cpe <- bsl * (ce1 + ce2 + ce3 + ce4)
cepop <- matrix(cpe, nrow = ni * nt * nd, ncol = 1)
# effect site concentrations
cepop <- cepop * dospop  #sample eta's for 1000 subjects
eta <- rnorm(ni, mean = 0, sd = sqrt(ome))
temp <- matrix(eta, nrow = nt, ncol = ni, byrow = T)
temp2 <- matrix(temp, nrow = ni * nt * nd, ncol = 1)
for(i in seq(4)) {
  pop[, i] <- temp2 + (cepop * em)/(cepop + ec5) + bb1 *
    (exp( - bl1 * tpop) - exp( - bl2 * tpop)) + bl[i]
```

```
}
pop <- exp(pop)/(1 + exp(pop))
pty1 <- matrix(1, nrow = nt, ncol = 5)
# calculate p(T!=t|Y=k), can be done here because no eta
# this calculates the survival ratio in each interval
#for each value of Y
# for people first allowed to quit at 1 hr
for(j in seq(5)) {
  for(i in seq(nt)) {
    if(tt[i] < 1) {
      lhaz <- 0
    }
    if(tt[i] == 1) {
      lhaz <- bh[j]
    }
    if(tt[i] > 1) {
      lh1 <- bh[j] * (tt[i] + th * (0.5 * tt[i] * tt[i] -
        tt[i]))
      lh2 <- bh[j] * (tt[i - 1] + th * (0.5 * tt[i - 1] *
        tt[i - 1] - tt[i - 1]))
      lhaz <- lh1 - lh2
    }
    pty1[i, j] <- exp( - lhaz)
  }
}
pty <- 1 * pty1
# calculate p(T!=t|Y), so this is the individual
# survival ratio given that subjects vector of Y
temp <- matrix(pty[, 1], nrow = ni * nt * nd, ncol = 1) *
  (1 - pop[, 1])
temp <- temp + matrix(pty[, 2], nrow = ni * nt * nd,
  ncol = 1) * (pop[, 1] - pop[, 2])
temp <- temp + matrix(pty[, 3], nrow = ni * nt * nd,
  ncol = 1) * (pop[, 2] - pop[, 3])
temp <- temp + matrix(pty[, 4], nrow = ni * nt * nd,
  ncol = 1) * (pop[, 3] - pop[, 4])
temp <- temp + matrix(pty[, 5], nrow = ni * nt * nd,
  ncol = 1) * (pop[, 4])
# now calculate the actual survival p(T>t|Y) as a
# product of the survival ratios
# move it up 1 because Y is still measured when
# people quit so p(T>=t|Y)
surpop <- matrix(temp, nrow = ni * nd, ncol = nt,
  byrow = T)
surv <- matrix(1, nrow = ni * nd, ncol = 1)
```

```
for(i in seq(nt)) {
  survrat <- surpop[, i]
  surpop[, i] <- surv
  surv <- surv * survrat
}
surpop <- t(surpop)
surpop <- matrix(surpop, nrow = ni * nd * nt, ncol = 1)
# pop and surpop (and dospop for selections) are the
#   important matrices
# generate response plot overlay
# P(Y>=k)
y1 <- matrix(0, nrow = nt, ncol = 1)
par(mfrow = c(2, 2), mar = c(5, 5, 3, 2))
for(k in dose) {
  plot(0, 0, type = "n", xlim = c(0, 12), ylim = c(0, 1),
    xlab = "time (hours)", ylab = "P(Y>=m)", lwd = 2)
  mtext("P(Y>=m|T>=t)", side = 4, line = 1)
  for(i in seq(4)) {
    yy <- matrix(pop[dospop == k, i], nrow = ni, ncol = nt,
      byrow = T)
    for(j in seq(nt)) {
      y1[j] <- mean(yy[, j])
    }
    lines(tt, y1, lwd = 1, lty = 1)
  }
  for(i in seq(4)) {
    yy1 <- pop[dospop == k, i] * surpop[dospop == k]
    yy1 <- matrix(yy1, nrow = ni, ncol = nt, byrow = T)
    yy2 <- matrix(surpop[dospop == k], nrow = ni,
      ncol = nt, byrow = T)
    for(j in seq(nt)) {
      y1[j] <- mean(yy1[, j])/mean(yy2[, j])
    }
    lines(tt, y1, lwd = 1, lty = 2)
  }
  title(paste("Dose = ", k, " mg: IR"), cex = 0.8)
}
legnam <- c("Unconditional probability",
  "conditional probability")
legend(locator(1), legend = legnam, lty = c(1, 2))
```

10.A.6   Code for Figure 10.6

```
# This file plots comparable efficacy of
# the 2*500 mg MR and 500mg IR formulations
ylir <- matrix(0, nrow = nt, ncol = 1)
ylcr <- matrix(0, nrow = nt, ncol = 1)
tt <- c(seq(0, 1.75, 0.25), seq(2, 12, 0.5))
plot(0, 0, type = "n", xlim = c(0, 12), ylim = c(0, 1),
  xlab = "time(hours)", ylab = "P(Y>=m)", lwd = 2)
# title(paste("Pain Relief Score >=", 2,
#    "for 500mgIR and 1000mgCR"),cex=0.8)
for(k in dose) {
  yyir <- matrix(popir[dospopir == 500, 2], nrow = ni,
    ncol = nt, byrow = T)
  yycr <- matrix(popcr[dospopcr == 1000, 2], nrow = ni,
    ncol = nt, byrow = T)
  for(j in seq(nt)) {
    ylir[j] <- mean(yyir[, j])
    ylcr[j] <- mean(yycr[, j])
  }
  lines(tt, ylir, lty = 1)
  lines(tt, ylcr, lty = 2, lwd = 2)
}
legnam <- c("500mg IR formulation",
  "2*500 mg MR formulation")
legend(4, 0.3, legend = legnam, lty = c(1, 2))
```

10.A.7   Code for Figures 10.7 and 10.8

```
# This file plots a 3-D image of the % patients with
# adequate PRL vs Time and Dose and contour plots
options(object.size = 6000000)
par(mfrow = c(1, 2), mai = c(3, 1, 1, 0.5))
y1 <- matrix(0, nrow = nt, ncol = 1)
y2 <- matrix(0, nrow = nt, ncol = 1)
y3 <- matrix(0, nrow = nt, ncol = 1)
z <- matrix(0, nrow = nt * 5, ncol = 1)
x <- matrix(0, nrow = nt * 5, ncol = 1)
y <- matrix(0, nrow = nt * 5, ncol = 1)
# time points for simulation
tt <- c(seq(0, 1.75, 0.25), seq(2, 12, 0.5))
nt <- length(tt)  # doses for simulation
```

```
dose <- c(0, 50, 100, 500, 1000, 1500)
# dose <- c(0,50.0,150.0,300.0,500.0,1000.0)
nd <- length(dose)
for(i in 2) {
  for(k in seq(nd)) {
    yy <- matrix(popir[dospopir == dose[k], i], nrow = ni,
      ncol = nt, byrow = T)
    for(j in seq(nt)) {
      indx <- j + (k - 1) * nt
# percentage of patients that have P(Y>=2)>0.75
      z[indx] <- (length(yy[yy[, j] > 0.75, j]) * 100)/ni
      x[indx] <- tt[j]
      y[indx] <- dose[k]
    }
  }
  popir.fit <- interp(x, y, z)
  persp(popir.fit, xlab = "time(hours)", ylab =
    "dose (mg)", zlab = "% patients", zlim = c(0, 100),
    lwd = 2, cex = 1)
}
for(i in 2) {
  for(k in seq(nd)) {
    yy <- matrix(popcr[dospopcr == dose[k], i], nrow = ni,
      ncol = nt, byrow = T)
    for(j in seq(nt)) {
      indx <- j + (k - 1) * nt
# percentage of patients that have P(Y>=2)>0.75
      z[indx] <- (length(yy[yy[, j] > 0.75, j]) * 100)/ni
      x[indx] <- tt[j]
      y[indx] <- dose[k]
    }
  }
  popcr.fit <- interp(x, y, z)
  persp(popcr.fit, xlab = "time(hours)",
    ylab = "dose (mg)", zlab = "% patients",
    zlim = c(0, 100), lwd = 2, cex = 1)
}

# or a contour plot
par(mfrow = c(1, 2), mai = c(3, 1, 1, 0.5))
  contour(popir.fit, xlab = "Time(hours)",
    ylab = "Dose (mg)", nlevels = 10)
```

Part 5:

Phase II and Phase III Clinical Trials

11

Power and Sample Size Calculations

Jürgen Bock

F. Hoffmann-La Roche AG, Basel, Switzerland

11.1 Why Use S-Plus for Sample Size Calculations?

Sample size calculations belong to the routine steps in the design of studies for drug development. The importance of the determination of the appropriate sample size for clinical studies is emphasized in the ICH Guidelines: *"The number of subjects in a clinical trial should always be large enough to provide a reliable answer to the question addressed."* A simple example may help to understand the basic problem.

Example 1. A pharmacodynamic study was conducted to investigate whether an antiepileptic drug reduced the number of electroencephalographic (EEG) spikes. EEG recording began 1 h before dosing and was continued for 1 h afterward; spike counts were performed at 2 min intervals. The aim was to determine whether in the majority of the patients the antiepileptic drug reduces the mean number of spikes per unit by at least half.

The mean number spikes_{BL} (from 30 intervals) before dosing (baseline) and the mean number spikes_T after dosing were calculated for each patient. Although counts are discretely distributed, the distribution of means from 30 observations has been assumed to be approximately continuous. The 10 mg dose was administered to $n = 9$ patients. The data for this example can be found in the data frame `spikes`. We want to know whether the mean spike counts are more than halved. For this purpose the spike counts after treatment were multiplied by 2, and the ratios $x = 2 * \text{spikes}_T / \text{spikes}_{BL}$ and logarithms $y = \log(x)$ calculated. This gives rise to expectation 1 for the ratios and 0 for the logarithms in case of halving.

Applying summary delivers the following lines for medians and means:

```
    spikes.BL        spikes.T            x                y
Median: 5.677 Median:2.4290 Median:0.5293 Median:-0.6361
 Mean: 8.871    Mean:3.0480    Mean:0.7062    Mean:-0.6609
```

The differences between the means and medians indicate highly skewed distributions of both the mean spike numbers and their ratios. However, for the logarithms of the ratios, the means and medians are roughly equal, their distributions being nearly symmetrical. One therefore can assume that the logarithms y in the population follow approximately a normal distribution with the expected value μ_y and variance σ_y^2. The ratios x must therefore be approximately log-normally distributed.

The null hypothesis that the population mean for y is zero means a halving of the mean spike counts after treatment. Because $y = \log(2 * \text{spikes.T}) - \log(\text{spikes.BL})$ is a vector of individual differences we can apply the paired t-test:

```
> t.test(spikes$y, alternative="less")$p.value
    0.02717707
```

For a predefined significance level of $\alpha = 0.05$ (one-sided) we would conclude that the mean difference of the log-transformed values is significantly smaller than zero. Or, equivalently, that the geometric mean of x is significantly smaller than 1. If the study would have been stopped earlier, say after patient 8, we would have obtained

```
> t.test(spikes$y[1:8], alternative="less")$p.value
    0.05325937
```

i.e., a nonsignificant result. According to

```
> mean(spikes$y)
    -0.6609389
> sqrt(var(spikes$y))
    0.8802705
> mean(spikes$y[1:8])
    -0.604568
> sqrt(var(spikes$y[1:8]))
    0.9235194
```

only a slight difference in means and standard deviations of y can be detected. The corresponding geometric means of x are $e^{-0.661} = 0.52$ and $e^{-0.605} = 0.55$.

A look at the t-test formula

$$t = \frac{y}{s_y}\sqrt{N}$$

reveals that even if the mean y and standard deviation s_y would stay constant, the t-statistic would change with the size N. The mean y denotes a mean difference here.

More generally, the means, mean differences, and standard deviations tend to their true values with increasing sample size according to the law of large numbers, whereas the term $\sqrt{N}$ increases monotonically. No matter how big the true population mean difference is, we obtain a significant result if the sample size used is sufficiently large. To be more precise we have to examine the probability of getting a significant t-test result. This is called **error probability** α in the case of the true mean difference being equal to zero or else **power**. Whilst α is prespecified before the start of the trial, the power depends on the true mean difference, the true standard deviation σ, and the sample size. We can neither change the true mean difference nor the true standard deviation, but the power will increase if we increase the sample size.

The difference found for a large sample size may be statistically significant, but of no clinical relevance. On the other hand, a clinically relevant difference may be overlooked if the sample size is too small, the study being "underpowered." Consequently, good clinical trial design requires specifying the **population mean difference** $\Delta = \mu_1 - \mu_2$ **to be detected** for the primary variable before the start of the study. Based on this, one then needs to calculate the corresponding sample size for the prespecified α-level and power, commonly using $\alpha = 0.05$ and $power = 0.80$.

In the case of a one-sided paired t-test the sample size can be calculated using the well-known approximation formula

$$ N \approx \frac{[t_{1-\alpha,N-1} + t_{1-\beta,N-1}]^2}{c^2} $$

with $c = \Delta/\sigma$, where σ denotes the true standard deviation of the pairwise differences. The quantiles $t_{1-\alpha,N-1}$ and $t_{1-\beta,N-1}$ of the t-distribution with $N - 1$ degrees of freedom can be computed from the S-PLUS function qt.

In the previous example we have asked whether the mean spike counts are at least halved. Any difference to be detected indicates a further decrease in mean spike counts, and $\Delta = \log(2) = 0.6931$ means a further halving since $e^{0.6931} = 2$. Planning another study that should detect a mean difference of $\Delta = 0.6931$ with power $1 - \beta = 0.80$, using a paired t-test at level $\alpha = 0.05$ (one-sided), our requirements are to be interpreted as follows: If the null hypothesis is true, i.e., if the mean spike counts are at most halved, then the risk of concluding that they are more than halved is only 0.05. If the mean spike counts are reduced to one quarter of the baseline counts, then we will detect it with probability 0.80.

For sample size calculations we must specify a value for the standard deviation, for example, $\sigma = 0.9$ mm Hg. We can compute the right-hand side of the formula above increasing N until the left-hand side exceeds it.

The function

```
> right.side <- function(N, delta, sigma=1, alpha=0.05,
  beta=0.2)
  { (sigma*(qt(1-alpha, N-1)+qt(1-beta, N-1))/delta)^2
}
```

provides

```
> right.side(N=10:15, delta=0.6931, sigma=0.9)
   12.44280 12.21485 12.03305 11.88471 11.76137 11.65722
```

For $N = 13$ the left-hand side is for the first time smaller than the right. Therefore a sample size of 13 should be used.

The crucial assumption made for the sample size calculations here is that of the variance. To get the "true" sample size we would need to know the "true" standard deviation σ. Usually σ is replaced by an estimate from a previous study with the same drug. In the worst case one uses the results of a similar study with a drug of the same class. Most often the inclusion and exclusion criteria, the treatments, the centers, as well as other factors, differ essentially between these, i.e., the target populations differ. The sample size is calculated under the assumption that the true standard deviation in the target population coincides with the estimate from the previous study.

If the previous study is small this estimate may be unreliable, hence it has been recommended to use an upper confidence limit instead of the standard deviation itself. This makes sample size calculations more conservative, but does not solve the problem of different target populations. The sample used to estimate σ is usually not drawn from the target population of the future study, i.e., the estimate and the confidence limits may be biased, leading to a biased estimate of the necessary sample size.

This problem cannot be solved with classical designs, but with sequential designs, where other assumptions may be critical. We must be aware that drawing statistical conclusions means making decisions under uncertainty and taking risks. On one hand, we must rely on our assumptions, but it is wise to vary these, and explore their influence on the power and sample size.

This is also required by the ICH Guidelines: "*It is important to investigate the sensitivity of the sample size to a variety of deviations from these assumptions and this may be facilitated by providing a range of sample sizes appropriate for a reasonable range of deviations from the assumptions.*" Generally this also includes different distributional assumptions, use of different tests, etc., which requires incorporating different sample size formulas. In a more narrow sense, we can investigate the sensitivity to deviations from the assumed design parameters, i.e., the argument values of a chosen sample size formula (e.g., Δ, σ, power$= 1 - \beta$ in the case of Student's t-test).

The S-PLUS library provides powerful graphical tools for presenting the results of such "sensitivity analyses." A set of functions to run these analyses has been developed in this context. Their application and usefulness will be demonstrated in the following sections.

Several commercial software packages for sample size calculations also provide graphical presentations of the functional dependency of the sample size or power on design parameters. Unfortunately, they often only allow to examine the dependency on a single parameter. To get more insight into the complex dependency structure, we will use the S-PLUS contour plots. This makes power and sample size calculations more transparent, and enables users to judge the risks involved.

The scope of the chapter does not go beyond sensitivity analyses for the t-tests and the corresponding confidence intervals, but includes the new concept of the "power of an estimator." This leads to "more acceptable" solutions for the old, but partially ignored, problem of determining the sample size for the estimation of parameters.

11.2   The Noncentral t-Distribution

Power and sample size calculations for normally distributed variables are based on the noncentral t-distribution. Whilst functions for the central t-distribution (i.e., for a zero noncentrality parameter) are provided by S-PLUS, the corresponding functions for the noncentral distribution are still missing. They could be linked as Fortran or C routines. It is very easy though to create an S-PLUS function for the noncentral t-distribution, because the more complicated subroutines for the beta distribution and log-gamma function are already available in S-PLUS.

The function ptnoncent can be found in the library for this book. It is called with the following arguments:

```
> ptnoncent(tx, df, nonc=0, itrmax=100, errmax=1.0e-6)
```

It calculates the probability of observing a value not larger than tx in the case of a t-distribution with noncentrality parameter nonc (default 0), i.e., the cumulative distribution function with:

required arguments
- tx, a vector of numerical values
- df, the vector of the corresponding degrees of freedom

optional arguments
- nonc, the vector of noncentrality parameter values
- itrmax, the allowed maximum number of iterations
- errmax, the numerical error bound

Example 2. For $tx = 2.042272$, i.e., the 0.975-quantile of the central t-distribution with $df = 30$ degrees of freedom, and the three values 0, 0.5, and 1 as the noncentrality parameter, code and output look as follows:

```
> ptnoncent(2.042272, 30, c(0,0.5,1))
     0.9750000 0.9299950 0.8393311
```

In the case of nonc $= 0$, i.e., the central t-distribution, we get - as expected - the value 0.975.

11.3   The Power of the t-Test

The power of a test is defined as the probability of rejecting the null hypothesis if the alternative hypothesis is true. Take, for example, the t-test with the test statistic

$$t = \frac{\bar{y}_1 - \bar{y}_2}{s_R}\sqrt{\frac{n_1 n_2}{n_1 + n_2}}$$

The arithmetic means $\bar{y}_1, \bar{y}_2$ are calculated from two independent random samples of sizes n_1, n_2, drawn from normally distributed populations with expectations μ_1 and μ_2 and a common variance σ^2. The variance is estimated by the degrees of freedom weighted mean $s_R^2 = [(n_1 - 1)s_1^2 + (n_2 - 1)s_2^2]/[N - 2]$ of the sample variances s_1^2 and s_2^2, where $N = n_1 + n_2$ is the total sample size. This is identical to the residual mean square of the one-way analysis of variance with group defined as a factor.

The null hypothesis $H_0 : \mu_1 = \mu_2$ will be rejected in favor of the alternative hypothesis $H_A : \mu_1 \neq \mu_2$ if $|t| > t_{1-\alpha/2, df}$, where $df = N - 2$ denotes the degrees of freedom and α the prespecified error probability. The critical value $t_K = t_{1-\alpha/2, df}$ can be calculated as a quantile qt$(1 - \alpha/2, df)$ of the central t-distribution. Equivalently, the p-value p <- 2*(1-pt(abs(t))) can be computed. To reject the null hypothesis, the p-value has to be smaller than 0.05.

If the alternative hypothesis is true, then the test statistic follows a noncentral t-distribution with $df = N - 2$ degrees of freedom, and the power is calculated as the probability $P(|t| > t_K)$ from

$$\text{ptnoncent}(-t_K, nonc, df) + 1 - \text{ptnoncent}(t_K, nonc, df)$$

The noncentrality parameter $nonc$ can be easily derived by replacing the arithmetic means in the test statistic by their expectations μ_1, μ_2, and s_R by σ:

$$nonc = \frac{\mu_1 - \mu_2}{\sigma}\sqrt{\frac{n_1 n_2}{n_1 + n_2}} = \frac{\Delta}{\sigma}\sqrt{\frac{n_1 n_2}{n_1 + n_2}}$$

where Δ denotes the prespecified difference to be detected. Usually the randomization ratio, i.e., the ratio $k = n_2/n_1$ of group sizes, is fixed in advance, leading to

$$\text{nonc} = c\sqrt{N}, \qquad c = \frac{\Delta}{\sigma}\frac{\sqrt{k}}{1+k}$$

The coefficient c of $\sqrt{N}$ depends mainly on the type of t-test. For a paired t-test we would get $c = \Delta/\sigma_d$, where Δ denotes the expectation of the paired differences and σ_d their true standard deviation. The degrees of freedom when comparing two independent groups are $N - 2$ and $N - 1$ for the paired t-test. In general, degrees of freedom are calculated by subtracting an integer df_{calc} from the total size N.

The power function of the t-test, t.power, is based on the above formulas. This function offers the options "paired," "independent," and "generic" for the type of t-test, making it flexible and easy to use. For the first two options the argument delta denotes the difference of the true means, and sigma is the standard deviation of the individual differences or the single observations respectively. In the "generic" case, delta has to be set equal to the coefficient c of $\sqrt{N}$, and sigma to 1. An application of the "generic" option can be found in the next section.

The power function is called by

```
> t.power(delta, n1, n2=n1, sigma=1, type="paired",
    alpha=0.05, sided=2, df.calc=1)
```

with the

required arguments
- delta, the vector of true differences
- n1, the vector of sample sizes (first group)

optional arguments
- n2, vector of sample sizes (second group)
- sigma, vector of standard deviations
- type, "paired", "independent", or "generic"
- alpha, vector of (type 1) error probabilities
- sided, 1 – one-sided, 2 – two-sided test
- df.calc, number to be subtracted from the total size N to obtain the degrees of freedom

N always stands for the total number of patients in the study. It is the same as n1 for the paired t-test and N=n1+n2 when comparing two independent groups of size n1 and n2, respectively.

The output of the function is a list containing the specified argument values, the total size, and the power vector.

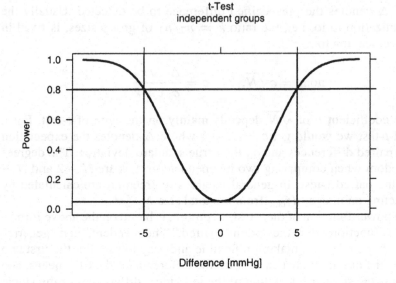

Figure 11.1. Power of the *t*-test as function of the difference of means.

Continuation of Example 1. Assuming $\sigma = 0.9$, we get different sizes of power for detecting a difference $\Delta = 0.6931$ with sample sizes 10, 11, 12, 13, 14 as follows:

```
> t.power(delta=0.6931,n1=10:14, sigma=0.9, sided=1)$power
    0.7263464 0.7676408 0.8033359 0.8340441 0.8603473
```

A power of 0.80 is first reached for the sample size $N = 12$, i.e., the exact sample size differs only by 1 from the approximate sample size calculated in the first subsection.

Example 3. In a study, $n_1 = 24$ patients are treated with a placebo, and $n_2 = 48$ patients with a drug that reduces blood pressure. The diastolic blood pressures of the two groups are compared using the two-sided *t*-test for independent samples at the level $\alpha = 0.05$, i.e., we test the null hypothesis $H_0 : \mu_1 = \mu_2$ versus the alternative hypothesis $H_A : \mu_1 \neq \mu_2$ for the population means μ_1, μ_2. Assuming that the standard deviation for diastolic blood pressure is $\sigma = 7$ mm Hg we can graphically display the power as a function of the difference $\Delta = \mu_2 - \mu_1$ mm Hg:

The code to create this figure reads:

```
pow <- t.power(delta=seq(-9,9,0.05), sigma=7,
    type="independent", n1=24, n2=48)
xyplot(pow$power~pow$delta,
```

```
ylim=c(0,1),ylab=list("Power",cex=0.6),
xlab=list("Difference [mm Hg]",cex=0.6),
main=list("t-Test\nindependent groups",cex=0.6),
scales=list(cex=0.6),
panel=function(x, y,...)
{
  lines(x,y, lwd=3)
  abline(h=c(0.05,0.80),lwd=2)
  abline(v=c(-5,5),lwd=2)
}
)
```

11.4   Therapeutic Equivalence Studies

For therapeutic equivalence studies, the mean effect μ_E of an experimental drug is compared with the mean effect μ_R of a reference drug. The objective here is to test whether the efficacy of the experimental drug is at least as good as that of the reference drug. More precisely, the hypotheses

$$H_0 : \mu_E \leq \delta\mu_R \quad \text{versus} \quad H_A : \mu_E > \delta\mu_R$$

or

$$H_0 : \mu_E - \delta\mu_R \leq 0 \quad \text{versus} \quad H_A : \mu_E - \delta\mu_R > 0$$

are tested for a given proportion $0 < \delta < 1$. Often $\delta = 0.8$ is chosen. For therapeutic equivalence studies, parallel group designs are preferred to the the crossover designs commonly used when investigating bioequivalence. For normally distributed efficacy variables using two independent groups, the test statistic

$$t = \frac{\bar{y}_E - \delta\bar{y}_R}{s_{TH}}$$

with

$$s_{TH} = s_R\sqrt{\frac{1}{n_E} + \frac{\delta^2}{n_R}}$$

can be applied. The residual standard deviation s_R is calculated as before by a one-way analysis of variance including group as a factor. The noncentrality parameter now becomes

$$\text{nonc} = \frac{\mu_E - \delta\mu_R}{\sigma\sqrt{1/n_E + \delta^2/n_R}}$$

If a ratio $n_E/n_R = k$ is fixed, then nonc $= c\sqrt{N}$ where $(N = n_E + n_R)$ with

$$c = \frac{\mu_E - \delta\mu_R}{\sigma\sqrt{1+k}\sqrt{1/k + \delta^2}}.$$

The power is calculated by

```
> t.power(delta, n1, n2 , sigma=1, type="generic",
  alpha=0.05, sided=1, df.calc=2)
```

for delta=c, n1=n_R, and n2=n_E. Sometimes α is replaced by $\alpha/2$ to stay conservative. This is not the same as testing for two-sided equivalence. Two-sided equivalence can be tested by using two one-sided tests simultaneously as described by Sasabuchi (1988). The power of this procedure has to be computed with numerical integration (Hauschke et.al. (1999)).

Example 4. Let us assume that the reference drug reduces the mean diastolic blood pressure by $\mu_R = 10$ mm Hg. We want to test whether the experimental drug lowers the diastolic blood pressure by at least $0.8 * 10 = 8$ mm Hg, i.e., only 2 mm Hg less than the reference drug. This means 80% therapeutic equivalence (i.e., $\delta = 0.80$). We will apply the t-test with $\alpha = 0.025$ for a parallel group study with $n_R = 100$ and $n_E = 200$ patients. What is the minimum power for this test, if the mean reduction in diastolic blood pressure of the experimental drug is greater than or equal to the mean of the reference medication ($\mu_E \geq \mu_R$)?

Assuming the same standard deviation $\sigma = 7$ mm Hg as used in the previous example, the power calculated for $\mu_E = \mu_R = 10$, $k = 2$, is

```
> k <- 200/100
> cc <- 10*0.2/7/sqrt(1+k)/sqrt(1/k+0.64)
> t.power(delta=cc, n1=100, n2=200, sigma=1,
  type="generic", alpha=0.025, sided=1, df.calc=2)$power
    0.7603294
```

It has to be noted that there is no link between the true ratio n2/n1 and k for the "generic" option. If the group sizes change, k and cc must be recalculated.

11.5   Calculation of Sample Sizes for t-Tests

When comparing means with a t-test, the sample size requirements normally are:

Find the minimum total sample size N to detect a given population mean difference Δ for the primary variable with a prespecified power $(1 - \beta)$ by an α-level (one- or two-sided) test assuming a standard deviation σ.

Δ is the **difference to be detected**. We want to find it with minimum probability $1 - \beta$ if the **true** difference is equal to or greater than Δ. Δ must not take the same value as the "minimum" clinically relevant difference. It is the clinically relevant difference in the context of the study.

As explained in the previous section, the degrees of freedom depend on the underlying model. Equal group sizes $n_1 = n_2 = N/2$ guarantee maximum power. Sometimes another ratio $k = n_2/n_1$ needs to be fixed in advance (e.g., to allow for a smaller placebo group).

The well-known formula (see, e.g., Bock (1998)):

$$N \approx \frac{[t_{1-\alpha,df} + t_{1-\beta,df}]^2}{c^2}$$

with the quantiles $t_{1-\alpha,df}$ and $t_{1-\beta,df}$ of the central t-distribution and

$$c = \frac{\Delta}{\sigma} \frac{\sqrt{k}}{1+k}$$

produces a very good approximation of the exact total sample size N for a one-sided t-test. In the two-sided case, α has to be replaced by $\alpha/2$. The exact sample size can be calculated using the "exact" power function t.power. One should be aware that "exact" does not mean "true," but "exactly computed." The sample size is only a function based on assumptions which could be wrong. For example, the specified standard deviation may deviate from the true value.

If the quantiles $u_{1-\alpha}$ and $u_{1-\beta}$ of the standard normal distribution are used instead of the t-quantiles, the calculated size decreases usually by no more than 2. Consequently, this has been used as a starting value within the frame of the function t.size which computes the exact sample size. The sample size is increased in steps of 1 until the required power power.stip is reached. Since t.size allows vector arguments, we are iterating for a vector of sample sizes. As soon as the desired power has been reached for an individual element it is not further increased. This is implemented by the code lines

```
n.add <- power<power.stip
nn <- nn + n.add
```

The elements of the logical vector power<power.stip are equal to 1 if the power is still too small and 0 otherwise. The stop criterion is sum(n.add)=0. Here the desired power has been reached for all components.

The function has to be called by

```
> t.size(delta, sigma=1, type="paired", power.stip=0.80,
    alpha=0.05, sided=2, df.calc=1, n2.over.n1=1)
```

with the

required arguments
   - delta, the vector of differences to be detected

optional arguments
   - sigma, vector of standard deviations
   - type, "paired", "independent" or "generic"
   - power.stip, vector of prespecified power
   - alpha, vector of (type 1) error probabilities
   - sided, 1 – one-sided, 2 – two-sided test
   - df.calc, number to be subtracted from the total size N to get the degrees of freedom
   - n2.over.n1 ratio of the sizes of the two groups involved

The output of the function is a list containing the specified argument values, the total sample size (n.total), the group sizes (n1,n2) in the case of independent groups, and the vector of actual power (power.act). This may exceed the prespecified power since we are looking for integer values when estimating sample sizes.

Continuation of Example 1. We get

```
> t.size(delta=0.6931, sigma=0.9, sided=1)[c(1:4,7)]
   $type:
   [1] "paired"

   $delta:
   [1] 0.6931

   $sigma:
   [1] 0.9

   $n.total:
   [1] 12

   $power.act:
   [1] 0.8033359
```

i.e., the sample size 12.

Continuation of Example 3. In a previous section we got a power of 0.80 for delta=5, sigma=7, n1=24, n2=48. The following two questions arise:

   - How does the sample size increase with the standard deviation?
   - How does the total size depend on the group size ratio?

To answer these questions we need to calculate the sample sizes for the same α and power, using all possible combinations of the standard deviation 7, 8, 9 and the ratio 1, 2, 3:

```
> x <- t.size(delta=5, sigma=rep(c(7,8,9), rep(3,3)),
  type="independent", df.calc=2,
  n2.over.n1=rep(c(1,2,3),3))
> data.frame(x$sigma, x$n2.over.n1, x$n.total,
  x$n1, x$n2)
```

x.sigma	x.n2.over.n1	x.n.total	x.n1	x.n2
7	1.000	64	32	32
7	2.000	72	24	48
7	2.909	86	22	64
8	1.000	84	42	42
8	2.000	93	31	62
8	2.964	111	28	83
9	1.000	104	52	52
9	2.000	117	39	78
9	2.971	139	35	104

Continuation of Example 4. To demonstrate 80% therapeutic equivalence with a power of at least 0.80, making the same assumptions as before, the total size N is

```
> t.size(delta=10*0.2/7/sqrt(3)/sqrt(1/2+0.64), sigma=1,
  type="generic", alpha=0.025, sided=1, df.calc=2,
  n2.over.n1=2)$n.total
  332
```

The complete output list gives the group size $n_R = 111$ for the reference group and $n_E = 121$ for the experimental group.

11.6   The t-Test for a Slope

The following regression analysis example demonstrates the wide scope of application for the new functions.

Example 5.   In an animal study the food absorption of an amino acid is examined. Two doses x_i (containing 0.08 and 20 g/day/kg livemass) are fed to animals measuring the excreted amounts y_i. Let's suppose that the data follows a linear regression model $y_i = \gamma_0 + \gamma_1 x_i + e_i$ ($i = 1, ..., N$) where the slope $\gamma_0 \approx 0$ describes the excreted proportion and $1 - \gamma_1$ the absorbed.

The aim of this study is to demonstrate that the absorption can be increased to more than 70% with a new food mixture (formulation), We therefore need to test the hypothesis $H_0 : \gamma_1 \geq 0.3$ versus $H_A : \gamma_1 < 0.3$. A true absorption proportion exceeding 80% (i.e., $\gamma_1 < 0.2$) should be detected with a power of 0.80. From a previous study we obtained a residual variance estimate of 0.00002.

Generally, the noncentrality parameter can easily be derived by replacing all random variables in the formula of the t-test with their expected values. For the linear regression model the slope γ_1 is compared to a given value γ using the test statistic

$$t = \frac{\gamma_1 - \gamma}{s_R}\sqrt{SQ_x}$$

where $\hat{\gamma}_1$ and s_R denote the estimators of the slope and the residual standard deviation with $N - 2$ degrees of freedom. The noncentrality parameter becomes

$$\text{nonc} = \frac{\Delta}{\sigma}\sqrt{SQ_x}$$

where $\Delta = \gamma_1 - \gamma$, σ denotes the true standard deviation of the responses, and $SQ_x = \sum(x - \bar{x})^2$.

Feeding three animals per dose group, we get SQ_x from

```
> SQX <- var(c(rep(0.08,3),rep(0.20,3)),SumSquares=T)
```

Replacing σ by the estimated standard deviation of $\sqrt{0.00002}$ from a previously conducted study, the noncentrality parameter becomes

```
> nonc <- (0.2-0.3)/sqrt(0.00002)*sqrt(SQX)
  -3.286335
```

The corresponding power is calculated using the "generic" option, delta $= nonc/\sqrt{N}$ for the total size $N = 6$ and n2=0.

```
> t.power(delta=nonc/sqrt(6), n1=6, n2=0, sigma=1,
  type="generic", alpha=0.05, sided=1, df.calc=2)$power
  0.8498736
```

Here we assume that delta multiplied by the square root of the total sample size N provides the noncentrality parameter.

N only occurs indirectly in this formula through SQ_x, i.e., the sum of squares of the chosen x-values (e.g., the doses used in the study). If the same measurement points are replicated then we can relate the total sample size to the number of replications. This is true if the same dose is fed to several animals.

Let's examine experimental plans which are composed by $r_p \geq 1$ repetitions of the same basic plan. Such a **basic plan** can be described by

$$\begin{bmatrix} x_1^* & x_2^* & \cdots & x_a^* \\ f_1 & f_2 & \cdots & f_a \end{bmatrix}$$

x_j^* denotes the a different measurement points (e.g., doses) - the **spectrum** - and f_j their frequencies - the **basic frequencies**. The relative frequencies f_j/b with $b = \sum f_j$ describe the weights of the points of the spectrum in the basic plan.

In the study, $r_p \times f_j$ observations are recorded at each measurement point x_j^*. The total sample size is $N = r_p \sum f_j = r_p \times b$. The sum of squares for the basic plan is

$$SQ_x^* = \sum_j f_j (x_j^* - \bar{x}^*)^2 i$$

with $\bar{x}^* = \sum_j f_j x_j^*/b$. Then $SQ_x = r_p \times SQ_x^*$, and hence

$$\text{nonc} = \frac{\Delta}{\sigma}\sqrt{r_p \times SQ_x^*} = \frac{\Delta}{\sigma\sqrt{b}}\sqrt{SQ_x^*}\sqrt{N} = c\sqrt{N}$$

with

$$c = \frac{\Delta}{\sigma\sqrt{b}}\sqrt{SQ_x^*}$$

Continuation of Example 5. Here the basic plan is

$$\begin{bmatrix} x_1^* & = & 0.08 & x_2^* & = & 0.20 \\ f_1 & = & 1 & f_2 & = & 1 \end{bmatrix}$$

The sum of squares for the basic plan SQ_x^* can be calculated as

```
> var(c(0.08,0.20), SumSquares=T)
  0.0072
```

With $b = 2$ we obtain $c^2 = 0.1^2 \times 0.0072/(0.00002 \times 2) = 1.8$. For $\alpha = 0.05$ and the prespecified power of 0.80 the total sample size is calculated by

```
> t.size(delta=sqrt(1.8), type="generic", sided =1,
  df.calc=2)$n.total
  6
```

The actual power from the output list of t.size is the same as before.

11.7   Sensitivity Analysis

As stated in the previous section, the vectorized function allows us to investigate how sensitive the sample size estimation is to assumption violations. This can be cumbersome. The question arises of how to present the results of such an analysis. In the clinical development process for a new drug it is crucial to choose the appropriate sample size for the pivotal Phase III studies. At this stage we usually have gained experience from the dose finding studies in Phase II. By then we should have a more or less reliable estimate for the variance of the primary variable. Having estimated the drug effects in Phase II makes it easier to choose a difference Δ to be detected. To get an idea of how the sample size and power changes with varying assumptions, we try to answer the following questions:

- What happens to the total sample size if we vary Δ and σ independently around the prespecified values by 20%?
- How does the power change if we vary the total sample size, the standardized difference Δ/σ, or both?
- How does the total size change if we vary the specified power, the standardized difference Δ/σ, or both?

To be able to answer these questions the following figures should be examined carefully.   They have been produced using

```
> t.size.plot(delta, sigma=1, type="paired",
    power.stip=0.8, alpha=0.05, sided=2, n2.over.n1=1,
    span=0.2, gridsize=30, ncuts=9)
```

The arguments of t.size.plot are the same as for t.size apart from the three new arguments describing the variation of the design parameters:

- span [(1-span)*parameter, (1+span)*parameter] is the variation interval of delta, sigma, or delta/sigma
- gridsize, number of supporting points per grid axis
- ncuts, number of cuts for the contour plot

Figure 11.2 displays the contours of the total sample size depending on Δ and σ. It gives an impression of how the sample size increases with the standard deviation or with decreasing Δ. Looking at Figure 11.3 one sees how the power is related to the total sample size and the standardized difference Δ/σ. This is useful for exploring the dependency of the power on the sample size for a fixed standardized difference. Figure 11.4 shows how the total sample size varies with the power and the standardized difference. This tells us what a higher power requirement would cost in terms of patient numbers.

The grid for the total sample size and the standardized difference Δ/σ in Figure 11.3 is created using the minimum and maximum values calculated

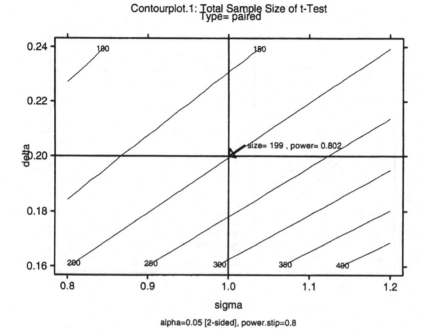

Figure 11.2. Sensitivity Analysis 1.

in Figure 11.2. The power ranges from 0.5 to 0.95 in Figure 11.4. The argument `gridsize` fixes the number of equally spaced supporting points within each of these intervals (as default 30 points).

Continuation of Example 3. So far we have calculated a total sample size of $N = 72$ and the group sizes $n_1 = 24$, $n_2 = 48$ for the chosen ratio $n_2/n_1 = 2$ starting with $\Delta = 5$ mm Hg, $\sigma = 7$ mm Hg and the prespecified power $1 - \beta = 0.80$. For the graph "Sensitivity Analysis 1" (Figure 11.2) Δ and σ are varied 20% around the prespecified values, i.e., between 4 and 6 or 5.6 and 8.4, respectively. The graph reveals that decreasing Δ from 5 to 4 mm Hg would require increasing the sample size from 72 to 110.

The minimum and maximum estimated total sample sizes are

```
> t.size(delta=c(6,4), sigma=c(5.6,8.4),
    type="independent", df.calc=2, n2.over.n1=2)$n.total
      33 159
```

For Figure 11.3 the total sample size N is varied between 33 and 159 with a standardized difference Δ/σ ranging from the minimum value 4/8.4=0.48 to the maximum value 6/5.6=1.07 of the first plot. The power is calculated

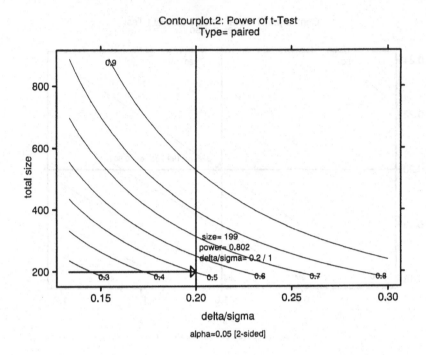

Figure 11.3. Sensitivity Analysis 2.

for the 900 grid points and the contour curves are drawn. Let's set the standardized difference to 5/7. Exploring the change in power for different sample sizes shows that by increasing the sample size to 100, the power exceeds 0.90.

Figure 11.4 shows the contour curves of the total sample size. The standardized difference has the same range as in the second plot and the power varies between 0.5 and 0.95. In each plot the starting point of the sensitivity analysis is indicated by an arrow.

11.8   Power and Sample Size for Confidence Intervals

As already explained in the first section of this chapter, statistical significance alone does not reveal whether the effect found (frequently defined as mean or mean difference) is of clinical relevance. The size of an effect can be estimated by confidence limits. These must be narrow enough to ensure a sufficiently precise estimate of the effect. For example, the confidence interval $[\overline{y} - HW, \overline{y} + HW]$ for the mean μ of a normal distribution with

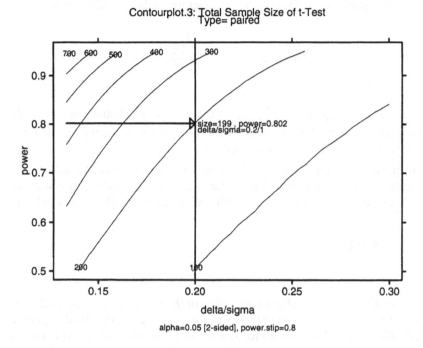

Figure 11.4. Sensitivity Analysis 3.

unknown variance has the half-width $HW = t^*s/\sqrt{N}$ with the quantile $t^* = t_{1-\alpha/2,df}$ of the central t-distribution with $df = N - 1$ degrees of freedom and the sample standard deviation s. Not only does the location vary, but also the width of the confidence interval. Only a sufficiently large sample size gives narrow limits with a high probability.

Statistical testing has dominated the clinical trial methodology for a long time. The problem of getting sufficiently precise estimates has been partially ignored, but received more and more attention during the last decade.

The simplest approach is to determine the sample size ensuring a pre-specified expected half-width. More precisely, $E(HW^2) < \Delta^2$ is required. This leads to the sample size formula

$$N_{HW} \geq \frac{t^2_{1-\alpha/2,df}}{c^2}$$

with $c = \Delta/\sigma$. Tukey (1953) requires that

$$P[HW < \Delta] = 1 - \beta$$

for a prespecified probability $1 - \beta$ leading to the sample size formula

$$N_{HW,\text{Tukey}} = \frac{\chi^2_{1-\beta,df}}{df} N_{HW}$$

with the $(1 - \beta)$-quantile $\chi^2_{1-\beta,df}$ of the central χ^2 distribution with df degrees of freedom. The function N.HW in the S-PLUS library of this book allows the calculation of these sample sizes.

Example 6. We want to estimate the population mean for the diastolic blood pressure by 0.95 confidence limits. Let's assume a standard deviation of $\sigma = 5$ mm Hg. An expected half-width of about 3 mm Hg is ensured for

```
> N.HW(3, 7)
    24
```

i.e., $N = 24$ patients need to be included in the study. The calculated half-width of the confidence interval will frequently exceed the fixed expected width. Since the observed width is randomly distributed a given bound will only be respected with a certain probability. If we require an observed half-width smaller than 3 mm Hg with probability $1 - \beta = 0.95$ we obtain

```
> N.HW(3, 7, T, 0.05)
    33
```

i.e., 33 patients are needed.

   Hsu (1998) computed the probability P_{HSU} that the observed half-width is restricted by Δ and, simultaneously, the true mean is covered by the confidence interval. Guilbaud (1997) derived narrow bounds for this probability allowing the computation of the sample size for the requirement $P_{HSU} \leq 1 - \beta$ using simple SAS statements.

   Usually it is desired to estimate the parameter with a prespecified precision, say $\pm\Delta$. This can only be achieved with a certain probability. Nevertheless, the mentioned requirements do not entirely reflect the expectation of keeping the estimates within the Δ-region of the true value. Also in the case of Hsu's requirement, it may happen that the true mean coincides with one of the confidence limits. In the worst case, the other limit deviates by 2Δ from the true mean. Bock (1998) has defined the **power of an estimator** as the probability that both confidence limits are found in the region of the true value, more precisely,

> *If a parameter ϑ is estimated by the confidence interval $[\vartheta_{low}, \vartheta_{upp}]$, then the power is defined as the probability that both confidence limits are included in a given precision interval $[\vartheta - \Delta_{low}, \vartheta + \Delta_{upp}]$ around the true value:*
>
> $$Power = P[\vartheta - \Delta_{low} < \vartheta_{low} < \vartheta_{upp} < \vartheta + \Delta_{upp}]$$

Specifying a power of 0.80 means that 80% of the calculated confidence limits will not deviate by more than the precision bounds from the true value. The power of an estimator compares to the power of an equivalence test. Equivalence is established if both confidence limits are included within given bounds.

We restrict ourselves to dealing with the estimation of the location parameter μ of a normal distribution. The location parameter could be, for example, the mean, the mean difference, a contrast, or a regression coefficient. If the estimator $\hat{\mu}$ has the standard error σ^*, the confidence limits are $\hat{\mu} \pm t^* s^*$, where s^* denotes the estimated standard error and t^* denotes the $(1 - \alpha/2)$-quantile of the t-distribution with the degrees of freedom of the standard error. The simplest case is the estimation of a population mean by the arithmetic mean ($\hat{\mu} = \bar{y}$, $s^* = s/\sqrt{N}$).

The power is given by an integral (see the Appendix for the derivation). It can be computed by the function

```
> tconf.power(delta, n1, n2=n1, sigma=1,
    type="paired", alpha=0.05, sided=2, df.calc=1)
```

found in the library. The arguments have the same naming convention as the function t.power. Interpretation is different though. delta now denotes the precision bound, not the difference to be detected. Dependent on the data structure the option type="paired" or "independent" is used. The argument sided=1 means using the lower $(1 - \alpha)$ confidence limit. The power here is the probability that the lower confidence limit exceeds $\mu - \Delta$, which can be computed directly from the noncentral t-distribution.

The function for the total sample size,

```
> tconf.size(delta, sigma=1, type="paired",
    power.stip=0.8, alpha=0.05, sided=2, df.calc=1,
    n2.over.n1=1)
```

has been written by replacing t.power in t.size with tconf.power. Finally, it has been entered into t.size.plot leading to

```
> tconf.size.plot(delta, sigma=1, type="paired",
    power.stip=0.8, alpha=0.05, sided=2,
    n2.over.n1=1, span=0.2, ncuts=9, gridsize=30)
```

This provides the same tools which are already available for the power analysis of the t-test.

Continuation of Example 1. For the paired t-test we have previously calculated the sample size of $N = 12$ to detect a difference $\Delta = \log(2) = 0.6931$ with a power of 0.80. A standard deviation $\sigma = 0.9$ has been assumed for the individual differences of logarithms y. The sample size for estimating the lower limit of the mean difference with precision $\Delta = 0.6931$ and a power of 0.80 is

```
> tconf.size(delta=0.6931, sigma=0.9, type="paired",
  power.stip=0.8, alpha=0.05, sided=1)$n.total
  12
```

i.e., the same as for the one-sided t-test. This is because the power numerically equals the power of the paired t-test for $\mu_y = \Delta$.

Continuation of Example 3. The function

```
> tconf.size(delta=5, sigma=7, type="independent",
  power.stip=0.8, alpha=0.05, sided=2,
  df.calc=2, n2.over.n1=2)
```

delivers the group sizes n1=32 and n2=64, whilst we obtained n1=24 and n2=48 for detecting a difference delta=5 with Student's t-test (refer to Section 11.5). The null hypothesis tested by Student's t-test is $H_0 : \mu_1 = \mu_2$ whereas the null hypothesis for equivalence testing is $H_0 : \mu_1 - \mu_2 \leq -\Delta$ or $\mu_1 - \mu_2 \geq \Delta$. It is rejected if both confidence limits are included in $[-\Delta, \Delta]$. The probability for inclusion, i.e., the power of the equivalence test for $\mu_1 = \mu_2$, is numerically the same as the power of the estimator. If we interpret equivalent as "almost equal" then the equivalence test could be regarded as a t-test for interchanged null and alternative hypotheses. Consequently, the sample sizes should be approximately the same as for the t-test if α and β are interchanged.

The command

```
> t.size(delta=5, sigma=7, type="independent",
  power.stip=0.975, alpha=0.20, sided=2,
  df.calc=2, n2.over.n1=2)
```

produces indeed the sizes n1=32 and n2=63.

Example 7. Drug interaction studies are performed to investigate whether the combined administration of two drugs A, B changes their kinetics, i.e., their blood plasma level curves over time. Either level for A and B could change. We will examine the levels of A only administering A and AB to two different groups of subjects (parallel groups). Crossover studies are dealt with in the next section.

Let's assume that the AUC (area under the concentration curve) follows a log-normal distribution. If A is administered, then $y = \log(AUC)$ has the mean μ_A. The mean of the combination is denoted by μ_{AB}, and we assume variance homogeneity for y, i.e., the same standard deviation σ in both groups. The coefficients of variation cv of the (untransformed) AUC and σ^2 are related by $\sigma^2 = \log(1 + cv^2)$.

The **interaction effect** is defined as the difference $\vartheta = \mu_{AB} - \mu_A$. Since we seldomly know in advance whether an interaction exists or not the testing of hypotheses would be inappropriate. It is not clear here whether to test for a difference between the group means or perform an equivalence test. A straightforward strategy is first to estimate the interaction effect with sufficient precision and power, and then judge the clinical relevance of the effect.

The exponentiation of the estimate $\vartheta = \overline{y}_{AB} - \overline{y}_A$ gives the ratio of the geometric means. By multiplying this ratio by 100 it becomes the percentage of the reference mean. When we allow μ_{AB} to deviate from μ_A by $80 - 125\%$, we require the same as the precision $\Delta = \log(1.25) = -\log(0.8) = 0.223$ for the estimator in the domain of log-transformed AUC. The variation coefficient for AUC often has been found to lie around 20%. Using this value in our next example we get a standard deviation of $\sigma = \sqrt{\log(1 + 0.2^2)} = 0.198$.

If we use 0.95 confidence limits, doubling the number of subjects in group AB (n2.over.n1=2) and specifying a power of 0.80, the sample sizes can be calculated from

```
> tconf.size(delta=0.223, sigma=0.198,
  type="independent", n2.over.n1=2)
```

This gives the group sizes $n_A = 14$, $n_{AB} = 27$, the total sample size $N = n_A + n_{AB} = 41$ and a power of 0.83. For equal group sizes we would get a smaller total sample size of $N = 36$ from

```
> tconf.size(delta=0.223, sigma=0.198,
  type="independent")$n.total
```

To explore the dependency on the assumptions we can use the function tconf.size.plot specifying:

```
> tconf.size.plot(delta=0.223, sigma=0.198,
  type="independent", n2.over.n1=2)
```

The corresponding graphs are not included here being similar to t-test plots apart from titles and labels.

11.9   Bioequivalence Studies

Most bioequivalence studies use a crossover design with two sequences T/R and R/T where T is the investigational and R the reference drug. These drugs are administered during two periods to two separate groups of individuals (sequence groups). As in drug interaction studies the log-transformed areas $y = \log(AUC)$ are often used as primary variables, assuming a normal distribution with mean μ_y and standard deviation σ_y. To avoid bias the period differences pd_{ij} of individuals j in sequence group i are used for estimation, deleting thereby the period effects, and getting the variance $\sigma_p^2 = 2\sigma_y^2$.

The difference τ of the treatment effects for drug T and drug R can be estimated using the arithmetic group means of the period differences by

$$\tau_p = \frac{\overline{pd}_{1.}}{2} - \frac{\overline{pd}_{2.}}{2}$$

from the two means $\overline{pd}_{1.}, \overline{pd}_{2.}$ of the sequence groups of sizes n_1 and n_2 (see, e.g., Jones and Kenward (1989)). The variance is then given as

$$V(\hat{\tau}_p) = \frac{\sigma_p^2}{4}\left(\frac{1}{n_1} + \frac{1}{n_2}\right)$$

Defining the half-period differences $pd_{ij}/2$ as new variables puts us into the scenario of a t-test with independent groups and the error variance $\sigma^2 = \sigma_p^2/4 = \sigma_y^2/2$.

According to the usual **inclusion rule** mean bioequivalence is established if both 0.90 confidence limits lie between $\pm\Delta$. Note that the confidence level 0.90 is used here, and not 0.95, making this rule algebraically equivalent to the **two one-sided tests procedure**. Guidelines to bioequivalence testing usually require the equivalence limit $\Delta = -\log(0.8) = \log(1.25) = 0.223$. Hence the total sample size can be computed with tconf.size specifying the parameters delta=0.223, sigma=$\sigma_y/\sqrt{2}$, type="independent", alpha=0.10.

Example 8. Assume again a variation coefficient of 20% for the AUC leading to $\sigma_y = 0.198$. For equal sequence group sizes we get

```
> tconf.size(delta=0.223, sigma=0.1987/sqrt(2),
   type="independent", alpha=0.1)$n.total
   16
```

Altogether 16 subjects have to be treated, 8 in each sequence group.

The power of estimators has been defined as the probability of observing confidence limits in a given region of the true mean or mean difference. For bioequivalence testing the power is defined as the probability of finding the confidence limits around zero. If the true mean difference is zero the two

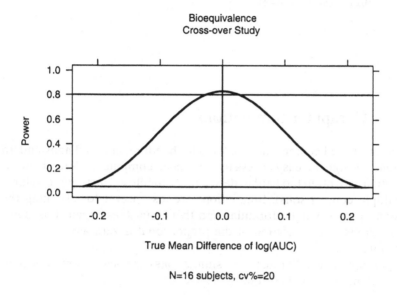

Figure 11.5. The power of mean bioequivalence testing.

definitions are the same, or else the integrand in tconf.power changes. Using the additional argument mean.dev allows the power (or sample size) calculation for bioequivalence studies in the case of nonzero mean differences.

Continuation of Example 8. The following code was used to create Figure 11.5 of the power for the two-sided tests procedure as a function of the true mean difference:

```
pow <- tconf.power(delta=0.223, n1=8, sigma=0.1987/
    sqrt(2), mean.dev=seq(-0.223, 0.223, length=100),
    type="independent", alpha=0.1)

xyplot(pow$power~pow$mean.dev,
    ylim=c(0, 1), ylab=list("Power", cex=0.6),
    xlab=list("True Mean Difference of log(AUC)", cex=0.6),
    main=list("Bioequivalence\nCross-over Study", cex=0.6),
    sub=list("N=16 subjects, cv=20", cex=0.6),
    scales=list(cex=0.6),
    panel=function(x, y, ...)
    {
        lines(x, y, lwd=3)
        abline(h=c(0.05, 0.80), lwd=2)
```

```
    abline(v=0, lwd=2)
  }
)
```

11.10   Computer Simulations

Simulations have become valuable tools for handling new and nonstandard situations through the easy access to high speed computers. Here, theoretical solutions are often too hard to derive (e.g., for closed testing procedures for multiple comparisons). Simulations are also useful for checking the precision of approximate formulas and their robustness against assumption violations (e.g., violation of the proportional hazard assumption for the log-rank test).

Simple approaches - "brute force" simulations - can be very time-consuming. Speed improvements can be achieved by:

- simulating the distribution of a sufficient statistic (e.g., $\bar{x}$, s^2 for t-test), and
- accelerating search algorithm by reuse of sampled values

Amaratunga (1999) proposed an accelerating algorithm called QuickSize where you need to iterate the following steps ($i = 1, \ldots, maxiter$), starting with sample size N_0:

- Let N_i be the current value of the sample size N. Randomly generate B pseudosamples of size N_i each and count how many of these lead to a rejection of the null hypothesis.
- Let $B_i(N_i)$ be the total number of pseudo-samples with the same size N_i drawn in all iterations up to and including the ith iteration. Suppose that $B_i^*(N_i)$ of them led to rejection of the null hypothesis. At this stage

$$P^* = \frac{B_i^*(N_i)}{B_i(N_i)}$$

serves as a "current" estimate of the power.

If $P^* < 1 - \beta$ increase sample size: $n_{i+1} = n_i + m_i$

If $P^* > 1 - \beta$ decrease sample size: $n_{i+1} = n_i - m_i$

Usually $m_i = 1$ will be used.

Example. The S-PLUS function QuickSize.Fisher for the calculation of the sample size for Fishers Exact Test has been provided by D. Amaratunga. It is included in the library of this book.

We get, for example,

```
> QuickSize.Fisher(niter=500, block=10, n=20, p1=0.1,
  p2=0.5, alpha=0.05, power=0.8, nmax=0, h=1)
      n nsig ntotal power
   18 18 1846    2340 0.789
   19 19 2013    2480 0.812
   20 20  124     150 0.827
   21 21   25      30 0.833
```

The first column gives the sample size, nsig denotes the number of significant tests, and ntotal denotes the total number of tests. The solution is n=19.

11.11   References

Amaratunga, D. (1999). Searching for the right sample size. *The American Statistician* **53**, 52–55.

Bock, J. (1998). *Bestimmung des Stichprobenumfangs für biologische Experimente und kontrollierte klinische Studien*. R. Oldenbourg Verlag, München.

Guilbaud, O. (1997). An approach to sample size determination for confidence intervals proposed by hsu. *Proceedings of the Biopharmaceutical Section of the American Statistical Association* pp. 179–184.

Hauschke, D., Kieser, M., Diletti, E., and Burke, M. (1999). Sample size determination for proving equivalence based on the ratio of two means for normally distributed data. *Statist. Med.* **18**, 93–105.

Hsu, J. (1998). Sample size computation for designing multiple comparison experiments. *Computational Statistics & Data Analysis* **7**, 79–91.

ICH Harmonized Tripartite Guideline: Statistical Principles for Clinical Trials (1998).

Jones, B., and Kenward, M. (1989). *Design and Analysis of Cross-Over Trials*. Chapman & Hall, London.

Sasabuchi, S. (1988). A multivariate one-sided test with composite hypotheses when the covariance matrix is completely unknown. *Memoirs of the Faculty of Science Kyushu University* **42**, 37–46.

Tukey, J. (1953). The problem of multiple comparisons. Unpublished manuscript cited by Hsu (1998).

11.A Appendix

11.A.1   The Power of the Confidence Interval for the Mean of a Normal Distribution

Because the confidence limits for the expectation of a normal distribution are symmetric, we only specify one precision bound $\Delta > 0$. We require that the lower confidence limit exceeds the lower precision bound

$$\mu - t^*s^* > \mu - \Delta \quad \Longleftrightarrow \quad -(\Delta - t^*s^*) < \mu - \mu$$

and simultaneously the upper confidence limit lies below the upper precision bound

$$\mu + t^*s^* < \mu + \Delta \quad \Longleftrightarrow \quad \mu - \mu < (\Delta - t^*s^*)$$

Hence the power is $P^* = P(|\hat{\mu} - \mu| < c(x))$ with the right hand side $c(x) = \Delta - t^*\sigma^*\sqrt{x}$ depending on $x = s^{*2}/\sigma^{*2}$. Since mean and variance estimators are independent for normally distributed variables, we first of all compute the conditional power for fixed x and then the expectation based on the density function of x:

$$g(x) \quad = \quad \kappa^\kappa e^{-\kappa x} x^{\kappa-1} / \Gamma(\kappa)$$

with $\kappa = df/2$. This leads to the integral

$$P^* \quad = \quad \int_0^\xi \left[2\Phi\left(\frac{c(x)}{\sigma^*}\right) - 1 \right] g(x)\, dx$$

with the cumulative distribution function Φ of the standard normal distribution and the upper limit $\xi = [\Delta/(\sigma^*t^*)]^2$.

11.A.2   List of Functions in the Library of the Book

The following functions are provided via the online (Web) library of this book:

- ptnoncent distribution function of the noncentral t-distribution
- t.power power of a t-test
- t.size total sample size of a t-test
- t.size.plot graphical sensitivity analysis of power and sample sizes for t-tests
- N.HW sample size calculations for confidence intervals according to Tukey
- tconf.power power of the confidence intervals (normal distribution)

- `tconf.size` total sample size for a confidence interval (normal distribution)
- `tconf.size.plot` graphical sensitivity analysis of power and sample sizes for confidence intervals
- `QuickSize.Fisher` simulation of the sample size for Fisher's exact test

12

Comparing Two Treatments in a Large Phase III Clinical Trial

Michaela Jahn
F. Hoffmann-La Roche AG, Basel, Switzerland

12.1 Introduction

This chapter outlines some of the analyses performed for a Phase III oncology trial. The trial was conducted by F. Hoffmann-La Roche and took place between 1996 and 1998. More than 600 cancer patients receiving either a new experimental therapy or a standard treatment were included in the trial. Besides a description of the data, the statistical analysis methods used for some of the analyses are detailed (with an emphasis on survival analysis). The applications of the methods using S-PLUS are outlined. The focus is on the analyses as specified in the trial protocol. Due to space limitation, not all analyses performed for the trial are mentioned and not all aspects of the trial are presented.

Section 12.2 gives an introduction to the data, the objectives, and the conduct of the trial. The trial data are introduced in Section 12.3, and the baseline characteristics of the patient population in the trial are presented in Section 12.4. The method for the analysis of the primary study parameter, as well as the results of the analysis, are presented in Section 12.5. In Section 12.6, the results of the analysis of the secondary study parameters are displayed. Since the S-PLUS manuals give a very good introduction to survival analysis, the explanation of the methodology is kept to a minimum. In Section 12.7, an example for the verification of the proportional hazard assumption is provided. Section 12.8 concludes with final remarks and a summary.

12.2 Trial Design

The trial described here is part of an international drug development for the treatment of advanced and/or metastatic colorectal cancer. Besides one Phase II trial, two open-label, multicenter, randomized Phase III studies comparing two treatments were conducted. One of the latter serves as the source of data.

The trial was set up to compare the experimental treatment (oral) to a standard therapy (intravenous infusion). After verification of inclusion and exclusion criteria, and recording of baseline information such as history of metastatic disease and demographic parameters, the patients were randomized to one of the two treatment arms. As an entry criterion the patients had to have measurable disease, implying that two dimensions of at least one lesion were measurable based primarily on CT-scans or X-rays. At baseline and approximately every 6 weeks during the 2-year trial, tumor assessments were performed during which the lesion sizes were recorded (in mm^2). Assessments performed after baseline were evaluated according to tumor response criteria set up by the World Health Organization (1979). Based on these criteria, patients were regarded as responders if a reduction in tumor size of 50% or more was observed between two visits more than 4 weeks apart. If all lesions disappeared, the patient was regarded as a complete responder (CR), otherwise as a partial responder (PR). In the case of an increase in tumor size by 25% or more, a patient was defined as having progression of disease (PD). Patients without response or progression were classified as being in stable disease (SD). Based on the response classification, the best overall response during the study was determined: a patient who responded but progressed afterward was counted as a responder. Similarly, for a patient with stable disease and progression afterward, the best overall response was SD. When disease progression was observed with the first tumor assessment after baseline, the best overall response was PD. For patients without post-baseline tumor assessment, the best overall response was missing. The response rate was defined to be the percentage of patients with a best overall response of either CR or PR.

With each tumor assessment the following information was captured: date of assessment, size of individual lesions, and response evaluation. Based on the date of the tumor assessment and the date of randomization, parameters like time to disease progression, time to response, or duration of response are available. In the case of death, the death date was recorded, such that information necessary for the evaluation of survival was captured. Besides information on efficacy, safety parameters such as adverse events and laboratory parameters were recorded.

The minimum observation period (treatment phase plus follow up) for a patient was set to be 7 months, meaning that patients who were recruited last into the trial had to be followed for at least 7 months before the trial was officially closed. If a patient exited the study without recording of disease progression (e.g., due to adverse events), tumor assessments and survival status were performed trimonthly to capture the date of disease progression and the date of death. For patients who progressed while on study, the survival status was recorded trimonthly as well as after a patient exited the study.

12.3 Data Overview

The trial was conducted in 59 centers. Each treatment arm included 301 patients (i.e., 301 patients received the standard treatment and 301 patients received the experimental treatment). Based on the information captured, the dataset studydata was created and forms the basis for the analyses. Explanations of the variables are provided in Appendix 12.A.1. For illustration the first five rows of the source dataset are shown:

```
> studydata[1:5,]
  Patient TimeToProgression ProgressionStatus SurvivalStatus
1    3001                43                 1              1
2    3002                89                 1              1
3    3003               535                 0              0
4    3004               134                 1              1
5    3005               146                 1              1

  Survival Age    Sex Randomization BestResponse MetSitesAtBaseline
1      292  62   male  Experimental           PD                  2
2      234  54   male  Experimental           SD                  2
3      535  72 female  Experimental           SD                  1
4      589  61   male  Experimental           SD                  1
5      547  37 female      Standard           SD                  2

  Country Exclusion AdvEvStatus TimeToAdvEv
1     AUS         1           0          70
2     AUS         1           1          14
3     AUS         1           1          59
4     AUS         1           0         196
5     AUS         1           1         113
```

For example, patient 3002 is a 54-year-old male patient, who received experimental treatment. He had two metastatic sites involved at baseline. His best overall response during the study was stable disease (SD). Progression was recorded 89 days after randomization. The patient died 234 days after randomization.

The summary function provides summary statistics for all of the variables in the dataset:

```
> summary(studydata)
     Patient     TimeToProgression ProgressionStatus
 Min.   :1001    Min.   :  5.0     Min.   :0.0000
 1st Qu.:1814    1st Qu.: 81.0     1st Qu.:1.0000
 Median :4006    Median :146.5     Median :1.0000
 Mean   :4291    Mean   :167.9     Mean   :0.8987
 3rd Qu.:6204    3rd Qu.:223.0     3rd Qu.:1.0000
 Max.   :9302    Max.   :625.0     Max.   :1.0000

 SurvivalStatus     Survival          Age            Sex
 Min.   :0.0000   Min.   :  9.0   Min.   :29.00   female:257
 1st Qu.:0.0000   1st Qu.:194.2   1st Qu.:56.00   male  :345
 Median :1.0000   Median :299.5   Median :64.00
 Mean   :0.5399   Mean   :300.3   Mean   :62.11
 3rd Qu.:1.0000   3rd Qu.:399.8   3rd Qu.:70.00
 Max.   :1.0000   Max.   :644.0   Max.   :86.00
                                  NA's   : 1.00

     Randomization   BestResponse  MetSitesAtBaseline
 Experimental:301       : 59       Min.   :0.0
 Standard    :301     CR: 14       1st Qu.:1.0
                      PD:110       Median :2.0
                      PR:120       Mean   :2.5
                      SD:299       3rd Qu.:3.0
                                   Max.   :8.0
```

```
     Country         Exclusion        AdvEvStatus        TimeToAdvEv
     GB:173         Min.:1.000        Min.:0.0000       Min.:   1.0
     I: 73      1st Qu.:1.000     1st Qu.:0.0000    1st Qu.:  54.0
     RU: 69       Median:1.000      Median:0.0000     Median:112.5
     F: 58          Mean:1.126        Mean:0.3205       Mean:127.5
     D: 55      3rd Qu.:1.000     3rd Qu.:1.0000    3rd Qu.:196.0
    AUS: 51         Max.:4.000        Max.:1.0000       Max.:421.0
 (Other):123                      NA's:6.0000       NA's:   6.0
```

To be able to use all of the variables in the dataset by their names (e.g., Age) instead having to explicitly use subset commands (e.g., studydata$Age), the data frame is attached using the command attach(studydata).

12.4   Baseline Characteristics

Patient characteristics such as sex, age, weight, cancer history, and number of metastatic sites at baseline are summarized and compared between the two treatment arms to check that the same patient populations are represented in each arm. As examples, the variables age (Age), sex (Sex), country (Country), and number of metastatic sites at baseline (MetSitesAtBaseline) are examined. A comparison of baseline parameters between treatment groups with continuous outcome, e.g., age, is performed using the by command:

```
> by(Age, Randomization, summary)
INDICES:Experimental
         x
   Min.:29.00
1st Qu.:55.00
Median:64.00
   Mean:61.94
3rd Qu.:70.00
   Max.:84.00
----------------------------------------------------------------
INDICES:Standard
         x
   Min.:36.00
1st Qu.:57.00
Median:63.50
   Mean:62.28
3rd Qu.:69.00
   Max.:86.00
   NA's: 1.00
```

The same command applied to the factor variable Sex results in the following display:

```
> by(Sex, Randomization, summary)
INDICES:Experimental
         x
 female:129
   male:172
-----------------------------------------------------
INDICES:Standard
         x
 female:128
   male:173
```

A combined evaluation of age and sex can be displayed graphically, using the function bwplot to generate a Trellis graph. A Trellis graph is an arrangement of multiple graphs of the same type, in which each graph is based on an underlying conditional variable. For example, in Figure 12.1, the two boxplots on the left-hand side of the figure show the age distribution for female patients in both treatment groups whereas on the right-hand side the age distribution of male patients is presented:

```
> bwplot(Randomization ~ Age | Sex,
      strip=function(...) strip.default(..., style= 1))
```

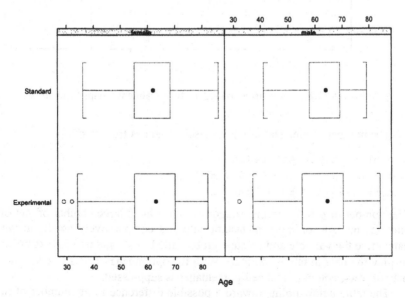

Figure 12.1. Boxplots of patient's age split by treatment and sex.

By setting the `strip` argument explicitly, the full strip label is colored in the background color and the text string for the current factor level is centered in it. A comparison of the groups "Standard" and "Experimental" shows that there is no obvious imbalance with respect to age and sex.

Next, we compare the age distribution for the two treatment groups with respect to different countries (Figure 12.2). For one center the country is not specified (see graph in the lower left corner), so the label is empty:

```
> bwplot(Randomization ~ Age | Country,
     strip=function(...) strip.default(..., style= 1))
```

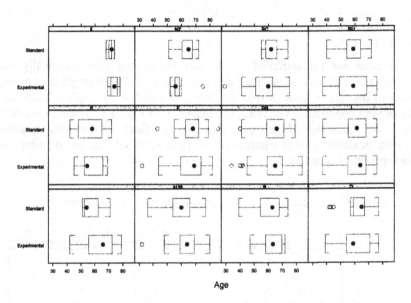

Figure 12.2. Age distribution for the two treatment groups by country.

The number of metastatic sites at baseline ranges from 0 to 8:

```
> table(MetSitesAtBaseline)
  0   1   2   3   4   5   6 7 8
  2 191 145 122 85 34 15 6 2
```

The number of patients in the categories indicating a large number of metastatic sites (5, 6, 7, 8, or more metastatic sites) is comparatively small, so we re-categorize the variable and create a group called "> 4" and use `crosstabs` to display the data in tabular format. With the option `margin=list()`, the display of row-, column-, and cell-percentages is suppressed.

The table below points toward a possible difference in the number of metastatic sites at baseline in the two treatment arms: In the experimental arm, more patients have only one or two metastatic sites involved, whereas in the standard group more patients have three or four sites at baseline involved. Since the number of metastatic sites is a known prognostic factor (patients with more sites involved do worse), this slight imbalance may bias the treatment effect. The impact is discussed later. In the table two points should be noted: Although the variable `Randomization` contains either "Experimental" or "Standard," S-PLUS is using abbreviations. The result of the chi-squared test, which is always provided together with the `crosstabs` output, is omitted here. Further analyses of other variables do not indicate an imbalance between the treatment groups with respect to baseline characteristics.

```
> x <- ifelse(MetSitesAtBaseline < 5, MetSitesAtBaseline, "> 4")
> crosstabs( ~ Randomization + x, margin=list())

Call:
crosstabs( ~ Randomization + x, margin = list())
602 cases in table
+--------+
|N       |
+--------+
Randomization|x
          |0       |1       |2       |3       |4       |>4      |RowTotl|
--------+--------+--------+--------+--------+--------+--------+--------+
Exprmnt|   2    |102     | 76     | 58     | 34     | 29     |301     |
       |        |        |        |        |        |        |0.5     |
--------+--------+--------+--------+--------+--------+--------+--------+
Standrd|   0    | 89     | 69     | 64     | 51     | 28     |301     |
       |        |        |        |        |        |        |0.5     |
--------+--------+--------+--------+--------+--------+--------+--------+
ColTotl|2       |191     |145     |122     |85      |57      |602     |
       |0.0033  |0.3173  |0.2409  |0.2027  |0.1412  |0.0947  |        |
--------+--------+--------+--------+--------+--------+--------+--------+
```

12.5   Primary Study Parameter

12.5.1   Methodology

The aim of the trial was to show at least equivalence between the two treatment arms in terms of response rate (percentage of patients with partial or complete response). For this study, at least equivalence is reached if the lower limit of the 95% confidence interval for the difference of the rates is not below −10%. Figure 12.3 illustrates the idea of the test procedure.

As an example, take a hypothetical observed response rate of 25% in the experimental arm and of 20% in the standard arm. The difference in response rates is 5% with a lower limit of the 95% confidence interval of −1.8%. In this case, the two treatments are regarded as at least equivalent.

The following primary hypothesis is tested:

$$H_0: RR_E \leq RR_S - 10\%$$

$$H_1: RR_E > RR_S - 10\%$$

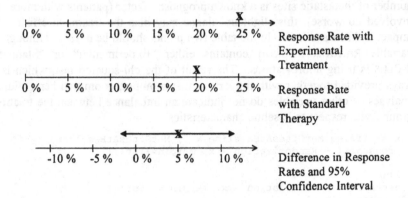

Figure 12.3. Test procedure for the primary efficacy parameter.

where RR_E and RR_S denote the response rates in the experimental and standard arms, respectively. The comparison is based on a one-sided chi-squared test with a correction according to Hauck–Anderson (Hauck and Anderson, 1986). The correction is used since the uncorrected chi-squared test is regarded as too liberal in terms of coverage probability. The significance level is 2.5%. The lower confidence limit for the difference of two rates according to Hauck and Anderson is defined as

$$P_E - P_S - CHA - u_{0.975}\sigma, \tag{12.1}$$

where $u_{0.975}$ is the 97.5 percentile of the standard normal distribution,

$$CHA = [2 \times \min(n_E, n_S)]^{-1} \tag{12.2}$$

is the correction, and

$$\sigma^2 = P_E \left(\frac{1 - P_E}{n_E - 1} \right) + P_S \left(\frac{1 - P_S}{n_S - 1} \right) \tag{12.3}$$

with

  P_E = proportion of patients responding in the experimental arm;
  P_S = proportion of patients responding in the standard arm;
  n_E = number of patients in the experimental arm; and
  n_S = number of patients in the standard arm.

In case the first test indicates at least equivalence between the two treatment arms, a further test is performed:

$$H_0 : RR_E = RR_S,$$

$$H_1 : RR_E \neq RR_S.$$

A chi-squared test with a Schouten continuity correction (similar to the correction according to Hauck and Andersen described above, see Schouten et al., 1980) is used at the 5% level. An adjustment for multiple testing is not needed since the test procedure is hierarchical. This can be seen in Figure 12.3: the comparison of the lower confidence limit to 0 is already included in the comparison to −10% (at least equivalence test).

A discussion of the population used in the analysis exceeds the scope of this chapter. Since the trial is an equivalence trial, the International Conference on Harmonization (ICH E 9, Step 4 (1998)) proposes the analysis to be based on the per-protocol (standard) population. To test for robustness of results, the analysis would be repeated on the intent-to-treat population (all randomized patients). For the purpose of this section, all presented analyses are performed on the all randomized population unless stated otherwise.

12.5.2   Results

For the evaluation of the primary efficacy parameter, the response rate, patients are categorized into responders and nonresponders. The responders are identified by partial or complete response (coded as "CR" and "PR"). The data are tabulated and the number of patients in each category is determined:

```
> x <- ifelse(BestResponse == "CR" | BestResponse=="PR",
    "Responder", "Non-Responder")
> crosstabs( ~ Randomization + x,
    margin= list("N/RowTotal" = setdiff(1, 2)))
Call:
crosstabs( ~ Randomization + x, margin = list("N/RowTotal" =
setdiff(1, 2)))
602 cases in table
+----------+
|N         |
|N/RowTotal|
+----------+
Randomization|x
        |Nn-Rspn|Respndr|RowTotl|
-------+-------+-------+-------+
Exprmnt|221    | 80    |301    |
       |0.73   |0.27   |0.5    |
-------+-------+-------+-------+
Standrd|247    | 54    |301    |
       |0.82   |0.18   |0.5    |
-------+-------+-------+-------+
ColTotl|468    |134    |602    |
       |0.78   |0.22   |       |
-------+-------+-------+-------+
```

M. Jahn

Out of 301 patients receiving the experimental treatment, 80 patients (27%) responded, whereas 54 patients (18%) responded to the standard therapy. Using the function ci.HauckAnderson given in Appendix 12.A.2, the lower confidence limit for the difference in rates is obtained as 1.85%. Therefore, at least equivalence is concluded:

```
> ci.HauckAnderson(80, 301, 54, 301, 0.05/2)
 Lower Limit Significance Level
     1.850696               0.025
```

Following the hierarchical test procedure, a test for a difference between treatments is given by the chi-squared test with a Schouten continuity correction. The function test.Schouten is provided in Appendix 12.A.3.

```
> test.Schouten(80, 301, 54, 301)
    0.01254856
```

With the response rates seen, the p-value for the difference is 0.0125, indicating a statistically significant difference in response rates in favor of the experimental treatment.

12.6 Secondary Study Parameters

12.6.1 Methodology

Several parameters are regarded as secondary parameters, including:

- Survival
- Time to progressive disease or death
- Time to first onset of adverse events

The first two parameters are related to the efficacy of the treatments, while the last one is a safety comparison of the two treatments. The results for the three parameters are displayed as graphs of the Kaplan–Meier estimates. The tests performed are regarded as exploratory. In the full analysis of the trial, parameters such as time to response and duration of response are evaluated as well.

Survival is expected to be similar in the two arms. Therefore, the goal is to show that with regard to survival the experimental treatment is at least equivalent compared to the standard therapy. Survival is defined as the time from randomization to death or the last date the patient is known to be alive (in the case of censoring). All death information, recorded during the study as well as during the trimonthly follow up evaluations, is used. For the test of at least equivalence, the upper limit of the 95% confidence interval of the relative risk is used. The relative risk is a measurement of the risk of dying in one treatment arm relative to the risk of dying in the other treatment arm. A relative risk of 1 indicates that the risk of dying is the same in both arms. The standard therapy is regarded as the base: a relative risk smaller than 1 indicates that the risk of dying is larger with the standard treatment. For the evaluation of the test, the upper 95% confidence limit of the relative risk is compared to the boundary 1.25. If the upper limit is smaller than 1.25, at least equivalence is concluded.

Time to disease progression is seen as a kind of surrogate marker for survival (Ellenberg and Hamilton, 1989). Disease progression indicates worsening of the disease which can be recorded earlier than death, allowing shorter clinical trial periods. An event is therefore defined as disease progression or death (sometimes without prior recording of disease progression). The reasoning for taking death without observed disease progression as an event is based on the following conservative assumption: if a death occurs without prior recording of progression, the tumor evaluations were not frequent enough to observe the progression.

Time to disease progression is defined as the time from randomization to first assessment of the progression or death. In the case of censoring, the last tumor assessment for which the patient has not progressed is used. A log-rank test is performed to analyze a possible difference between treatments. As with the survival analysis, all information available is taken into account. This includes the information from the trimonthly follow up evaluations.

Besides the efficacy, the safety (e.g., adverse effects, nondesired side effects, etc.) of a treatment is of equal importance. A treatment which has comparable efficacy to another treatment, but fewer adverse effects or less severe adverse effects, may be regarded as favorable. Several methods may be used to analyze a possible difference in safety. Since the two treatments under investigation are assumed to differ with regard to safety, a safety comparison was planned in the protocol. For the comparison, the incidence rates of the most common adverse effects was selected for each treatment arm. For the standard treatment, the selection was based on the literature. For the experimental treatment, experience from Phase II and other Phase I trials was used. The following types of adverse effects, which the investigator regarded as related to treatment and severe or life-threatening, were selected: diarrhea, stomatitis, nausea, vomiting, alopecia, hand–foot syndrome, and leukopenia. For convenience, reference is always made to this special class of adverse events unless explicitly stated otherwise.

The primary evaluation is done by comparing the time from randomization to the onset of adverse events using a log-rank test. Each patient experiencing at least one of the effects is regarded as having an event in the analysis. In the case of more than one event for one patient, the first event is used. For example, if a patient experiences a severe diarrhea on day 30 and a life-threatening stomatitis on day 38, the patient is regarded as experiencing an event at day 30. In the case of censoring, the time from randomization to study exit is analyzed. The secondary evaluation is based on the frequency of patients with at least one adverse event. The comparison between the two treatments is done using a chi-squared test with Schouten correction. This analysis is based on the safety population defined as all patients receiving at least one dose of trial medication.

12.6.2   Results

In this section, the results for the secondary study parameters as outlined above are displayed. Graphs of the Kaplan–Meier estimates are presented for each parameter.

Survival

The following statements create a graphical display of the Kaplan–Meier estimates (Figure 12.4):

```
> x <- survfit(Surv(Survival, SurvivalStatus) ~
      Randomization, data = studydata, na.action = na.omit)
> plot(x, lty = c(1, 3), ylab = "Estimated Probability",
      xlab = "Days", mark.time = F)
> title("Kaplan-Meier Estimates\nof Survival")
> key(text = list(c("Experimental", "Standard"), adj = 0),
      line = list(lty = c(1, 3)))
```

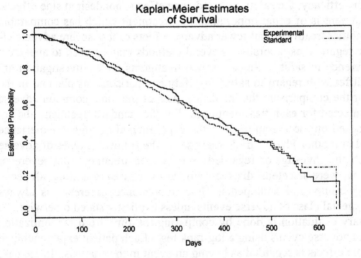

Figure 12.4. Survival curves for the two treatment arms.

The summary statistics are as follows:

```
> x
          *** Nonparametric Survival ***
Call: survfit(formula = Surv(Survival, SurvivalStatus) ~
Randomization, data = studydata, na.action = na.omit)
```

	n	events	mean	se(mean)	median	0.95LCL	0.95UCL
Randomization= Experimental	301	163	382	12.5	396	349	452
Randomization= Standard	301	162	382	12.7	364	337	415

For 54% of the patients (325 out of 602 patients), death is recorded during the study or the regular follow up. The median survival is 396 days for the experimental treatment and 364 days for the standard therapy.

The survival curves for the two treatment arms are very similar. With regard to survival the goal is to show at least equivalence of the experimental treatment compared to the standard therapy. For the calculation the treatment code needs to be a factor variable. This is done by creating the variable rnd. Using the standard therapy as base, the relative risk is 0.977 and the related 95% confidence interval ranges from 0.786 to 1.21:

```
> rnd <- ifelse(Randomization == "Experimental", 1, 0)
> summary(
    coxph(formula = Surv(Survival, SurvivalStatus) ~ rnd))
Call:
coxph(formula = (Surv(Survival, SurvivalStatus) ~ rnd))

  n= 602

       coef exp(coef) se(coef)      z    p
rnd -0.0232     0.977    0.111 -0.209 0.83

       exp(coef) exp(-coef) lower .95 upper .95
rnd      0.977      1.02      0.786     1.21

Rsquare= 0    (max possible= 0.998 )
Likelihood ratio test= 0.04  on 1 df,   p=0.835
Wald test             = 0.04  on 1 df,   p=0.835
Efficient score test = 0.04  on 1 df,   p=0.835
```

Using the treatment as only a factor in the model, the risk of dying is slightly smaller for patients who received experimental treatment. Since the upper limit of the confidence interval is smaller than 1.25, at least equivalence between the two treatment arms is concluded.

Patients without disease progression or death were to be treated for at least 7 months (with an exception for patients with a complete tumor response). Afterward, a regular follow up on survival and disease progression (if not recorded whilst on study) was performed. The first censoring with respect to survival can be expected after 7 months, except for patients lost to follow up. For patients lost to follow up, no further information on survival or disease progression is available. The censoring pattern is reflected in the graph of the Kaplan–Meier estimates (Figure 12.5), in which censoring is indicated by a "+" sign. The censoring pattern seems to be similar in the two treatment arms. Before day 210 (about 7 months), a total of 27 patients are censored in the survival analysis, indicating a good adherence to the trial protocol:

```
> x <- survfit(Surv(Survival, SurvivalStatus)~
   Randomization)
> plot(x, lty = c(1, 3), ylab = "Estimated Probability",
   xlab = "Days")
> key(text = list(c("Experimental", "Standard"), adj = 0),
   line = list(lty = c(1,3)))
> title("Kaplan-Meier Estimates\nof Survival")
```

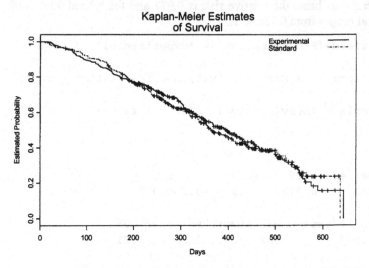

Figure 12.5. Survival curves with censoring marked by "+".

Time to Disease Progression

Time to disease progression is correlated with survival. Because of more events
(patients with progressive disease who did not die) and earlier events (progressive disease is recorded before death), a steeper drop in the Kaplan–Meier
curves can be expected in comparison to the survival curves. This is depicted in
Figure 12.6.

```
> x <- survfit(Surv(TimeToProgression, ProgressionStatus)~
   Randomization)
> plot(x, lty= c(1, 3), ylab = "Estimated Probability",
   xlab = "Days", mark.time=F)
> key(text = list(c("Experimental", "Standard"), adj=0),
   line=list(lty = c(1,3)))
> title(paste("Kaplan-Meier Estimates",
   "of Time to Disease Progression or Death", sep="\n"))
```

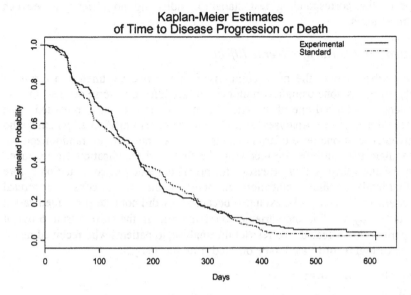

Figure 12.6. Time to disease progression or death.

```
> x
Call: survfit(formula = Surv(TimeToProgression, ProgressionStatus)
~ Randomization)

                     n events mean se(mean) median 0.95LCL 0.95UCL
Randomization=
      Experimental 301   269  185    8.46     160     134     169
Randomization=
          Standard 301   272  175    7.86     144     123     169

> survdiff(Surv(TimeToProgression, ProgressionStatus) ~
    Randomization)
Call:
survdiff(formula=Surv(TimeToProgression, ProgressionStatus) ~
Randomization)

                              N Observed Expected (O-E)^2/E (O-E)^2/V
Randomization=Experimental 301    269      274     0.0999    0.206
    Randomization=Standard 301    272      267     0.1027    0.206

Chisq= 0.2  on 1 degrees of freedom, p= 0.65
```

As seen in the analysis of survival, about half the patients died during the study or the trimonthly follow-up period. For all death cases, disease progression is recorded before death. In addition, about 110 patients in each arm progressed during the study. In the experimental arm, 32 patients are censored in comparison to 29 patients in the standard arm, because neither death nor disease progression occurred. The median time to disease progression is 160 days for patients treated with the experimental drug and 144 days for patients receiving standard therapy. The p-value for the log-rank test is 0.65 (with 0.2 being the value of the corresponding test statistic), indicating no difference between treatment arms.

Time to First Onset of Adverse Effects

As described above, the main comparison of the two treatments in terms of safety profile is done using a certain class of predefined adverse events. In case a patient experienced one of the selected adverse effects, this is regarded as an event in the analysis. Analyses on safety data are performed on all patients who received at least one dose of trial medication. Patients who are randomized but discontinue the study before receiving the first study medication are excluded from the analysis of safety. Reasons for premature study discontinuation before the first intake of study medication can include withdrawal of consent or refusal of treatment. Patients to be excluded because they did not take study medication are marked by a "4" in the variable Exclusion. In the current trial, 6 out of 602 patients are affected. To restrict the analysis to patients who received medication, certain records need to be omitted from the analysis:

```
> which <- Exclusion != 4
> table(which)
  FALSE TRUE
      6  596
> x <- survfit(
    Surv(TimeToAdvEv[which], AdvEvStatus[which]) ~
    Randomization[which])
> plot(x, lty = c(1, 3), ylab = "Estimated Probability",
    xlab = "Days", mark.time = F, conf.int = T)
> key(text = list(c("Experimental", "Standard"), adj=0),
    line = list(lty = c(1,3)))
> title(paste("Time to First Onset of Adverse Events",
    "of Pre-Specified Class", sep="\n"))
```

The Kaplan–Meier estimates for the time to first onset of selected adverse events are displayed in Figure 12.7 together with the 95% confidence intervals. The curves are well separated. Even the confidence limits are not overlapping up to day 100. This is reflected in the p-value of the log-rank test of 0.00841 (applying the function survdiff), indicating a statistically significant difference between the two treatments in favor of the experimental drug. The number of events in each treatment arm is different: 85 patients receiving experimental treatment experienced at least one of the adverse events of the prespecified group, whereas 106 patients receiving standard therapy experienced at least one of the adverse events. The secondary comparison results in a p-value of 0.08153 (function test.Schouten). The higher p-value is probably due to the reduced power of the chi-squared test compared to the log-rank test. This analysis is of course influenced by the choice of selected adverse events. However, since the specification of the analysis is already done in the protocol (prior to study start), only limited criticism would be valid.

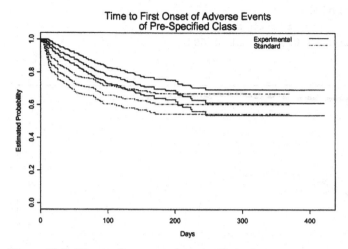

Figure 12.7. Time to first onset of prespecified adverse events with confidence intervals.

In summary, for the secondary parameters survival is regarded as at least equivalent. No difference in time to disease progression is visible. The safety profiles show a significant difference in favor of the experimental treatment.

12.7   Proportional Hazard Assumption and Subgroup Analysis

The test for at least equivalence in the analysis of survival is based on the methodology of the Cox regression model and thus on the proportional hazards assumption (the likelihood of dying is proportional between the two treatment arms at all time points). This assumption is evaluated by using Schoenfeld residuals, which have an expected value of zero in large samples. A spline smooth (Venables and Ripley, 1997) of the residuals over survival times should result in a flat curve if proportional hazards holds. Additionally, the slope of the smoothing is tested for a nonsignificant deviation from 0. With the following commands a Cox regression and an analysis of residuals are performed:

```
> x <- coxph(formula = Surv(Survival, SurvivalStatus) ~
    Randomization)
> plot(cox.zph(x))
```

The plot of the Schoenfeld residuals for the survival data is shown in Figure 12.8. The fitted line does not clearly deviate from 0. The test mentioned above for significant slope results in a p-value of 0.691. Therefore proportional hazards seems justifiable:

```
> print(cox.zph(x))
                         rho chisq      p
   Randomization -0.0221 0.158 0.691
```

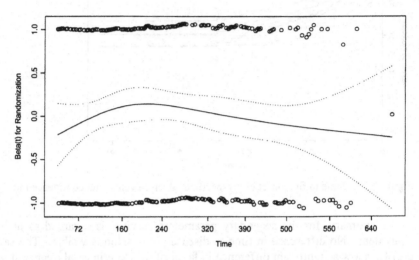

Figure 12.8. Rescaled Schoenfeld residuals with a smooth fit.

A slight imbalance at baseline with regard to number of metastatic sites was seen. To assess a possible bias in the analysis, the impact of the number of metastatic sites at baseline on survival is evaluated. With Kaplan–Meier estimates modeling in terms of using additional stratification variables (e.g., patients with one or two metastatic sites involved versus patients with more than two sites) estimating the survival function is possible. The advantage is that the resulting curves are intuitive and easy to interpret. In contrast, Cox regression allows "real" modeling. In the following both approaches are pursued, first in terms of separate plots of the Kaplan–Meier estimates and second performing the Cox regression.

We create a graph of the survival curve for patients with one or two metastatic sites at baseline and another for patients with more than two metastatic sites at baseline (Figure 12.9):

```
> x <- survfit(Surv(Survival[MetSitesAtBaseline <= 2],
    SurvivalStatus[MetSitesAtBaseline <= 2]) ~
    Randomization[MetSitesAtBaseline <= 2])
> y <- survfit(Surv(Survival[MetSitesAtBaseline > 2],
    SurvivalStatus[MetSitesAtBaseline > 2]) ~
    Randomization[MetSitesAtBaseline > 2])
> par(mfrow=c(1,2))
> plot(x, lty = c(1, 3), ylab = "Estimated Probability",
    xlab = "Days", mark.time = F)
```

```
> key(text = list(c("Experimental", "Standard"), adj = 0),
    line = list(lty = c(1, 3)))
> title("Survival - Number of Metastatic Sites <= 2")
> plot(y, lty = c(1, 3), ylab = "Estimated Probability",
    xlab = "Days", mark.time = F)
> key(text = list(c("Experimental", "Standard"), adj = 0),
    line = list(lty = c(1, 3)))
> title("Survival - Number of Metastatic Sites > 2")
```

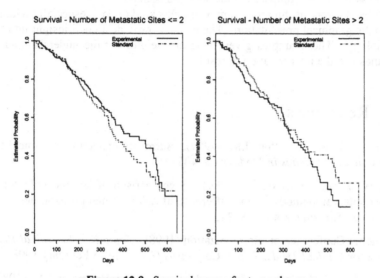

Figure 12.9. Survival curves for two subgroups.

Although differences can be seen, the survival curves of the two treatments can be regarded as similar within each subgroup. The second approach is to perform Cox regression analysis to model the impact of the number of metastatic sites at baseline. Two models are fitted: one containing only the treatment information and the other containing treatment and a number of metastatic sites as factors. The magnitude of the difference between the two models is tested to assess the "gain" by adding the factor "number of metastatic sites at baseline":

```
> c1<-coxph(Surv(Survival, SurvivalStatus) ~ Randomization)
> c2<-coxph(Surv(Survival, SurvivalStatus) ~
    Randomization + MetSitesAtBaseline)
> pchisq(-2 * (c1$loglik[2] - c2$loglik[2]), df = 1)
[1] 0.9984804
```

With a p-value of 0.998, there is no significant difference between the two models, indicating that the number of metastatic sites at baseline does not improve the fit. Although slight imbalances are seen, the survival is mostly explained by the impact of the treatment received by the patients.

12.8 Summary

The trial shows a statistically significant difference in response rates favoring the experimental treatment. Survival can be regarded as at least equivalent and time to disease progression is not statistically significantly different between the treatment arms. The safety comparison shows that the difference in time to the first onset of adverse effects is statistically significant in favor of the experimental treatment. Altogether, the experimental treatment shows improved efficacy and safety when compared to the standard therapy.

S-PLUS is a very useful tool in analyzing this kind of data, especially with respect to the graphical presentation. The manuals provide a good insight into the methodology. The examples given there are helpful for the understanding of commands and the interpretation of results.

12.9 References

Ellenberg, S.S. and Hamilton, J.M. (1989). Surrogate endpoints in clinical trials: Cancer. *Statistics in Medicine* **8**, 405–413.

Hauck, W.W. and Anderson, S. (1986). A comparison of large-sample confidence interval methods for the difference of two binomial probabilities. *The American Statistician* **40**, 318–322.

International Conference on Harmonization (1998). *Topic E 9: Statistical Principles for Clinical Trials, Step 4, Consensus Guideline*. 5 February 1998.

Schouten, H.J.A., Molenaar, I.W., Strik, R. van, and Boomsma, A. (1980). Comparing two independent binomial proportions by a modified chi square test. *Biometrical Journal* **22**, 241–248.

Venables, W.N. and Ripley, B.D. (1997). *Modern Applied Statistics with S-PLUS*, 2nd ed. Springer-Verlag, New York.

World Health Organization–Geneva (1979). *WHO Handbook for Reporting Results of Cancer Treatment*.

12.A. Appendix

12.A.1 Variables Used in the Text

Patient	Patient number as used in study
Age	Age at time of randomization
Sex	Male/Female
Randomization	Experimental—experimental treatment
	Standard—standard therapy

Country	Country in which the patient is randomized
MetSitesAtBaseline	Number of metastatic sites at baseline
Exclusion	1—patients in the per protocol analysis
	4—patients randomized but without treatment
BestResponse	CR (complete response)
	PR (partial response)
	SD (stable disease)
	PD (progressive disease)
TimeToProgression	Time from randomization to disease progression or death
ProgressionStatus	0—no disease progression or death
	1—disease progression or death
Survival	Time from randomization to death
SurvivalStatus	0—alive
	1—dead
TimeToAdvEv	Time from randomization to first onset of preselected treatment-related grade 3 or 4 adverse events
AdvEvStatus	0—no experience of preselected adverse events
	1—experience of at least one pre-selected adverse events

12.A.2 Hauck–Anderson Modified Chi-Squared Test

```
ci.HauckAnderson <- function(resp.a, n.a, resp.b, n.b,
  alpha = 0.025)
{
# Chi-square test with Hauck-Anderson modification
# Input:
# resp.a, resp.b: responders in group a and b
# n.a, n.b: number of patients in group a and b
# alpha: significance level
#
# Returns:
# lower confidence limit
# ---
  resprate.a <- resp.a/n.a
  resprate.b <- resp.b/n.b
  cha <- 1/(2 * min(n.a, n.b))
  sigma <- sqrt((resprate.a * (1 - resprate.a))/(n.a - 1) +
    (resprate.b * (1 - resprate.b))/(n.b - 1))
  lowerlimit <- resprate.a - resprate.b - cha -
    sigma * qnorm(1 - alpha)
```

```
result <- c(lowerlimit * 100, alpha)
names(result) <- c("Lower Limit", "Significance Level")
return(result)
}
```

12.A.3 Schouten Modified Chi-Squared Test

```
test.Schouten <- function(resp.a, n.a, resp.b, n.b)
{
# Chi-square test with Schouten modification
# Input:
# resp.a, resp.b: responders in group a and b
# n.a, n.b: number of patients in group a and b
# alpha: significance level
#
# Returns:
# p value
# ---
  resprate.a <- resp.a/n.a
  resprate.b <- resp.b/n.b
  nom <- abs(resprate.a - resprate.b) -
    1/(2 * max(n.a, n.b))
  p0 <- (n.a * resprate.a + n.b * resprate.b)/(n.a + n.b)
  se2 <- ((p0 * (1 - p0) * (1/n.a + 1/n.b)) *
    (n.a + n.b))/(n.a + n.b - 1)
  chi <- (nom * nom)/se2
  p <- 1 - pchisq(chi, 1)
  return(p)
}
```

13

Analysis of Variance: A Comparison Between SAS and S-PLUS

Melvin Olson

Allergan, Inc., Irvine, CA, USA

13.1   Introduction

The purpose of this chapter is to show how to perform a simple analysis of variance (ANOVA) model in S-PLUS. Many practicing statisticians in the pharmaceutical industry think of S-PLUS as a nice exploratory tool, but continue to use SAS to run their confirmatory models. The point to be made here is not that there is necessarily anything wrong with that practice, but rather that an alternative exists. In fact, not only does S-PLUS have excellent facilities for confirmatory statistics and modeling, but when combined with its unsurpassed graphics and programming capabilities, it is a very attractive package indeed.

Since so many industry statisticians are familiar with running ANOVA models in SAS, this chapter will show how to run one in S-PLUS, making a direct comparison between the two packages. In this way, you can see exactly how to run things in S-PLUS that you are already used to doing in SAS. In the process, you will gain an understanding of how S-PLUS treats ANOVA models and appreciate some of the advantages it has to offer.

The focus of this chapter is a little different from many others as it is not designed to explain the statistical technique itself, but rather to illustrate its implementation in S-PLUS. While it will be assumed that the reader is familiar with ANOVA models, a short review will be provided in Section 13.2. The remaining sections will cover the topics of the example dataset, summary statistics, fitting ANOVA models, contrasts, and a summary.

13.2   ANOVA Models

The classical ANOVA model is used in settings where there is a normally distributed outcome and a discrete predictor variable. The assumption is made that the subpopulations (groups), defined by the discrete levels of the predictor variable, all come from the same underlying population with the possible exception of a location shift (i.e., the groups share the same variance but have possibly different means). The question of interest is whether or not the means from the various groups are the same.

Given that the question of interest involves means, it appears that "analysis of variance" is a misnomer. The key to this curiosity is that the test statistic, for whether or not the means are the same, is in the form of an F-test. As such, a ratio is formed between the "variance" associated with the various group means and that associated with the null hypothesis of equality of group means. In this way, "variances" are analyzed to investigate a hypothesis concerning means.

The simplest ANOVA model, the one-way ANOVA model, can be written in the following form:

$$Y_{ij} = \mu + \alpha_i + \varepsilon_{ij}$$

$$\text{where } \mu = \text{overall mean}$$

$$\alpha_i = \text{effect of level } i, i = 1, \ldots, I; \sum_i \alpha_i = 0$$

$$\varepsilon_{ij} = \text{residual error of the } j^{\text{th}} \text{ individual}$$
$$\text{in the } i^{\text{th}} \text{ group, } j = 1, \ldots, n_i$$

$$Y_{ij} = \text{outcome of the } j^{\text{th}} \text{ individual}$$
$$\text{in the } i^{\text{th}} \text{ group}$$

$$\varepsilon_{ij} \sim N(0, \sigma^2)$$

$$Y_{ij} \sim N(\mu, \sigma^2)(\text{under } H_0)$$

The residuals are assumed to be normally distributed with mean zero and variance σ^2. Since the α_i's are considered to be fixed effects, Y is also normally distributed with variance σ^2. With this notation, the null and alternative hypotheses are written as

$$H_0 : \quad \alpha_1 = \ldots = \alpha_I = 0$$

$$H_A : \quad \text{not all } \alpha_i\text{'s equal}$$

The implication of these hypotheses on the outcome is that

$$\text{under } H_0 : \quad Y_{ij} \sim N(\mu, \sigma^2)$$

$$\text{under } H_A : \quad Y_{ij} \sim N(\mu_i, \sigma^2)$$

With this formulation, the objective of the ANOVA model is to determine if there is evidence that each level of the predictor warrants its own mean or if the groups share a common mean.

One of the most important things to note about the formulation of the model is the necessity for a restriction, $\sum_i \alpha_i = 0$. Without this restriction, the model would be overparameterized and could not be fit. If we tried to include an indicator variable for each level of α_i, call it X_i, then the indicator for the mean, call it X_μ, is a linear combination of the others defined by $X_\mu = X_1 + \ldots + X_I$. In such cases the fit of the model fails.

Since we have included a parameter for the overall mean (μ), we need to impose a restriction, such as $\sum_i \alpha_i = 0$, in order for the model to be fit. The question becomes how to implement this restriction in practice. The approach taken by SAS is to set $\alpha_I = 0$ and estimate the other α_i's.

The way in which a statistical package actually implements a restriction on the α_i's may seem like an unimportant technical detail, but the interpretation of all of the estimated effects depends on it. A comparison of the interpretation of estimated model effects is given in Table 13.1 in the context of the example dataset used in the first sections of this chapter.

The S-PLUS approach to estimating the parameters differs from the SAS approach. The contrast type associated with a particular variable is used to fit the model in a reparameterized form. A simple function can be used to recover the original model parameters, no matter what the contrast type.

Here ends the quick review of the theory of ANOVA. The discussion and examples used in the rest of this chapter build on the basic model shown here.

13.3   Example Dataset

One of the S-PLUS datasets will be used to illustrate the ideas discussed in the rest of the chapter. The dataset is called drug.mult and concerns an experiment in which six subjects were given a treatment at baseline and at weekly intervals for three successive weeks. The variables recorded are subject, gender (F/M), $Y.1$, $Y.2$, $Y.3$, and $Y.4$, where $Y.1$ contains the outcome of interest at baseline and $Y.2$–$Y.4$ contain the outcome of interest for weeks 1–3, respectively.

The first obstacle encountered, in both SAS and S-PLUS, is that the data need to be reformatted to run ANOVA models with a single response variable per row. The first four rows of the original dataset are printed below to show that there are 3 response variables per row instead of the single response variable required for ANOVA models:

```
> drug.mult[1:4,]
      subject gender  Y.1  Y.2  Y.3  Y.4
           S1      F 75.9 74.3 80.0 78.9
           S2      F 78.3 75.5 79.6 79.2
           S3      F 80.3 78.2 80.4 76.2
           S4      M 80.7 77.2 82.0 83.8
```

There are many ways around this problem in SAS (including copying the dataset for each value of time, renaming variables, and setting the datasets together), but that will not be shown here as the data will be transformed in S-PLUS and then exported for use in SAS. The code for one way of transforming the data in S-PLUS is given below:

S-PLUS code for transforming data

```
> subj <- factor(rep(drug.mult$subject, 4))
> sex <- factor(rep(drug.mult$gender, 4))
> time <- ordered(rep(0:3, rep(6, length(0:3))))
> response <- c(drug.mult$Y.1, ..., drug.mult$Y.4)
> newdrug <- data.frame(response, subj, sex, time)
> newdrug[1:4,]
      response subj sex time
          75.9   S1   F    0
          78.3   S2   F    0
          80.3   S3   F    0
          80.7   S4   M    0
```

The variables subj and sex have been defined as factors because they are discrete and have no inherent ordering. The variable time is also discrete (could be considered as continuous), but has an inherent ordering so it has been given the class ordered.

The new dataset, newdrug, is now ready to be exported and the first four rows have been printed above for inspection. Under Windows, this is done most easily with **File-Export Data**. The easiest file type to choose for this purpose is SAS (*.sd2), but a simple text file (ASCII) could also be used. Under UNIX, either of the functions exportData or write.table could be used.

13.4   Summary Statistics

The first thing to do with our new dataset is get a quick picture of what the data look like. Obviously, this can be done in many ways, including generating summary statistics and simple graphical displays. A small sample of such techniques will be shown in this section.

In SAS, the two most used ways of generating summary statistics are PROC MEANS and PROC UNIVARIATE. The syntax for both procedures is shown below with the output for PROC MEANS only. Suffice it to say that PROC MEANS has more succinct output and shows the sample size (n), mean, standard deviation, minimum, and maximum by default. PROC UNIVARIATE shows more statistics, especially ones related to percentiles:

SAS code for summary statistics

```
proc means data=newdrug;
    class sex;
    var response;
proc sort data=newdrug;
    by time;
proc univariate data=newdrug plot;
    by time;
    var response;
```

Another convenient feature of PROC UNIVARIATE, when used with a by statement and the plot option, is that boxplots of the response are generated for each grouping of the by variable (in our example, time). The boxplots are simple line printer drawings that give a quick idea of trends that may exist in the data, but are too crude to be used in a report. The syntax of producing boxplots from PROC UNIVARIATE has been shown in "SAS code for summary statistics," but the output has been omitted due to its length:

SAS output for summary statistics (PROC MEANS only)

```
SEX Obs  N       Mean    Std Dev     Minimum    Maximum
--------------------------------------------------------
F    12 12 78.0666667 2.0786068 74.3000000 80.4000000
M    12 12 81.2500000 2.3368588 77.2000000 86.4000000
--------------------------------------------------------
```

PROC MEANS can be used with either a class or a by statement, the only real difference being in the layout of the output dataset (not generated in this example).

Summary statistics in S-PLUS can be generated either by the individual functions (mean, stdev, etc.) or by using the summary function. The summary function gives the quartiles, minimum, maximum, and mean for continuous variables and for factor (and ordered) variables, it lists the distinct levels along with a count of their occurrence. Summary statistics can also be produced for each level of a factor variable using the by function:

S-PLUS code and output for summary statistics

```
> summary(newdrug)
       response    subj   sex   time
     Min.:74.30    S1:4   F:12  0:6
   1st Qu.:78.27   S2:4   M:12  1:6
   Median:80.05    S3:4         2:6
     Mean:79.66    S4:4         3:6
   3rd Qu.:81.18   S5:4
     Max.:86.40    S6:4
```

```
> by(newdrug, newdrug$sex, summary)
   newdrug$sex:F
         response      subj    sex     time
         Min.:74.30    S1:4    F:12    0:3
      1st Qu.:76.12    S2:4    M: 0    1:3
      Median:78.60     S3:4            2:3
        Mean:78.07     S4:0            3:3
      3rd Qu.:79.70    S5:0
        Max.:80.40     S6:0

-------------------------------------------

   newdrug$sex:M
         response      subj    sex     time
         Min.:77.20    S1:0    F: 0    0:3
      1st Qu.:80.25    S2:0    M:12    1:3
      Median:81.25     S3:0            2:3
        Mean:81.25     S4:4            3:3
      3rd Qu.:81.93    S5:4
        Max.:86.40     S6:4
```

The by function operates by specifying the dataset name, the factor variable, and the function to be calculated. The equivalent in SAS is the by statement available in most of the statistical procedures, and/or the class statement in certain procedures.

Another useful way of discovering what information is contained in the data is to create simple graphics such as scatterplots, histograms, and box-plots. If you have the graphics module for SAS, it is possible to create publication quality graphics in SAS using such procedures as GPLOT and GCHART. Highly specialized graphs can always be obtained by using PROC ANNOTATE to manually draw the graph from scratch or add to an existing one. For example, boxplots can be drawn in SAS using PROC GPLOT, the problem being that the width of the box is much too narrow and cannot be scaled. PROC ANNOTATE could be used to draw the boxes manually to the desired specifications given the time and patience. As the syntax of saving SAS graphs is machine-dependent, more will not be covered on the topic of SAS graphs.

Graphics in S-PLUS are simple, powerful, and informative. An example that puts several types of graphs together on one graphics sheet is shown in Figure 13.1, a one-line equivalent of PROC GREPLAY in SAS. The command par(mfrow=c(2, 2)) defines two rows and two columns of graphs:

S-PLUS commands for exploratory graphs

```
> par(mfrow=c(2, 2))
> plot(newdrug$time, newdrug$response)
> plot(newdrug$sex, newdrug$response)
> hist(newdrug$response)
```

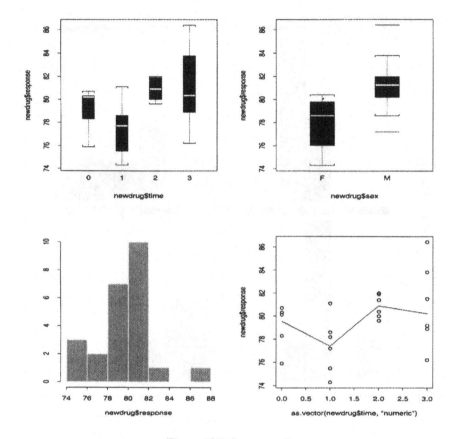

Figure 13.1. Summary plots.

```
> plot(as.vector(newdrug$time, "numeric"),
  newdrug$response)
> lines(lowess(as.vector(newdrug$time, "numeric"),
  newdrug$response))
```

The default plot type involving time and sex is boxplot because the variables were defined as factor and ordered, respectively. From the boxplots we can already see that there is an initial drop in response at week 1 followed by an increase and that males have a higher response than females. The histogram shows the overall distribution of the response variable. To obtain a standard scatterplot with factor and ordered variables, it is necessary to coerce them into numeric format using the as.vector function. A simple scatterplot smooth can then be added to the plot using the lines and lowess functions. Note that elements may be added to the current

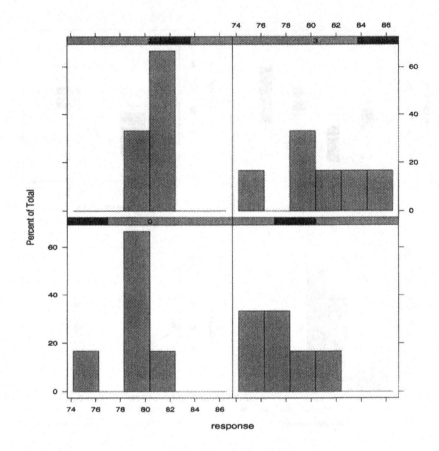

Figure 13.2. Trellis graphs

graph until the next graph is begun.

Another powerful visual tool available in S-PLUS is the so-called Trellis graph. Conditional graphs are plotted in a matrix layout with uniform axes making for easy comparisons across levels of the conditioning variable(s). The easiest explanation comes from looking at the example code below and the graph in Figure 13.2:

S-PLUS code for a Trellis graph

```
> histogram(~response|time, data=newdrug)
```

The tilde (~) is used in the specification of statistical models and has more obvious function in two-dimensional Trellis graphs such as xyplot and dotplot. The Trellis version of boxplots (bwplot) uses the one-dimensional syntax above. The variable(s) to the right of the vertical bar

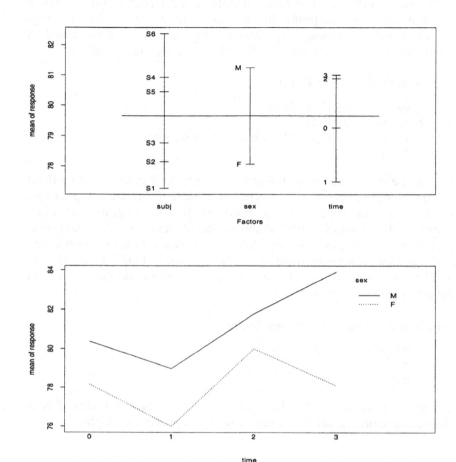

Figure 13.3. Design plots.

specifies the conditioning to be made. One panel is drawn for each level of the conditioning variable(s), in our example, four. The lowest level of the conditioning variable(s) is put in the lower left-hand panel and the highest level in the upper right-hand. It is possible to see that after the baseline time point, the response shifts from left to right (lower value to higher) over time.

There are two special plot functions in S-PLUS that are particularly useful for performing ANOVA models: plot.design and interaction.plot. The first function, plot.design, finds all factor (and ordered) variables and plots the mean of the response variable at each level of the factors. The response variable can be specified explicitly as well as a function to be used other than the mean (median, etc.).

The `interaction.plot` function plots the mean profile of the response variable over time, one profile line for each level of sex (as shown in Figure 13.3). We can immediately address the question of whether or not an interaction between time and sex exists (i.e., are the lines parallel?):

S-PLUS code for ANOVA plots

```
> plot.design(newdrug)
> attach(newdrug)
> interaction.plot(time, sex, response)
> detach()
```

One of the most convenient features of graphs in S-PLUS is saving them for insertion into documents. In Windows, create the desired graph, copy, and paste into the document of your choice. To save the graph into a file, click on **File-Export Graph** and specify the file name, location, and type. S-PLUS can save graphs in many different file types. One that works well with MS Word is the type Windows metafile (*.wmf). In UNIX (also works in Windows), specify a driver, create the graph, and close the graph as in the small example below:

S-PLUS code for saving a graph to file

```
> postscript("filename")
> plot(x, y)
> dev.off()
```

The words in *italics* have to be replaced by names and variables of your choice. Incidentally, MS Word can also import postscript graphs.

13.5   Fitting ANOVA Models

In this section, the syntax for running ANOVA models in both SAS and S-PLUS will be shown, and various aspects of options and interpretation of the output will be discussed. The models fit to the dataset in this section are not necessarily ones that should be used in practice, but they are used in an illustrative manner to compare and contrast ANOVA models.

13.5.1   One-Way ANOVA Models

The simplest form of an ANOVA model is the one-way model. The full specification of this model was shown earlier in Section 13.2. The effect of sex on the outcome will now be examined.

SAS code for one-way ANOVA

```
proc mixed data=newdrug;
    class sex;
    model response=sex / s;
    lsmeans sex / cl;
```

PROC MIXED is used here in preference to PROC GLM because it offers more flexibility with regard to saving intermediate and final results. The variable sex is defined as a class variable. The model option s requests that the solution to the fixed effects portion of the model be printed. The default is that only the summary of the model fit is printed.

In the SAS output below, we see that the effect of sex on the response variable is statistically significant ($p = 0.0019$).

SAS output for one-way ANOVA

```
              Solution for Fixed Effects
    Effect      SEX      Estimate       Std Error     DF
    INTERCEPT           81.25000000    0.63840671     22
    SEX          F      -3.18333333    0.90284343     22
    SEX          M       0.00000000        .          .


    Effect      SEX          t   Pr > |t|
    INTERCEPT           127.27    0.0001
    SEX          F       -3.53    0.0019
    SEX          M          .        .


          Tests of Fixed Effects
    Source      NDF    DDF    Type III F   Pr > F
    SEX          1      22       12.43     0.0019


              Least Squares Means
    Effect  SEX      LSMEAN       Std Error     DF        t
    SEX      F    78.06666667    0.63840671     22    122.28
    SEX      M    81.25000000    0.63840671     22    127.27
    Effect  SEX   Pr > |t|  Alpha      Lower     Upper
    SEX      F     0.0001    0.05     76.7427   79.3906
    SEX      M     0.0001    0.05     79.9260   82.5740
```

Since F comes before M alphabetically, this is the level of sex that has been estimated and the level M has been set to zero with missing values for standard error, degrees of freedom, t-statistic, and p-value.

To understand how the model was fit, it is necessary to look at the estimated parameters and, for clarity, the least squares means (lsmeans) which estimates effects given the model. SAS has fit an intercept of 81.25, a female effect of -3.18, and has set the male effect to zero. None of these

Table 13.1. Parameter estimation and mean response.

	SAS	S-Plus
Constraint	$\alpha_I = 0$	$\sum_i \alpha_i = 0$
Overall mean	$\mu + \frac{1}{2}\alpha_1$	μ
Treatment difference	α_1	$2\alpha_1$
Women	$\mu + \alpha_1$	$\mu + \alpha_1$
Men	μ	$\mu - \alpha_1$

values correspond to parameters of the model shown in Section 13.2. The female effect was obtained by adding a dummy variable to the model with a coding of 1 for F and 0 for M. The intercept used was simply a dummy variable of all 1's. To see the relationship between the model parameters and the SAS output, we look at the lsmeans. The estimated response for males is 81.25 which, according to our model, should be an estimate of $\mu + \alpha_2$ (remembering that with SAS, $\alpha_I = \alpha_2 = 0$). A little thought, and perhaps some algebra as well, will lead to the relationships found in Table 13.1. Note that in SAS model fitting, the intercept is not an estimate of the overall mean.

The approach S-Plus takes to fitting ANOVA models is different to that taken by SAS. The overall result is, of course, the same as that obtained by SAS. However, since the model estimates from S-Plus differ from those in SAS, experienced SAS users may think that the results from S-Plus are incorrect because it is not clear what they refer to. The goal of the rest of this section is two-fold; to show the S-Plus syntax for fitting a one-way ANOVA model and to explain how the model estimates were obtained:

S-Plus code and output for one-way ANOVA

```
> nd.aov.sex <- aov(response~sex, data=newdrug)
> summary(nd.aov.sex)
             Df Sum of Sq  Mean Sq  F Value       Pr(F)
      sex     1   60.8017 60.80167 12.43195 0.001902097
Residuals    22  107.5967  4.89076
> coef(nd.aov.sex)
    (Intercept)      sex
       79.65833 1.591667
> dummy.coef(nd.aov.sex)
    $"(Intercept)":
    (Intercept)
       79.65833

    $sex:
           F        M
   -1.591667 1.591667
> hist(resid(nd.aov.sex))
```

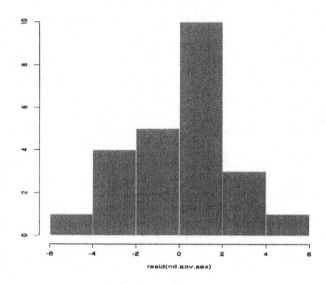

Figure 13.4. Histogram of residuals.

```
> plot(nd.aov.sex)
> nd.lsmeans.sex <- multicomp(nd.aov.sex, comparisons=
  "none", method="lsd", error.type="cwe")
> nd.lsmeans.sex
    95 % non-simultaneous confidence intervals for
    specified linear combinations,
    by the Fisher LSD method

    critical point: 2.0739
    response variable: response

    intervals excluding 0 are flagged by '****'

      Estimate Std.Error Lower Bound Upper Bound
    F    78.1     0.638        76.7        79.4 ****
    M    81.3     0.638        79.9        82.6 ****
> plot(nd.lsmeans.sex, href=80)
```

The idea behind the S-PLUS way of fitting statistical models is to save the fit into an object and use various standard functions for extracting certain pieces of information. The advantage to this approach is that the standard functions are the same for all modeling procedures. Thus, summary is used to produce an ANOVA table, coef prints the estimated coefficients from

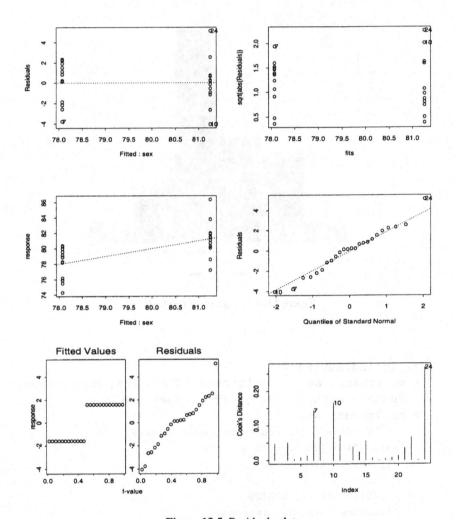

Figure 13.5. Residuals plots.

the fitted model, dummy.coef prints the estimated model coefficients in a nice format (and for all levels), contrasts shows how the dummy variable for sex was coded, and resid extracts the residuals which can be easily displayed with the hist function (see Figure 13.4).

The plot function is intelligent in that it realizes that nd.aov.sex is an ANOVA model fit and produces six diagnostic graphs chosen for this type of model (see Figure 13.5). The multicomp function with the options used above gives the lsmeans and 95% confidence intervals (alternatively, model.tables can be used). The plot function recognizes that the vari-

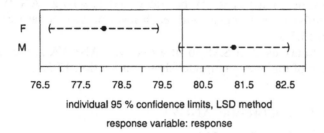

Figure 13.6. LSMeans plot.

able is of type `multicomp` and produces a suitable graph (see Figure 13.6).

Notice, from the output of the `contrasts` function, that S-PLUS does not use the 0/1 coding of SAS. The variable `sex` was defined to be of class factor which has, by default, a Helmert contrast system. Hence, `sex` is given the coding of -1 for Female and 1 for Male. The estimated coefficients of the two dummy variables used to fit the model, `intercept` and `sex` (see `coef` output) are 79.66 and 1.59. The function `dummy.coef` is then used to relate the dummy variable coefficients back to the μ, α_1, and α_2 of the theoretical model. What we see is that the parameterization used by S-PLUS produces effect estimates that match the parameters of the model presented in Section 13.2 and is, in this sense, more natural than that used by SAS. Note that the parameterization used in the S-PLUS model uses orthogonal contrasts while those used by SAS are not (more on this topic later in Section 13.6). Also note that the nice correspondence between fitted effects and model effects, as shown in Table 13.1, will not always hold in S-PLUS (depends on the model), and that it is always important to check the coding being used through the use of the `contrasts` function.

One of the built-in S-PLUS contrast systems is called `treatment`. This system is similar to that used in SAS except that in S-PLUS the first level is set to 0. Hence, if the levels were to be recoded to put the last level as the first, the `treatment` contrasts could be used to give the same results as SAS.

13.5.2   Two-Way ANOVA Models

A second factor variable of interest is added to create a two-way ANOVA model. The added complexity with this model is that a term for the interaction between the two factor variables may be included. There are two other factor variables in our dataset to be considered, `subj` and `time`. It only makes sense to perform a two-way ANOVA on factors that are crossed (i.e., each level of the first factor variable appears with each level of the second). There is at least one measurement of each sex at each time point and thus, the factor `time` can be added to the two-way ANOVA. The levels of `subj` do not appear with each level of `sex`, a case considered in Section 13.5.3.

The syntax used in SAS to fit a two-way ANOVA model differs little from that of the one-way model shown in Section 13.5.1:

SAS code for a two-way ANOVA

```
*Without interaction;
proc mixed data=newdrug;
    class sex time;
    model response=sex time / s;
    lsmeans sex time / cl;
*With interaction, output omitted;
    proc mixed data=newdrug;
    class sex time;
    model response=sex time sex*time / s;
    lsmeans sex time sex*time / cl;
```

An interaction term in SAS is made by inserting an asterisk between each of two or more terms. The S-PLUS syntax for fitting a two-way ANOVA is also similar to its one-way counterpart:

S-PLUS code and output for a two-way ANOVA

```
> nd.aov.both <- aov(response~sex+time, data=newdrug)
> summary(nd.aov.both)
            Df Sum of Sq Mean Sq F Value      Pr(F)
       sex   1  60.80167 60.80167 19.75149 0.000278263
      time   3  49.10833 16.36944  5.31763 0.007861779
 Residuals  19  58.48833  3.07833
> nd.aov.inter <- aov(response~sex+time+sex:time,
  data=newdrug)
> summary(nd.aov.inter)
            Df Sum of Sq Mean Sq F Value    Pr(F)
       sex   1  60.80167 60.80167 22.26827 0.0002317
      time   3  49.10833 16.36944  5.99522 0.0061389
  sex:time   3  14.80167  4.93389  1.80701 0.1864698
 Residuals  16  43.68667  2.73042
```

The functions used earlier to extract information from the fit of a model (`coef`, `dummy.coef`, etc.) work here as well. The only comment to be made is that the main effect of time is significant ($p = 0.0079$), whereas the interaction between sex and time is not ($p = 0.1865$).

13.5.3   Split-Plot ANOVA Models

It was noted in the discussion of two-way ANOVA models that the variables of sex and subj are not crossed. The variable subj is said to be "nested" within sex because certain levels of subj (S1, S2, and S3) appear only in one level of sex (F) and others (S4, S5, and S6) appear only in the other level of sex (M). The design of such an experiment fits into the framework of a split-plot model.

The subplot variable (time) typically has levels that are independent of one another. In our particular dataset, the measurements made at the different time points are related to one another within a subject. The good news is that a split-plot model sets up a compound symmetry variance structure that has one variance along the diagonal and one covariance term in all of the off-diagonal elements of the variance-covariance matrix. Thus, this type of model is appropriate for our dataset if the correlation (covariance) between two time points is the same as between any other two. Although this variance structure may seem implausible, it is quite often sufficient for such models in practice.

The code for a split-plot model in SAS is given for both PROC GLM and PROC MIXED. The advantage of using PROC GLM is that the expected mean squares are printed so that a check can be made that the correct error term is used for the test:

SAS code for a split-plot model

```
proc glm data=newdrug;
    class sex time subj;
    model response=sex time sex*time subj(sex);
    random subj(sex) / test;
proc mixed data=newdrug;
    class sex time subj;
    model response=sex time sex*time / s;
    repeated / subject=subj type=cs;
```

The SAS manual offers two approaches to the split-plot syntax to be used in PROC MIXED (SAS (1997, pp. 637/38)), both of which need to be modified to match our situation. The repeated solution appears in the code above and the appropiate random statement is random subj(sex).

To save space, only partial output from SAS is shown. The important thing to note is that sex is tested with the subj(sex) error term whereas time is tested with the residual error term (actually time*subj(sex) interaction). The four degrees of freedom for the subj(sex) term come from counting the degrees of freedom for subj within each level of sex (two for F and two for M):

SAS output for a split-plot model

```
General Linear Models Procedure
Source       Type III Expected Mean Square
SEX          Var(Error) + 4 Var(SUBJ(SEX))
                        + Q(SEX,SEX*TIME)
TIME         Var(Error) + Q(TIME,SEX*TIME)
SEX*TIME     Var(Error) + Q(SEX*TIME)
SUBJ(SEX)    Var(Error) + 4 Var(SUBJ(SEX))

Source: SEX *
Error: MS(SUBJ(SEX))
                      Denominator   Denominator
  DF  Type III MS          DF            MS F Value Pr > F
   1 60.801666667          4 3.1466666667 19.3226 0.0117
Source: TIME *
Error: MS(Error)
                      Denominator   Denominator
  DF  Type III MS          DF            MS F Value Pr > F
   3 16.369444444         12 2.5916666667  6.3162 0.0081
* - This test assumes one or more other fixed effects
    are zero.
Source: SEX*TIME
Error: MS(Error)
                      Denominator   Denominator
  DF  Type III MS          DF            MS F Value Pr > F
   3 4.9338888889         12 2.5916666667  1.9038 0.1829
Source: SUBJ(SEX)
Error: MS(Error)
                      Denominator   Denominator
  DF  Type III MS          DF            MS F Value Pr > F
   4 3.1466666667         12 2.5916666667  1.2141 0.3552
```

The S-PLUS syntax for a split-plot model differs little from that of the one- and two-way ANOVA models. The Error function is added to the model specification to indicate that another component of variance is being added for subj. The asterisk in S-PLUS notation is equivalent to sex + time + sex:time in S-PLUS and sex|time in SAS.

S-PLUS code and output for split-plot ANOVA

```
> nd.split.plot <- aov(response~sex*time+Error(subj),
  data=newdrug)
```

```
> summary(nd.split.plot)
    Error: subj
               Df Sum of Sq Mean Sq  F Value    Pr(F)
        sex    1  60.80167 60.80167 19.32256  0.01173
   Residuals   4  12.58667  3.14667

    Error: Within
               Df Sum of Sq  Mean Sq  F Value      Pr(F)
       time    3  49.10833 16.36944 6.316184  0.0081378
   sex:time    3  14.80167  4.93389 1.903751  0.1828514
   Residuals  12  31.10000  2.59167
```

The S-PLUS output is separated into two ANOVA tables, one for the between-subject error term (labeled Error: subj) and one for the within subject error term (labeled Error: Within). The fixed effects of sex, time, and sex:time appear in the appropriate ANOVA table according to which error term is used for the test of that effect. We see that the effect of sex is significant ($p = 0.0117$) as is that of time ($p = 0.0081$), results that were perhaps evident from the interaction plot.

It should be emphasized that this was the first statistical model fit to the data that was perhaps realistic. As stated earlier, the compound symmetry variance structure inherent to the split-plot model assumes that, within subjects, the correlation between time points are equal. A simple visual check can be done by computing the correlation matrix over time. Two other approaches are available; a Geisser–Greenhouse adjustment to the degrees of freedom could be used, or, a mixed effects model as in Laird and Ware (1982) could be used to compare the validity of various variance structures.

One additional note should be made about the variable time. In all these models, time has been treated as a factor variable but it could also be used as a continuous variable. As a continuous variable, time would only have one degree of freedom and would represent the linear effect of time. The interpretation of the effect of time certainly changes depending on whether it is coded as a factor or as a continuous variable.

13.6   Contrasts

As we have seen in the previous sections, unless you are simply searching for p-values, it is important to know the contrasts used to fit a model in order to be able to interpret the results. This section is devoted to explaining how to work with contrasts in S-PLUS and the equivalent in SAS.

To get a feeling for how contrasts work, it is instructive to work with an example where one of the factor variables has at least three levels. For this purpose, a simple dataset is constructed below and exported to SAS. The seed for the random number generator has been fixed so that the data and results can be reproduced. The simulation assumes three treatment groups: a placebo (P) and two doses of the treatment (T1 and T2). The response is normally distributed for all groups with means of 0, 2, and 3, and standard deviations of 3, 3, and 5 for the groups of P, T1, and T2, respectively. The sample size is 50 per group:

S-PLUS code for generating example data

```
> set.seed(7027)
> response <- c(rnorm(50, 0, 3), rnorm(50, 2, 3),
  rnorm(50, 3, 5))
> response <- round(response, 2)
> treat.f <- factor(c(rep("P", 50), rep("T1", 50),
  rep("T2", 50)))
> example <- data.frame(response, treat.f)
> write.table(example, file="filename", sep=" ",
  dimnames.write=F)
```

The code for writing the dataset to an ASCII file has been included for reference (write.table function).

To work with contrasts in SAS it is necessary to know the order that SAS has assigned to the levels of the factor variable. When in doubt, look at the first part of the SAS output where the values of the levels are printed in the order in which SAS uses them:

SAS code for contrasts

```
proc glm data=example;
     class treat_f;
     model response=treat_f;
     lsmeans treat_f / cl;
     contrast 'P-T1' treat_f 1 -1 0;
     contrast 'P-T2' treat_f 1 0 -1;
     contrast 'T1-T2' treat_f 0 1 -1;
```

Note that PROC MIXED can be used with exactly the same syntax as above by substituting "mixed" for "glm." PROC GLM has been used here because the output of the contrasts shows the sums of squares associated with each of the pairwise contrasts:

SAS output for contrasts

```
General Linear Models Procedure
Dependent Variable: RESPONSE
Contrast DF  Contrast SS  Mean Square F Value Pr > F
P - T1    1   77.19379600  77.19379600    6.42 0.0123
P - T2    1  127.66740100 127.66740100   10.62 0.0014
T1 - T2   1    6.31516900   6.31516900    0.53 0.4698
```

We see from the SAS output that the mean response in the placebo group is different from that in both of the treatment groups, but that the mean response does not differ between the two treatment groups (T1–T2). It is also evident that the sums of squares from the contrasts (77.19, 127.67, and 6.32, respectively) sum to more than the sum of squares for the overall treatment effect (140.78). This is partially because three single degree of freedom contrasts have been specified whereas the overall treatment effect has only two. However, if we choose the first two contrasts, the total sums of squares are 77.19 + 127.67 = 204.86 which is still greater that that of the overall treatment effect.

The contrasts used for pairwise comparisons, the type most often used in the pharmaceutical industry, are simply not orthogonal as seen in our example. (It should be noted that orthogonal contrasts can be constructed in SAS, but can be tricky with complicated designs.) Multiply the first 2 rows of the contrast coefficients elementwise, and sum them ($1*1+(-1)*0+0*(-1) = 1$). If the two contrasts had been orthogonal, the sum would have been zero. This point is the key to understanding contrasts in S-PLUS.

Before we continue with contrasts, it is worth mentioning that we would normally be penalized for making so many contrasts by making some form of a multiple comparison adjustment (e.g., Bonferroni). However, in the special case of 3 treatment groups we may adopt the following conditional approach to testing. Assuming that T2 represents the high-dose group and T1 the low-dose group, the comparison of interest is normally between placebo and T2. Specify that this contrast is of primary interest. If this contrast is significant, then the other two contrasts come "free of charge" (i.e., no adjustment to α is necessary). If the primary contrast is not significant, do not calculate the other contrasts. If the primary contrast is not significant in this framework, there is no reason to expect that the others will be; see Koch and Gansky (1996).

There are two solutions to performing pairwise (nonorthogonal) contrasts in S-PLUS. The first and much easier solution is to use the multicomp function. Not only can the multicomp function be used to perform multiple comparisons (Bonferroni, Tukey, Dunnett, Scheffé, etc.), but it can also be used to perform pairwise comparison contrasts (with no adjustment) as shown below. The second solution involves recoding the treatment groups to change their order of appearance and then using either user-defined or built-in contrast matrices (more on this approach later).

An example of the multicomp approach is shown here:

S-PLUS code and output for lsmeans and pairwise comparisons

```
> ex.lsmeans <- multicomp(ex.helmert, method="lsd",
  error.type="cwe", comparisons="none")
> par(mfrow=c(2, 1))
```

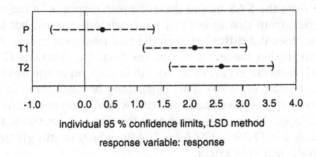

Figure 13.7. LSMean plots.

```
> plot(ex.lsmeans)
> ex.pair.lsd <- multicomp(ex.helmert, method="lsd",
  error.type="cwe")
> ex.pair.lsd
     95 % non-simultaneous confidence intervals for
     specified linear combinations,
     by the Fisher LSD method

     critical point: 1.9762
     response variable: response

     intervals excluding 0 are flagged by '****'

          Estimate Std.Error Lower Bound Upper Bound
     P-T1   -1.760    0.693      -3.13      -0.387 ****
     P-T2   -2.260    0.693      -3.63      -0.889 ****
     T1-T2  -0.503    0.693      -1.87       0.868
> plot(ex.pair.lsd)
```

The output includes the point estimates, standard errors, and confidence intervals for each of the three pairwise comparisons. Plots of the lsmeans and pairwise comparison contrasts are shown in Figures 13.7 and 13.8, respectively.

An indicator is used in the output to show which of the contrasts do not contain zero, but no *p*-values are given. The *p*-value should not be necessary given the confidence interval, but as it will be asked for anyway, a trick for calculating the *p*-values is shown below.

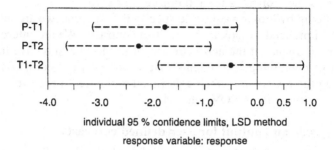

Figure 13.8. Pairwise comparison plots.

S-PLUS code and output for calculating p-values

```
> names(ex.pair.lsd)
    "table"        "alpha"       "error.type" "method"
    "crit.point"  "lmat"        "Srank"      "simsize"
    "ylabel"      "call"        "lmcall"
> ex.pair.lsd$table
          estimate    stderr       lower        upper
    P-T1   -1.7572  0.6934783  -3.127675  -0.3867251
    P-T2   -2.2598  0.6934783  -3.630275  -0.8893251
    T1-T2  -0.5026  0.6934783  -1.873075   0.8678749

> t.stats <- ex.pair.lsd$table[,1]/ex.pair.lsd$table[,2]
> t.stats
         P-T1        P-T2        T1-T2
    -2.533893  -3.258646  -0.7247523

> t.pvals <- 2*(1-pt(abs(t.stats),
  ex.helmert$df.residual))
> t.pvals
         P-T1          P-T2       T1-T2
    0.01232792  0.001390651  0.469756
```

Most intermediate and/or final results are stored with the fit of a statistical model and many other functions. The names function is used to see what components have been stored with a variable. By finding the appropriate components, it is easy to calculate the t-statistics and p-values. SAS has chosen to use F-values for the contrasts (pairwise comparisons) but the p-values are the same as those from S-PLUS. In the case of 1 numerator degree of freedom F-tests, $F = t^2$.

Contrasts in S-PLUS are regulated by contrast functions that specify the type of contrast to be used. User-defined contrasts can easily be constructed or one of four built-in S-PLUS contrasts can be used; contr.helmert, contr.poly, contr.sum, and contr.treatment.

We begin by looking at how to specify the contrasts we are interested in, and then proceed to look at the built-in contrasts. We are interested in the two comparisons of the active treatment to placebo and have to construct a matrix containing coefficients that specifies these comparisons. After creating the matrix, it has to be assigned to the treatment variable and can then be used in our ANOVA model:

S-PLUS code and output for user-defined contrasts

```
> contr.placebo <- matrix(c(-1,1,0,-1,0,1), nrow=3)
> contrasts(example$treat.f) <- contr.placebo(3)
> contrasts(example$treat.f)
       [,1] [,2]
   P    -1   -1
   T1    1    0
   T2    0    1
> ex.placebo <- aov(response~treat.f, data=example)
> summary(ex.placebo, split=list(treat.f=
  list(c1=1, c2=2)))
               Df Sum of Sq  Mean Sq  F Value       Pr(F)
    treat.f     2   140.784 70.39212 5.854884  0.00357648
 treat.f: c1    1    77.194 77.19380 6.420615  0.01232792
 treat.f: c2    1    63.590 63.59045 5.289153  0.02286634
  Residuals   147  1767.352 12.02280
```

The first column of the contrast matrix defines the comparison of the low dose to placebo and the second column defines the comparison of the high dose to placebo. We have split the output for treat.f into two pieces, c1 and c2, such that c1 contains the results of the contrast corresponding to the first column of the contrast matrix and c2 to those from the second column. The first contrast of our user-defined contrast matrix (contr.placebo) yields a sum of squares equal to 77.194, the same as SAS. The second contrast appears as though it is a comparison of "T2-P" and should give a sum of squares of 127.667, whereas the result from S-PLUS is 63.590. This is not what we were expecting! Note that the sum of the two contrast sums of squares (77.194 + 63.590) is equal to the sum squares of the main effect of treat.f of 140.784.

Note the following: with three levels of the factor of interest, the first contrast gives the expected result. The second contrast, on the other hand, gives the minimum of the true sum of squares and the remainder of the total sum of squares not accounted for by the first contrast. The implication is that you cannot rely on this method to produce results you might be expecting unless you are working with orthogonal contrasts.

To be on the safe side, it is a good idea to redefine the contrast matrix (by switching the order of the columns) such that the second contrast from the original definition is now estimated correctly.

To put the information into practice, we need to redefine the contrast matrix and redo the ANOVA model to get the results below:

S-PLUS code and output for user-defined contrasts – continued

```
> contr.t2.p <- matrix(c(-1,0,1,-1,1,0), nrow=3)
> contrasts(example$treat.f) <- contr.t2.p(3)
> contrasts(example$treat.f)
        [,1] [,2]
   P    -1   -1
   T1    0    1
   T2    1    0
> ex.t2.p <- aov(response~treat.f, data=example)
> summary(ex.t2.p, split=list(treat.f=list(c1=1, c2=2)))
```

	Df	Sum of Sq	Mean Sq	F Value	Pr(F)
treat.f	2	140.784	70.3921	5.85488	0.0035765
treat.f: c1	1	127.667	127.6674	10.61877	0.0013907
treat.f: c2	1	13.117	13.1168	1.09100	0.2979650
Residuals	147	1767.352	12.0228		

From this model, we see that the sum of squares for the contrast "T2-P," this time represented by c1, is 127.667, exactly what we got from SAS. Using this approach, we can specify any contrast we are interested in, whether or not it is orthogonal.

There are two default contrasts, Helmert (contr.helmert) for factor variables and polynomial (contr.poly) for ordered variables. The contrast matrices for these defaults, as well as the other built-in matrices, are printed below for reference:

S-PLUS built-in contrast matrices

```
> contrasts(example$treat.f) <- contr.helmert(3)
> contrasts(example$treat.f)
        [,1] [,2]
   P    -1   -1
   T1    1   -1
   T2    0    2
> contrasts(example$treat.f) <- contr.poly(3)
> contrasts(example$treat.f)
              .L          .Q
   P  -0.7071068   0.4082483
   T1  0.0000000  -0.8164966
   T2  0.7071068   0.4082483
> contrasts(example$treat.f) <- contr.sum(3)
```

```
> contrasts(example$treat.f)
      [,1] [,2]
   P    1    0
   T1   0    1
   T2  -1   -1
> contrasts(example$treat.f) <- contr.treatment(3)
> contrasts(example$treat.f)
       2 3
   P   0 0
   T1  1 0
   T2  0 1
```

The Helmert contrasts are orthogonal and it so happens that the first column gives one of our comparisons of interest "T1-P." The second column of the Helmert contrasts compares T2 to the average of T1 and P, and has little meaning in this context. The polynomial contrasts are also orthogonal and the first column gives our other comparison of interest "T2-P" (the actual value of the coeffecients in the matrix are relative and effect the estimate, not the p-value). The fact that the first column of this contrast is also the linear contrast can be ignored in our setting. As well as being the quadratic contrast, the second column of the polynomial contrasts compares T1 to the average of P and T2 in our setting. The contrast function contr.sum is close to what we would like to use, except that the position of P and T2 are reversed. As defined, the first column gives us "T2-P" and the second column compares T1 to T2. The columns of the treatment contrasts, in addition to not being orthogonal to one another, do not even define proper contrasts and their interpretation must be made with care.

In this section, we have looked at how contrasts can be used in both SAS and S-PLUS. We have also seen that S-PLUS prefers to deal with orthogonal contrasts, whereas in the pharmaceutical industry, nonorthogonal contrasts are often of main interest. Easy solutions for obtaining non-orthogonal contrasts exist in S-PLUS, but rely on knowledge of contrasts and their implication on the fit of an ANOVA model. This knowledge is taught in standard ANOVA courses, but is quickly forgotten. Pairwise contrasts can easily be constructed in S-PLUS using the multicomp function.

13.7   Summary

A review of the main concepts in ANOVA has been given so that a comparison can be made between the use of SAS and the use of S-PLUS for fitting these models. SAS has been the standard for statistical software used in the pharmaceutical industry for decades. Thanks to the gradual acceptance of other software packages by regulatory agencies, more and more statisticians are beginning to use other packages, especially S-PLUS, at least for certain types of analyses.

ANOVA models comprise a large percentage of the statistical models used by industry statisticians, performed mostly in SAS through years of familiarity. A first attempt at running ANOVA models in S-PLUS probably resulted in frustration because the model parameters "were not the same as SAS so they must be wrong."

This chapter has shown through syntax and output comparisons that SAS and S-PLUS can give the same results. The assumptions have been reviewed and the interpretations clarified so that confusion with ANOVA models in S-PLUS should no longer exist.

Related topics of lsmeans and contrasts have been compared between the two software packages. In addition, several special graphical functions for ANOVA models in S-PLUS have been shown.

After reading this chapter, you should be able to obtain from S-PLUS the results that you are used to seeing in SAS. In short, there's no reason not to use S-PLUS for ANOVA models, and a few reasons you should.

13.8   References

Koch, G. G., and Gansky, S. A. (1996). Statistical considerations for multiplicity in confirmatory protocols. *Drug Information Journal* **30**, 523–534.

Laird, N. M., and Ware, J. H. (1982). Random-effects for longitudinal data. *Biometrics* **38**, 963–974.

SAS (1997). *SAS/STAT® Software: Changes and Enhancements through Release 6.12*. SAS Institute Documentation, SAS Institute.

14

Permutation Tests for Phase III Clinical Trials

Vance Berger
U.S. Food and Drug Administration, Washington, DC, USA

Anastasia Ivanova
University of North Carolina, Chapel Hill, NC, USA

The opinions and assertions expressed in this chapter are the private views of the authors. No endorsement by the Food and Drug Administration or the University of North Carolina at Chapel Hill is intended or should be inferred.

14.1 Introduction

The general goal of clinical research is to determine which medical interventions are most beneficial for which patient populations. There are typically four phases of drug development. In this chapter we focus on Phase III clinical trials, which are comparative in nature, and constitute the definitive step in the evaluation of a new treatment, which can be a drug, vaccine, device, or screening test. The purpose of a Phase III trial is to determine the safety and efficacy of the new treatment relative to a standard treatment or placebo.

If a new treatment appears to be effective, then either the new treatment actually is effective, or else a Type I error (see the Glossary in Appendix 14.A.1) may have occurred in which an ineffective treatment is mistakenly found to be effective. To attribute the observed treatment effect to the treatment, one needs to control the Type I error rate, which should be done in two ways. First, steps need to be taken to ensure that biases do not inflate the Type I error rate. Second, an exact test needs to be used so that the actual Type I error rate is no larger than the nominal Type I error rate. The double-blind randomized clinical trial is the design of choice for minimizing the effect of biases and providing the strongest evidence with which to establish the efficacy of a treatment (Chalmers et al., 1983; Vickers et al., 1997; Ederer, 1975; Sacks et al., 1982; Simon, 1982).

With the random allocation of patients to treatment groups and the masking of treatment identities from both patients and investigators, the biases that can invalidate a one-arm or nonrandomized study would be expected to affect all treatment groups comparably. In fact, treatments have been reported to be effective based on nonrandomized studies, only to be found ineffective, or even harmful, by subsequent randomized clinical trials (Vickers et al., 1997; Placebo Working Group, 1996). Even in a randomized clinical trial, however, selection bias can invalidate the results. Consequently, steps need to be taken to minimize the effect of selection bias. See Berger and Exner (1999) for strategies for controlling the effects of selection bias. Once biases have been minimized, the key remaining issue is preservation of the Type I error rate.

14.1.1   Preservation of the Type I Error Rate

Before one can control the Type I error rate, one must define it. Consider the p-value obtained from answering the question: What is the probability of finding a result such as the observed result, or a more extreme one, if one were to sample randomly from a normal distribution? This p-value has little to do with the extent to which the observed result is extreme among the other results that could have been obtained from the experiment actually being conducted, because the experiment actually conducted did not involve random sampling from a normal distribution.

Put differently, the Type I error rate of interest concerns not the probability of a Type I error computed under a sampling model (e.g., the normal distribution) that was NOT used in the conduct of the trial (Ludbrook and Dudley, 1998). Rather, interest centers on the probability of a Type I error computed under a randomization model that mimics the random allocation of patients to treatment groups that was used during the conduct of the study. The only way to preserve this Type I error rate is to enumerate the permutation sample space or the set of outcomes that could have occurred in the trial, to find the null probability of each based on its likelihood of occurrence in the study, and to perform a design-based permutation test (Berger, 2000).

The p-value of interest, then, is obtained from answering the question: What is the probability of finding a result as extreme, or a more extreme one, as the observed result if one were to repeat hypothetically the study, under the assumption that the active treatment is no better than the control treatment? We will consider only the case of a 1:1 allocation to each treatment group without blocking (except that we do allow for stratification in Section 14.4). Permutation tests may be conservative, meaning that their actual size may be less than the nominal size of α. However, if the permutation test reflects the design (including blocking when blocking was used in the design), then it will guarantee validity. Validity can be expressed as strict preservation of the Type I error rate, or a null distribution of the p-value that is stochastically no smaller than uniform on [0, 1], or for any $0 < k < 1$, $P\{p < k\} \le k$ (Edgington, 1995). In this chapter, we will discuss the mechanics of conducting a permutation test in S-PLUS.

14.1.2   Ensuring Good Power to Detect All Types of Meaningful Treatment Effects

Once the permutation sample space is constructed (see Sections 14.2.3, 14.3.3, and 14.4.3), the next step is to determine the test statistic. Subject to preserving the Type I error rate, one would like the best chance for identifying effective treatments, so statistical power becomes the objective. Except for the simplest problems (technically, when there is monotone likelihood ratio in some statistic (Berger, 1998)), there is no uniformly most powerful test, because each type of treatment effect may potentially be best detected by a different locally most powerful test. In practice, the type of treatment effect would not be known until completion of the study. For robustness, then, we propose that the alternative hypothesis be made explicit as the set of all types of treatment effects that are worth detecting, as opposed to the more common formulation which is only the single type of treatment effect that is considered most likely.

It is not always obvious how to delineate what exactly constitutes a treatment effect that is worth detecting, so this is a nontrivial but important step. With a broad alternative hypothesis, one would like to use a test that is constructed to offer good power over the entire alternative hypothesis. For any outcome in the permutation sample space, an extreme region can be constructed as the set of outcomes as extreme or more extreme (offering more evidence against the null hypothesis in favor of the alternative of interest) as the one in question. The p-value for a given outcome is then the null probability of its extreme region. Extremity is measured by the value of the test statistic.

14.1.3   Organization of the Chapter and of a Typical Section

In Section 14.2, we consider a single binary response variable. In Section 14.3, we present a single 2×3 contingency table, with two rows (one per treatment group) and three ordered columns (one per outcome level). The methods extend naturally, however, to problems with more than three columns. In Section 14.4, we present a pair of 2×2 contingency tables, as would result from a study with a single binary endpoint when the analysis calls for stratification. Again, the methods extend naturally to problems in which there are more than two tables.

For each problem we will first present a real data example. As our work is of a proprietary nature, these examples will be drawn from the literature. This should not give the impression that the problems we treat are in any way contrived. Rather, these problems are encountered frequently in actual studies.

After an example is presented, we will briefly discuss some (but not necessarily all) of the tests that are currently used for the data structure in question. Some of these tests rely on unrealistic or inappropriate assumptions, such as treating an ordered categorical variable as binary or continuous, or assuming a common odds ratio across strata for a stratified analysis of a binary endpoint. We prefer to use permutation tests that require no assumptions (Berger, 2000). To this end, we will discuss the degrees of freedom and the permutation sample space (which is independent of the choice of the particular test). The alternative hypothesis and notion of extremity will then be developed. Based on this, we will develop specific permutation tests. Finally, we apply these tests to the example that was presented at the beginning of this section. Some of the permutation tests we develop have not appeared in the literature previously and are introduced here as novel tests. It is beyond the scope of this chapter to provide the theory behind these tests, but much of this theory can be found in Berger (1998) and Berger et al. (1998).

14.2   A Single 2×2 Contingency Table

14.2.1   Example Presented

Ondansetron (OND) was compared to OND + dexamethasone + chlorpromazine (ODC) for the prevention of nausea and vomiting associated with multiple-day cisplatin therapy (Fox et al., 1993). There were 10/22 antiemetic responses in the OND arm, versus 19/22 responses in the ODC arm. With a single binary endpoint, the data are best presented as a 2×2 contingency table. The two rows correspond to the two treatment groups (by convention, we display the control group first, then the investigational treatment), and the two columns correspond to the two outcome levels (by convention, we display the better one last).

Input of the data in S-PLUS:

```
> ond <- c(12, 10)
> odc <- c(3, 19)
> d <- rbind(ond, odc)
> d
     [,1]  [,2]
ond   12    10
odc    3    19
```

Table 14.1. Antiemetic response data after 2 days (Fox et al., 1993).

	Level of response attained		
	None	Some	Total
OND	12	10	$N_1 = 22$
ODC	3	19	$N_2 = 22$
Total	15	29	$N = 44$

14.2.2   Common Tests

Pearson's chi-square test is frequently used but suffers from the drawback that its actual size (significance level) depends on the cell probabilities, and can exceed the nominal significance level (typically 0.05 in Phase III clinical trials). Upton (1982) presented 22 different tests for analyzing 2×2 contingency tables, and did not seem to be particularly fond of Fisher's exact test. However, Upton (1992) later reversed his position and endorsed Fisher's exact test because only Fisher's exact test preserves the conditional (on the margins) Type I error rate. Were one to construct a permutation test based on the chi-square test statistic, then the resulting test would be equivalent to the two-sided Fisher's exact test.

14.2.3   Degrees of Freedom and the Permutation Sample Space

With fixed margins, a 2×2 contingency table has only one degree of freedom. For example, if we know the upper-left cell count, then we can calculate the remaining cell counts by subtraction. Consequently, we can refer to a 2×2 contingency table in the sample space by its upper-left cell count. The permutation sample space is indexed by the upper-left cell count, and is generated by letting this upper-left cell count vary over its range.

14.2.4   Alternative Hypothesis and Definition of Extremity

Letting $P(OND)$ and $P(ODC)$ be the population response rates of OND and ODC, respectively, we consider the one-sided alternative hypothesis $P(ODC) > P(OND)$. As the upper-left cell count is the number of control patients who did not respond, larger numbers are more extreme than smaller numbers. Linearly ordering the sample space by the upper-left cell count provides the same ordering on the sample space as the difference in observed response rates, $p(ODC) - p(OND)$. While it is this difference in response rates, and not the upper-left cell count, that should be presented in the study report as the more interpretable estimate of the magnitude of the treatment effect, basing the analysis on the upper-left cell count sets the stage for the analyses to be presented in the forthcoming sections.

14.2.5   Specific Tests

We will pursue only Fisher's exact test, which is based on enumerating all the 2×2 contingency tables with the same margins as the observed 2×2 contingency table. These tables are then sorted by extremity (the upper-left cell count). The p-value is the null probability of the extreme region.

14.2.6   Example Worked

We present the data in Table 14.1 succinctly as (12, 10; 3, 19). The upper-left cell, observed to be 12, could have been any integer between 0 and 15. This gives a permutation sample space with 16 distinct points, all with row margins (22, 22) and column margins (15, 29). Not all of these 16 points are equally likely (each corresponds to a possibly different number of the $\binom{44}{22}$ ways to pick 22 patients to receive OND). The 16 possible 2×2 contingency tables are listed in Table 14.2. In addition, we list corresponding hypergeometric null probabilities, cumulative null probabilities, the mid-p-value (Berry and Armitage, 1995), a designation indicating whether or not each table is in the rejection region (at $\alpha = 0.05$), a designation indicating whether or not each table is in the extreme region, and the chi-square p-value.

The built-in S-PLUS function dhyper can be used to generate probabilities in column 2 of Table 14.2. For example, the first row of the table is obtained from dhyper(0,22,22,15), the second from dhyper(1,22,22,15), etc. The entire second column, a vector of null probabilities, nullprob, of all the tables in the reference set, can be obtained by

```
> nullprob <- dhyper(0:15, 22, 22, 15)
```

Note that the null probability for (0, 22; 15, 7) is the same as the null probability for (15, 7; 0, 22). Such symmetry is observed because the hypergeometric null probability depends on the 2×2 contingency table only through its unordered set of cell counts, and not on the location in the table of these cell counts. The cumulative probability in the third column can be calculated by calling either of the built-in S-PLUS functions cumsum or phyper. Since the most extreme table is (15, 7; 0, 22), the cumulative probability is calculated starting from the end and then gets inverted:

```
> cumprob <- cumsum(nullprob[16:1])[16:1]
> cumprob <- phyper(15:0, 22, 22, 15)
```

The mid-p-value in column four is a standard p-value (cumulative probability) minus half the null probability of the table:

```
> mid.p.value <- cumprob - nullprob/2
```

The one-sided p-value for Fisher's exact test is the null probability of the one-sided extreme region. For observed table (12, 10; 3, 19), the one-sided extreme region consists of the last four tables listed, with null probability 0.0048625. The two-sided p-value for Fisher's test can be obtained by applying the built-in S-PLUS function fisher.test. For example, the two-sided p-value for observed table (12, 10; 3, 19) is simply twice the one-sided p-value, or 0.0097:

```
> fisher.test(d)$p.value
```

Table 14.2. Dichotomized antiemetic response data after 2 days (Fox et al., 1993).

Hypothetical Table	Null Prob	Cum Prob (Fisher's exact test p-value, one-sided)	Mid-p-value	Reject region at $\alpha = 0.05$	One-sided extreme region for observed outcome (12,10;3,19)	χ^2 p-value, one-sided
(0, 22; 15, 7)	0.000001	1.000000	1.000000	Fail to reject:	Less extreme than observed table (12,10;3,19)	0.999996
(1, 21; 14, 8)	0.000031	0.999999	0.999984			0.999932
(2, 20; 13, 9)	0.000500	0.999969	0.999719	This		0.999265
(3, 19; 12, 10)	0.004331	0.999469	0.997303	region has null		0.994525
(4, 18; 11, 11)	0.022444	0.995138	0.983915	prob 0.073		0.971820
(5, 17; 10, 12)	0.074067	0.972693	0.935660			0.898342
(6, 16; 9, 13)	0.161427	0.898626	0.817913			0.737638
(7, 15; 8, 14)	0.237199	0.737199	0.618600			0.500000
(8, 14; 7, 15)	0.237199	0.500000	0.381400			0.500000
(9, 13; 6, 16)	0.161427	0.262801	0.182087			0.262362
(10, 12; 5, 17)	0.074067	0.101374	0.064340			0.101658
(11, 11; 4, 18)	0.022444	0.027307	0.016085	Reject:		0.028180
(12, 10; 3, 19)	0.004331	0.004862	0.002697	This region	As or more extreme than observed table (12,10;3,19)	0.005475
(13, 9; 2, 20)	0.000500	0.000531	0.000281	has null		0.000735
(14, 8; 1, 21)	0.000031	0.000031	0.000016	prob 0.027		0.000068
(15, 7; 0, 22)	0.000001	0.000001	0.000000			0.000004
	1.000000					

The chi-square test, which is inherently two-sided, is made one-sided by cutting the p-value in half for the last eight tables, and using one minus half the p-value for the remaining tables. For example, the p-value in the last column and the last row of the table is obtained from

```
> chisq.test(matrix(c(15, 7, 0, 22), nrow = 2,
    byrow = T))$p.value/2
```

The p-value in the last column and the first row of the table is obtained from

```
> 1 - chisq.test(matrix(c(0, 22, 15, 7),
    nrow = 2, byrow = T))$p.value/2
```

Fisher's exact test is often criticized as being overly conservative. For the dataset presented, the one-sided 0.05 critical region consists of the last five entries of Table 14.2. The actual significance level is the null probability of the critical region, or 0.027, which is obviously less than the intended 0.05. However, what is rarely appreciated is the fact that the chi-square test can also be conservative (as well as anticonservative). For the dataset presented, the chi-square test has the same critical region and, consequently, the same actual significance level, 0.027, as Fisher's exact test. In fact, because the chi-square p-value exceeds the Fisher p-value for each outcome in the critical region, one might say that the chi-square test is even more conservative than Fisher's exact test in the following sense. Had the nominal significance level been 0.027, Fisher's exact test would not be conservative, while the chi-square test would have actual significance level 0.004862. The same phenomenon is observed for any other attainable significance levels, including 0.101374 and 0.004862.

Figure 14.1 illustrates the null distribution of the upper-left cell count, C_{11} (which stands for the 1, 1 cell), as in the second column of Table 14.1. Figure 14.1 is obtained using a built-in function barplot(). The x-coordinates of the bar centers are saved in the vector x:

```
> x <- barplot(nullprob, xlab = "C11", ylab = "pdf(C11)")
> text(x, nullprob + 0.0055, round(nullprob, digits=3),
    cex = 1)                    # label the tops of the bars
> text(x, -0.01, (0:15))        # label horizontal axis
```

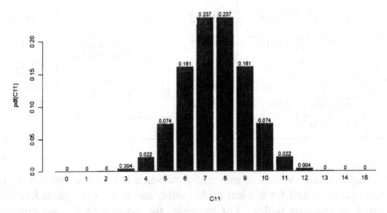

Figure 14.1. Null probabilities of each table in the sample space, as a function of the upper-left count C_{11}. Row margins are $N_1 = N_2 = 22$, column margins are $T_1 = 15$, $T_2 = 29$.

The numbers above the bars were rounded to three digits. Hence, zeros above the bars indicate that the null probability is ≤ 0.0005.

Figure 14.2 illustrates the cumulative distribution function (CDF) of the upper-left cell count, as in the third column of Table 14.1. This was obtained by

```
> plot((0:15), cumprob, xlab = "C11",
    ylab = "cdf(C11)", lab = c(15, 5, 0))
> lines(c(-1, 10.7), c(0.05, .05))
```

```
> text(-1.6, 0.05, "alpha=0.05", cex = .7)
> lines(c(-1, 12), c(0.005, .005), lty = 3)
> text(-1.6, 0.005, "p=0.005", cex = .7)
> lines(10:11, cumprob[11:12], lty = 3)
> lines(c(12, 12), c(-1, 0.005))    # vertical line observed
> lines(c(10.7, 10.7), c(-.1, 0.05))    # vertical line
> text(c(0:15), -0.11, c(0:15))    # label horizontal axis
> text(-1.2, seq(0.2, 1, 0.2), seq(0.2, 1,0.2))
> # label vertical axis
```

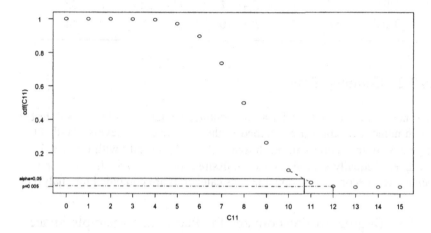

Figure 14.2. Cumulative distribution function of the upper-left cell count C_{11}.

14.3   A Single 2×3 Ordered Contingency Table

We now consider a 2×3 contingency table with two treatment groups and three ordered outcome levels.

14.3.1   Example Presented

For the data presented in Section 14.2, each patient was classified as a responder or a nonresponder only. In actuality, the response of each responder was further classified as either complete or partial. The full 2×3 contingency table appears in Table 14.3. Notice that if the partial response and complete response categories are combined, then the result is the 2×2 contingency table (12, 10; 3, 19) as given in Section 14.2. Unlike as in Section 14.2, though, there are now numerous reasonable tests that preserve the conditional Type I error rate, and generally there is no most powerful test among these. Let $C_1 = (C_{11}, C_{12}, C_{13})$ and $C_2 = (C_{21}, C_{22}, C_{23})$ be the pair of multinomial random vectors with cell probabilities

$\pi_1 = (\pi_{11}, \pi_{12}, \pi_{13})$ and $\pi_2 = (\pi_{21}, \pi_{22}, \pi_{23})$. The row margins are fixed by design (product multinomial sampling).

Table 14.3. Antiemetic response data after 2 days (Fox et al., 1993).

	Level of response			
	None	Partial	Complete	Total
OND	$C_{11} = 12$	$C_{12} = 3$	$C_{13} = 7$	$N_1 = 22$
ODC	$C_{21} = 3$	$C_{22} = 7$	$C_{23} = 12$	$N_2 = 22$
Total	$T_1 = 15$	$T_2 = 10$	$T_3 = 19$	$N = 44$

14.3.2   Common Tests

The most common analysis for a 2×3 contingency table is a linear rank test, for which numerical scores are assigned to the three response levels. Without loss of generality the scores can be chosen as $(0, v, 1)$, usually with $0 \le v \le 1$. For example, if equally spaced scores are desired, then $v = 0.5$. If $v = 0$ or $v = 1$, then the test becomes a binomial test, as categories are combined for analysis.

14.3.3   Degrees of Freedom and the Permutation Sample Space

Under the strong null hypothesis that the outcome depends on the patient only, and not on the treatment group, the column margins $\mathbf{T} = (T_1, T_2, T_3)$ are fixed. With fixed margins, all cell counts can be calculated by subtraction with knowledge of only C_{11} and C_{12}. The sample space is then two-dimensional (two degrees of freedom), and can be enumerated by letting C_{11} and C_{12} range independently over their respective potential ranges ($C_{11} \le \min\{T_1, N_1\}$ and $C_{12} \le \min\{T_2, N_1\}$), calculating the remaining four cell counts by subtraction, and deleting entries with negative cell counts. For example, the permutation sample space Ω for the data in Table 14.3 has 164 points and is obtained as follows:

```
> col.m <- c(c11 + c21, c12 + c22, c13 + c23)
> row.m <- c(c11 + c12 + c13, c21 + c22 + c23)
> v1 <- rep(0:col.m[1], col.m[2] + 1)
> v2 <- rep(0:col.m[2], rep(col.m[1] + 1, col.m[2] + 1))
> v3 <- row.m[1] - v1 - v2
> w1 <- col.m[1] - v1
> w2 <- col.m[2] - v2
> w3 <- col.m[3] - v3
> Omega <- cbind(v1, v2, v3, w1, w2, w3)
> Omega <- Omega[v3 >=0 & w3 >=0, ]
```

To plot the permutation sample space we use

```
> plot(Omega[ ,1], Omega[ ,2])
```

14.3.4   Alternative Hypothesis and Definition of Extremity

We share the view of McKinlay and Marceau (1999) that interventions are most beneficial to public health when they affect the entire population (of patients exposed to it), rather than just the sickest patients. As such, we do not consider a formulation couched in terms of either the complete response rate or the overall (complete or partial) response rate to be sufficient. Rather, we look for a shift of the entire ODC distribution to the right (better response outcomes) relative to the OND distribution. This leads us to test for independence against the one-sided alternative hypothesis that the ODC response distribution is stochastically larger than the OND distribution:

$$H_0: \quad \pi_1 = \pi_2$$

$$H_A: \quad \pi_{11} \geq \pi_{21}, \quad \pi_{11} + \pi_{12} \geq \pi_{21} + \pi_{22}, \quad \pi_1 \neq \pi_2.$$

The test statistic for a linear rank test is the difference in means, which is equivalent to using $C_{11} + (1 - v)C_{12}$, rejecting H_0 for large values of this test statistic. As v is varied, the extreme region for each resulting linear rank test can be found by plotting the sample space (C_{11} against C_{12}) and then simply drawing a line through the observed outcome, with the slope of the line determined by v as $1/(v-1)$ if v is not 1, or a vertical line if $v = 1$. If $v = 1$, then the test rejects H_0 for large values of C_{11} or, equivalently, for large values of $C_{22} + C_{23}$. This test is equivalent to Fisher's exact test on combined response categories discussed in Section 14.2. If $v = 0$, then the test rejects H_0 for large values of $C_{11} + C_{12}$ or, equivalently, for large values of C_{23}. This is equivalent to Fisher's exact test based on the complete response rate.

14.3.5   Specific Tests

In addition to the three most common linear rank tests ($v = 0, 0.5, 1$), we will also consider three nonlinear rank tests. These are the Smirnov test (Berger et al., 1998), which is based on the largest difference between the empirical CDFs and has piecewise linear boundaries of the rejection and extreme regions, the convex hull test (Berger et al., 1998), which is based on directional convex hull peeling of the permutation sample space, and the adaptive test (Berger, 1998), which essentially picks v based on the observed data, then adjusts so as to preserve the overall Type I error rate.

14.3.6   Example Worked

Figure 14.3 shows the extreme regions of the most common linear rank tests (v = 0, 0.5, 1), plus the Smirnov test, and can be obtained by

```
> ttestpv(12, 3, 7, 3, 7, 12)
```

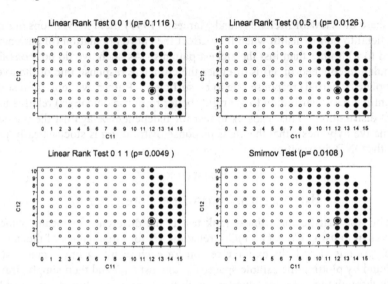

Figure 14.3. Extreme regions for three linear rank tests and the Smirnov test for (12, 3, 7; 3, 7, 12) with the observed data point circled.

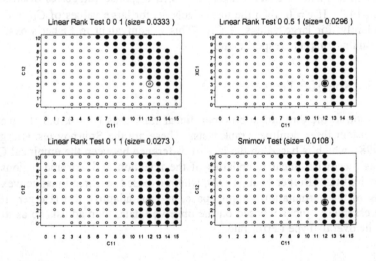

Figure 14.4. Rejection regions for three linear rank tests and the Smirnov test for (12, 3, 7; 3, 7, 12) with the observed data point circled.

The function `ttestpv` is written in S-PLUS by the authors and is available on the companion web page for this book, as are the other functions used in this chapter. The sum of the null probabilities of all the points in the extreme region is equal to a p-value, hence the name of the S-PLUS function. We find p-values of 0.1116 ($v = 0$), 0.0126 ($v = 0.5$), 0.0049 ($v = 1$), and 0.0108 (Smirnov). Figure 14.4 shows the rejection regions ($\alpha = 0.05$) of these same tests, and can be obtained using the function `ttestsize` written by the authors:

```
> ttestsize(12, 3, 7, 3, 7, 12)
```

The sum of the null probabilities of the points in the rejection region is the actual size of the test. Figure 14.5 plots the p-value of the linear rank test with middle score v against v, with v ranging in the unit interval (rather than over the entire real line). See Kimeldorf et al. (1992) for details regarding the interpretation of this type of figure. We see that $p < 0.05$ if $0 < v \leq 1$, and $p < 0.025$ if $0.25 < v \leq 1$. The smallest attainable p-value of 0.0048 resulting from a linear rank test with $0 \leq v \leq 1$ is attained when $0.87 \leq v < 1$. Note that this is consistent with the fact that the MP test, to detect the alternative for which the population cell probabilities match the observed ones, would be the linear rank test with middle score equal to 1.04 (Berger et al., 1998).

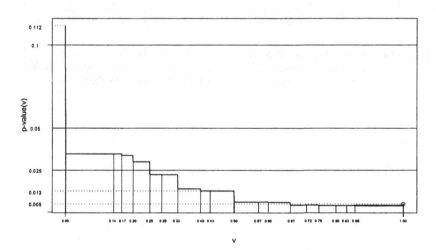

Figure 14.5. The p-value as a function of v in the unit interval.

Figure 14.6 is produced by the function `chull.test` written in S-PLUS by the authors:

```
> chull.test(12, 3, 7, 3, 7, 12)
```

The figure shows the derivation of the convex hull test. Berger et al. (1998) showed that for a test to not be severely biased, it would need to reject the null

362   V. Berger and A. Ivanova

hypothesis on each of the three points marked with number 12. This is the idea
behind the convex hull test. The number 12 was chosen arbitrarily, as all that
matters is the order induced on the sample space by the test statistic. The three
points marked 12 are the most extreme points, as they are the directed extreme
points of the sample space (Berger, 1998; Berger et al., 1998). Once these three

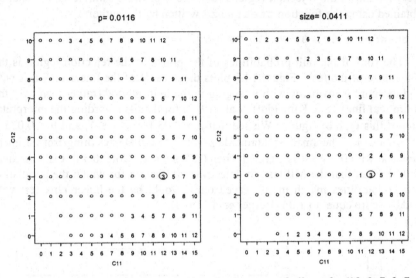

Figure 14.6. Extreme and rejection regions for the convex hull test for (12, 3, 7; 3, 7,
12) with the observed data point circled. Points that belong to the region are marked
with the values of the test statistic (somewhat arbitrary) assigned by the test.

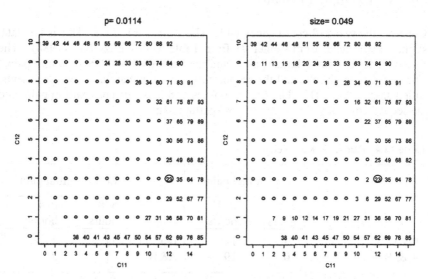

Figure 14.7. Extreme and rejection regions for the adaptive test for (12, 3, 7; 3, 7, 12) with the observed data point circled. Points that belong to the region are marked with the values of the test statistic (somewhat arbitrary) assigned by the test.

points are peeled into the critical region, the points that become directed extreme points of the remaining subset of the sample space are most extreme, and are peeled into the critical region next. These are the points marked 11. The convex hull peeling continues until just before the critical region has a null probability that exceeds the nominal significance level. The first plot gives the extreme region ($p = 0.0116$) and the second plot gives the rejection region at $\alpha = 0.05$ (with conservatism, the actual significance level is $\alpha = 0.0411$).

The results of the adaptive test (Berger, 1998) are shown in Figure 14.7, produced using the following command:

```
> adaptive.test(12, 3, 7, 3, 7, 12)
```

The test yields $p = 0.0098$. The adaptive test can be performed using the function adaptive.test, written by the authors, and is based on the smallest p-value attainable, for any given point in the sample space, as all linear rank tests are considered.

14.4   A Pair of Stratified 2×2 Contingency Tables

In this section, we analyze a pair of 2×2 tables, resulting from a study with a single binary endpoint when the analysis calls for stratification (e.g., by gender).

14.4.1   Example Presented

Consider the two 2×2 contingency tables shown in Table 14.4. This is the data, slightly rearranged, from Center #1 from Table 3 of Kuritz et al. (1988). The first table is for "some initial pain," and the second is for "lots of initial pain." The first row is placebo and the second row is a combination of two treatments, called simply "A & B." The first column is 5–8 hours of post-partum pain, and the second column is 0–4 hours of post-partum pain.

Table 14.4. Pain data, Kuritz et al. (1988).

	Some initial pain			Lots of initial pain	
	5–8 hours	0–4 hours		5–8 hours	0–4 hours
Placebo	4	14	Placebo	11	15
A & B	0	19	A&B	1	25

14.4.2   Common Tests

Some of the common tests for this problem are presented by Miller (1980). The Mantel–Haenszel test is based on a ratio, with the sum of the upper-left cell counts minus their null expectations as the numerator and the square root of the sum of their variances as the denominator. This test is typically not a permutation test, but there is no reason why a permutation test cannot be conducted using this same test statistic. However, because the null expectations and variances depend on the data through only the margins, which do not vary with different permutations, great simplification is possible, and the test can be conducted by simply using the sum of the upper-left cell counts as the test statistic. A sign test is also possible, in which each table contributes only through the sign of the estimated log-odds ratio (positive or negative, with zeros removed prior to analysis). This method loses substantial information by ignoring the magnitude of effect in each table. With only two tables, as we consider, the lowest attainable p-value would be $0.5 \times 0.5 = 0.25$. Finally, it is possible to combine individual p-values, one per table, with Fisher's method.

14.4.3   Degrees of Freedom and the Permutation Sample Space

With fixed margins, each of the 2×2 tables has one degree of freedom. The two tables together then have two degrees of freedom. As the upper-left cell counts of the two tables vary independently, the sample space can be generated as a product space, with each upper-left cell count varying as in Section 14.2.3.

14.4.4   Alternative Hypothesis and Definition of Extremity

As in Section 14.2, large counts in the upper-left cells are indicative of a treatment effect in the direction of interest. The most extreme outcomes are then those with the largest upper-left cell counts in each table. It is clear, then, that one outcome (pair of tables) is more extreme than another if the first has a larger upper-left cell count than the other in each table. However, it is unclear how to compare different outcomes (pairs of tables) for which the upper-left cell count is larger for one table and the other upper-left cell count is larger for the other table (e.g., the upper-left cell count is larger for table 1 in pair 1 but larger for table 2 in pair 2). Consequently, only a partial ordering on the sample space is obvious, and many complete orderings on the sample space will seem intuitive and will not violate the partial ordering. We are interested in detecting any type of positive treatment effect, which means that we want good power any time each odds ratio (θ_1 in the first table, θ_2 in the second table) exceeds or equals 1 and at least one of them strictly exceeds 1. This corresponds to requiring a benefit in at least one strata, and no harm in either strata, and is formalized as follows:

$$H_0: \theta_1 = \theta_2 = 1$$

$$H_A: \max(\theta_1, \theta_2) > 1 \; ; \; \min(\theta_1, \theta_2) \geq 1$$

14.4.5 Specific Tests

The test statistic used by the permutation version of the Mantel–Haenszel test is linear in the cell counts. Consequently, we will refer to this as the linear rank test. We will also develop two nonlinear rank tests, the Smirnov test (based on the larger of the two between-group success rates and 0) and the convex hull test (based on directional convex hull peeling of the permutation sample space). These two nonlinear rank tests are introduced here as novel tests, but are close analogues to the tests bearing the same names presented in Section 14.3.

14.4.6 Example Worked

Using notation we introduced in Section 14.2, the observed tables are $(x_{11}, x_{12};$ $x_{21}, x_{22}) = (4, 14; 0, 19)$ and $(y_{11}, y_{12}; y_{21}, y_{22}) = (11, 15; 1, 25)$. The permutation sample space, Ω, has $5 \times 13 = 65$ points, so for brevity we again rely on the Ω-plot (Figure 14.8) instead of a tabulation. The Ω-plot is obtained by

```
> minx <- min(x11 + x12, x11 + x21)
> miny <- min(y11 + y12, y11 + y21)
> v1 <- rep(0:minx, miny + 1)
> v2 <- as.vector(t(matrix(rep(0:miny, minx+1),
    ncol = (minx + 1))))
> Omega <- cbind(v1, v2)
```

The observed point, $\{(4, 14; 0, 19), (11, 15; 1, 25)\}$, is quite extreme, falling just below the most extreme point, which lies at the upper-right corner of Γ, $\{(4, 14; 0, 9), (12, 14; 0, 26)\}$. The p-values are $p = 0.0001$ (linear rank test), $p = 0.001$ (Smirnov test), and $p = 0.0001$ (convex hull test). In fact, the extreme regions for the linear rank and convex hull tests coincide. However, the linear rank test is much more conservative than either the Smirnov or the convex hull tests, with actual significance level 0.0229, compared to 0.0487 and 0.0446, respectively. In fact, the linear rank test critical region is a proper subset of the convex hull test critical region. The graphs in Figures 14.8, 14.9, and 14.10 were obtained by using the functions MH2x2.test, Smirnov2x2.test, and chull2x2.test written in S-PLUS by the authors.

```
> MH2x2.test(4, 14, 0, 19, 11, 15, 1, 25)
> Smirnov2x2.test(4, 14, 0, 19, 11, 15, 1, 25)
> chull2x2.test(4, 14, 0, 19, 11, 15, 1, 25)
```

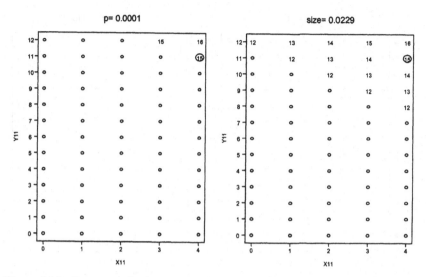

Figure 14.8. Extreme and rejection regions for the Mantel–Haenszel test for (4, 14, 0, 19); (11, 15, 1, 25) with the observed data point circled. Points that belong to the region are marked with the values of the test statistic.

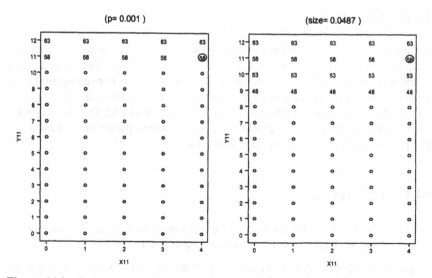

Figure 14.9. Extreme and rejection regions for the Smirnov test for (4, 14, 0, 19); (11, 15, 1, 25) with the observed data point circled. Points that belong to the region are marked with the values of the test statistic.

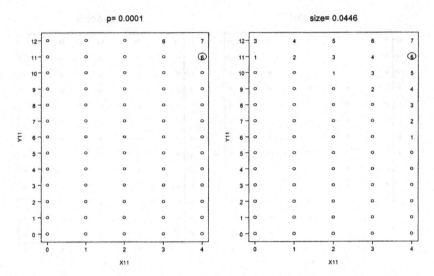

Figure 14.10. Extreme and rejection regions for the convex hull test for (4, 14, 0, 19); (11, 15, 1, 25) with the observed data point circled. Points that belong to the region are marked with the values of the test statistic (somewhat arbitrary).

14.5   Discussion

We have illustrated the computation of permutation tests, some old and some new, in S-PLUS. When the study design is a randomized clinical trial, these permutation tests strictly preserve the Type I error rate of between-group comparisons, while approximate or asymptotic tests may not. Because it is so critical that the Type I error rate be preserved in clinical trials, we feel that permutation tests should play a larger role in drug development, particularly in between-group analyses in pivotal Phase III clinical trials.

14.6   References

Berger, V.W. (1998). Admissibility of exact conditional tests of stochastic order. *Journal of Statistical Planning and Inference* **66**(1), 39–50.

Berger, V.W. (2000). Pros and cons of permutation tests. *Statistics in Medicine* **19**, 1319–1328.

Berger, V.W. and Exner, D.V. (1999). Detecting selection bias in randomized clinical trials. *Controlled Clinical Trials* **20**, 319–327.

Berger, V.W., Permutt, T., and Ivanova, A. (1998). The convex hull test for ordered categorical data. *Biometrics* **54**(4), 1541–1550.

Berger, V.W. and Sackrowitz, H. (1997). Improving tests for superior treatment in contingency tables. *Journal of the American Statistical Association* **92**, 700–705.

Berry, G. and Armitage, P. (1995). Mid-P confidence intervals: A brief overview. *The Statistician* **44**(4), 417–423.

Chalmers, T.C., Celano, P., Sacks, H.S., and Smith, H. (1983). Bias in treatment assignment in controlled clinical trials. *The New England Journal of Medicine* **309**, 1358–1361.

Ederer, F. (1975). Why do we need controls? Why do we need to randomize? *American Journal of Ophthalmology* **76**, 758–762.

Edgington, E.S. (1995). *Randomization Tests*, 3rd ed. Marcel Dekker, New York.

Fox, S.M., Einhorn, L.H., Cox, E, Powell, N., and Abdy, A. (1993). Ondansetron versus ondansetron, dexamethasone, and chlorpromazine in the prevention of nausea and vomiting associated with multiple-day cisplatin chemotherapy. *Journal of Clinical Oncology* **11**, 2391–2395.

Kimeldorf, G., Sampson, A.R., and Whitaker, L.R. (1992). Min and max scorings for two-sample ordinal data. *Journal of the American Statistical Association* **87**, 241–247.

Kuritz, S.J., Landis, J.R., and Koch, G.G. (1988). A general overview of Mantel-Haenszel methods. *Annual Review of Public Health* **9**, 123–160.

Ludbrook, J. and Dudley, H. (1998). Why permutation tests are superior to t and F tests in biomedical research. *The American Statistician* **52**(2), 127–132.

McKinlay, J. B. and Marceau, L. D. (1999). A tale of three tails. *The American Journal of Public Health* **89**(3), 295–298.

Miller, R.G. (1980). Combining 2×2 contingency tables. In: Miller R.G., Efron B., Brown B.W., and Moses L.E., eds. *Biostatistics Casebook.* Wiley, New York.

Permutt, T. and Hebel, J.R. (1989). Simultaneous-equation estimation in a clinical trial of the effect of smoking on birth weight. *Biometrics* **45**, 619–622.

Placebo Working Group (1996). Placebo effects and research in alternative and conventional medicine. *Chinese Journal of Integrated Medicine* **2**(2), 141–148.

Sacks, H., Chalmers, T.C., and Smith, H. (1982). Randomized versus historical controls for clinical trials. *The American Journal of Medicine* **72**, 233–240.

Simon, R. (1982). Randomized clinical trials and research strategy. *Cancer Treatment Reports* **66**(5), 1083–1087.

Upton, G.J.G. (1982). A comparison of alternative tests for the 2×2 comparative trial. *Journal of the Royal Statistical Society* **A 145**(1), 86–105.

Upton, G.J.G. (1992). Fisher's exact test. *Journal of the Royal Statistical Society* **A 155**(3), 395–402.

Vickers, A., Cassileth, B., Ernst, E., Fisher, P., Goldman, P., Jonas, W., Lewith, G., Schulz, K., and Silagy, C. (1997). How should we research unconventional therapies? *International Journal of Technology Assessment in Health Care* **13**(1), 111–121.

14.A. Appendix

14.A.1 Glossary of Hypothesis Testing Terms

This section presents a glossary terms used in describing hypothesis tests.

The *sample space* is the set of possible outcomes that might obtain, with positive probability, from a designed experiment.

A hypothesis is *simple* if its truth leads to a unique probability distribution on the sample space, and is *composite* if its truth leads to two or more probability distributions on the sample space.

The *decision space* consists of the outcomes {reject the null hypothesis, fail to reject the null hypothesis}.

A *hypothesis test* is a mapping from the sample space to the decision space.

The *critical (rejection) region* of a hypothesis test is the set of points in the sample space that result in the decision to reject the null hypothesis.

A *Type I error* is the error of rejecting a true null hypothesis.

The *Type I error rate* is the probability of committing a Type I error. This probability can be computed under either a sampling paradigm (which is not applicable to a clinical trial because clinical trials do not generally involve a random sample of patients) or a permutation paradigm (resulting in a permutation test that is applicable to clinical trials).

The *nominal level of significance*, α, is the largest probability one is willing to allow of a Type I error.

The *actual level of significance* of a hypothesis test is the null probability (maximized, if the null hypothesis is composite) of the critical region.

A hypothesis test is *conservative* if its actual significance level never exceeds α, but is at times less than α. A hypothesis test is *anticonservative* if its actual significance level may exceed α.

A *type II error* is the error of failing to reject a false null hypothesis.

The *power* of a hypothesis test to detect a given alternative is the probability, computed under that alternative, of the critical region of the hypothesis test.

One outcome is more *extreme* than another if it provides more evidence against the null hypothesis in favor of the alternative hypothesis than the other.

The *test statistic* is a computable function of the observed data that serves as a means of sorting the sample space by extremity. That is, points with larger values of the test statistic are more extreme than points with smaller values.

The *extreme region* is the (data-dependent) set of points in the sample space that are as extreme or more extreme than the observed point.

The *p*-value is the null probability (maximized, if the null hypothesis is composite) of the extreme region. When a hypothesis test is not based on a test statistic (such as the improved hypothesis tests of Berger and Sackrowitz, 1997), then neither an extreme region nor a *p*-value can be defined unambiguously.

A *significance test* is a mapping from the sample space to the unit interval, yielding *p*-values. A single significance test generates an entire family of nested hypothesis tests, one for each α-level, as "reject the null hypothesis if and only if $p \leq \alpha$." That is, the critical region of a level-α hypothesis test generated by a significance test is the set of points of the sample space for which $p \leq \alpha$.

A *permutation test* is a hypothesis test that is based on the sample space constructed by considering the set of treatment allocations that were possible, with assigned probabilities, based on the conduct of the actual study.

14.A.2   S-PLUS Code for Plotting Extreme Regions of Linear Rank Tests (v=0, 0.5, 1) and the Smirnov Test

Below is the S-PLUS code for the function `ttestpv`. Code for the other functions discussed in this chapter is available on the companion web site.

```
ttestpv <- function(x1, x2, x3, y1, y2, y3) {
# plots extreme regions for
# t-tests with v=0, 0.5, 1, and the Smirnov test
# calcualting corresponding p-values
# for 2x3 contingency table
#   x1,x2,x3
#   y1,y2,y3
# with ordered outcome levels
#
# constructing the reference set,
# enumerating all the tables with
# given col.m (column margin), and row.m (row margin)
  col.m <- c(x1 + y1, x2 + y2, x3 + y3)
  row.m <- c(x1 + x2 + x3, y1 + y2 + y3)
  v1 <- rep(0:col.m[1], col.m[2] + 1)
  v2 <- rep(0:col.m[2], rep(col.m[1] + 1, col.m[2] + 1))
  v3 <- row.m[1] - v1 - v2
  w1 <- col.m[1] - v1
  w2 <- col.m[2] - v2
  w3 <- col.m[3] - v3
  Omega <- cbind(v1, v2, v3, w1, w2, w3)
```

```
Omega <- Omega[v3 >= 0 & w3 >= 0, ]
  if(!is.matrix(Omega)) Omega <- t(Omega)   #
# calculating the null probability of each table
  if((x1 + y1) == 0)
    d1 <- 1
  else d1 <- dbinom(Omega[, 1], Omega[, 1] + Omega[, 4],
    0.5)
  if((x2 + y2) == 0)
    d2 <- 1
  else d2 <- dbinom(Omega[, 2], Omega[, 2] + Omega[, 5],
    0.5)
  if((x3 + y3) == 0)
    d3 <- 1
  else d3 <- dbinom(Omega[, 3], Omega[, 3] + Omega[, 6],
    0.5)
  nprob <- (d1 * d2 * d3)/
    dbinom(x1 + x2 + x3, x1 + x2 + x3 + y1 + y2 + y3, 0.5)
  scores1.test <- Omega[, 3]
  scores2.test <- Omega[, 2] + Omega[, 3]
  scores3.test <- Omega[, 1] + 2 * Omega[, 2] +
    3 * Omega[, 3]
  Smirnov <- pmax(Omega[, 1], (Omega[, 1] + Omega[, 2]) -
    row.m[1]/sum(row.m) * col.m[2], 0)
  observed <- as.numeric(Omega[, 1] == x1 &
    Omega[, 2] == x2)
  Omega <- cbind(Omega, nprob, Smirnov, scores1.test,
    scores2.test, scores3.test, observed)   #
# calculating p-value
  scores1.region <- Omega[scores1.test <= x3, ]
  if(!is.matrix(scores1.region))
    scores1.region <- matrix(scores1.region, nrow = 1)
  scores1.p <- sum(scores1.region[, 7])
  scores2.region <- Omega[scores2.test <= x2 + x3, ]
  if(!is.matrix(scores2.region))
    scores2.region <- matrix(scores2.region, nrow = 1)
  scores2.p <- sum(scores2.region[, 7])
  scores3.region <- Omega[scores3.test <=
    x1 + 2 * x2 + 3 * x3, ]
  if(!is.matrix(scores3.region))
    scores3.region <- matrix(scores3.region, nrow = 1)
  scores3.p <- sum(scores3.region[, 7])
  Smirnov.region <- Omega[Smirnov >=
    pmax(x1, x1 + x2 - row.m[1]/sum(row.m) * col.m[2], 0),
    ]
```

```
  if(!is.matrix(Smirnov.region))
    Smirnov.region <- matrix(Smirnov.region, nrow = 1)
  Smirnov.p <- sum(Smirnov.region[, 7])   #
# plotting the rejection regions for
# t-test with v=0, .5, and 1 and Smirnov test
  minx <- min(Omega[, 1])
  maxx <- max(Omega[, 1])
  miny <- min(Omega[, 2])
  maxy <- max(Omega[, 2])
  par(oma = c(2, 0, 0, 0), mfcol = c(2, 2))
  plot(Omega[, 1], Omega[, 2], xlab = "C11", type = "p",
    ylab = "C12",
    lab = c(maxx - minx + 1, maxy - miny + 1, 0)) #
# label x
  text(minx:maxx, miny - 0.1 * (maxy - miny), minx:maxx) #
# label y
  text(minx - 0.1 * (maxx - minx), miny:maxy, miny:maxy)
  points(scores1.region[, 1], scores1.region[, 2],
    type = "p", pch = 16, cex = 1.2)   #
# mark points in extreme region
  points(x1, x2, type = "p", pch = 1, cex = 2)   #
# mark the observed point
  title(main = paste("Linear Rank Test 0 0 1", "(p=",
    round(scores1.p, digits = 4), ")"))
  plot(Omega[, 1], Omega[, 2], xlab = "C11", type = "p",
    ylab = "C12",
    lab = c(maxx - minx + 1, maxy - miny + 1, 0)) #
# label x
  text(minx:maxx, miny - 0.1 * (maxy - miny), minx:maxx)   #
# label y
  text(minx - 0.1 * (maxx - minx), miny:maxy, miny:maxy)
  points(scores2.region[, 1], scores2.region[, 2],
    type = "p", pch = 16, cex = 1.2)
  points(x1, x2, type = "p", pch = 1, cex = 2)
  title(main = paste("Linear Rank Test 0 1 1", "(p=",
    round(scores2.p, digits = 4), ")"))
  plot(Omega[, 1], Omega[, 2], xlab = "C11", type = "p",
    ylab = "C12",
    lab = c(maxx - minx + 1, maxy - miny + 1, 0)) #
# label x
  text(minx:maxx, miny - 0.1 * (maxy - miny), minx:maxx)   #
# label y
  text(minx - 0.1 * (maxx - minx), miny:maxy, miny:maxy)
  points(scores3.region[, 1], scores3.region[, 2],
    type = "p", pch = 16, cex = 1.2)
```

```
points(x1, x2, type = "p", pch = 1, cex = 2)
title(main = paste("Linear Rank Test 0 0.5 1", "(p=",
  round(scores3.p, digits = 4), ")"))
plot(Omega[, 1], Omega[, 2], xlab = "C11", type = "p",
  ylab = "C12",
  lab = c(maxx - minx + 1, maxy - miny + 1, 0)) #
# label x
text(minx:maxx, miny - 0.1 * (maxy - miny), minx:maxx) #
# label y
text(minx - 0.1 * (maxx - minx), miny:maxy, miny:maxy)
points(Smirnov.region[, 1], Smirnov.region[, 2],
  type = "p", pch = 16, cex = 1.2)
points(x1, x2, type = "p", pch = 1, cex = 2)
title(main = paste("Smirnov Test", "(p=",
  round(Smirnov.p, digits = 4), ")"))
mtext(outer = T, paste("Data=", x1, ",", x2, ",", x3,
  ",", y1, ",", y2, ",", y3), side = 1, cex = 1.5)
c(scores1.p, scores2.p, scores3.p, Smirnov.p)
}
```

15

Sample Size Reestimation

Wenping Wang
Pharsight Corporation, Cary, NC, USA

Andreas Krause
Novartis Pharma AG, Basel, Switzerland

15.1   Introduction

The number of subjects is an important design parameter in clinical trials. The key information when planning the sample size is the postulated effect and its variation. The effect size may come from prior trials, from literature review, or quite often from the best guess by the investigator.

The estimation of the effect size when developing a protocol is usually an approximation. Even with prior studies, the assumptions made for the effect size may not be satisfied for the current trial. This could be the result of the subtle shift of the patient population (different stage, different inclusion/exclusion criteria), the variation of the medical practice of the participating investigators, the difference of treatment plan, or random variations from trial to trial. Consequently, the planned sample size may not always address the scientific questions of interest with adequate power. One of Murphy's Laws vividly captures the need for the sample size reestimation in clinical trials: "There is never enough time to do a job right the first time, but there is always time to do it again."

The work by Gould and Shih (1992) is among the earliest efforts of dealing with the problem of adjusting the sample size at the interim stage. Other references on this topic include Wittes and Brittan (1990), Shih (1992), Shih (1993), and Shih and Gould (1995). Gould and Shih (1992) studied the problem of samples from two univariate Normal distributions with different means and common variance. A distinguished feature of Gould and Shih's method is that the sample size is reevaluated at the interim stage without unblinding the treatment assignment. By keeping the blinding of clinical trials, the type I error is under control; and more importantly, the integrity of the trials is preserved. The ability of maintaining blinding makes the approach appealing to investigators to apply it for Phase II and Phase III trials.

In this chapter we formulate the problem as a finite mixture of populations. It is noted that this is essentially the same formulation as Gould and Shih (1992), although not explicitly stated in their paper. With this formulation the EM algorithm of Gould and Shih (1992) is generalized to the K-sample Normal distributions with unknown common variance. The EM approach is further extended to a Gibbs sampling setup for assessing parameter variation. All algorithms are implemented in S-PLUS and applied to simulated data.

15.2   The General Formulation of the Sample Size Reestimation Problem

The sample size reestimation problem can be regarded as a special case of the finite mixture of distributions. The finite mixture model assumes that the data $x_1, \ldots, x_n$ are independent observations from a mixture density with K components:

$$p(x \mid \pi, \phi, \eta) = \pi_1 f(x; \phi_1, \eta) + \ldots + \pi_K f(x; \phi_K, \eta)$$

where $\pi = (\pi_1, \ldots, \pi_K)$ are the nonnegative mixture proportions that sum to unity; $\phi = (\phi_1, \ldots, \phi_K)$ are component specific parameters; and η is a component invariant parameter. Some key references on finite mixture models include Diebolt and Robert (1994), Robert (1996), Escobar and West (1995), and Richardson and Green (1997). For a comprehensive review of the current development, see Stephens (1997).

The finite mixture model formulation is equivalent to a more convenient missing data formulation of Gould and Shih (1992). The missing data formulation assumes that each datum x_i arises from one unknown component z_i of the mixture. Therefore, the observed data are $x_1, \ldots, x_n$, and the complete data are $(x_1, z_1), \ldots, (x_n, z_n)$. The missing treatment identification variables $z_1, \ldots, z_n$ are assumed to be a realization of independently distributed discrete random variables with probability mass function

$$p(Z_i = k \mid \pi, \phi, \eta) = \pi_k$$

with $i = 1, \ldots, n, k = 1, \ldots, K$. Conditional on the Z's, $x_1, \ldots, x_n$ are assumed to be independent observations from the densities

$$p(x_i \mid Z_i = k, \pi, \phi, \eta) = f(x_i; \phi_k, \eta), \ i = 1, \ldots, n$$

With the sample size reestimation problem we usually assume that the mixture proportions are equal at the interim stage, the component specific parameters are different populations means, and the component invariant parameter is the unknown variance.

15.3   The Maximum Likelihood
Estimation Approach

We develop in the following the general idea of the EM algorithm, derive the Gould–Shih algorithm for two treatment groups, and extend it to K groups.

15.3.1   The EM Algorithm

The EM (Expectation Maximization) algorithm was introduced to derive maximum likelihood (ML) estimates in the presence of missing data (Dempster et.al. (1977)). Under mild conditions, the EM algorithm ensures that ML estimates are derived iteratively. EM consists of two steps, the E step and the M step. The E step imputes the missing data by replacing them with their expected values, and the M step derives maximum likelihood parameter estimates based on the full dataset (the observed and the imputed data). Alternating these two steps results in a convergence towards the ML parameter estimates.

15.3.2   Gould and Shih's EM Algorithm

In the current context the assignment of each patient to the groups placebo and treatment is unknown and can be treated as a missing data problem. Gould and Shih (1992) presented the EM algorithm for the reestimation of sample sizes in clinical trials.

Consider the case that the measurements of the response are from two univariate Normal distributions with different means and unknown common variance. Suppose that an equal number of patients per treatment arm, say $n/2$, are available at the interim stage. Denote x_i as the observation of the primary endpoint upon which the sample size calculation was performed and denote z_i as the treatment assignment indicator for subject i. Hence for each patient we have a pair of data $(x_i, z_i), i = 1, \ldots, n$. Without unblinding the treatment identification, the treatment indicators $z_1, \ldots, z_n$ are missing at random according to the taxonomy of missingness of Rubin (1976). An EM algorithm (Dempster et al. (1977)) was proposed by Gould and Shih (1992) to estimate the common variance σ^2.

The complete-data log likelihood function (ignoring a constant) is

$$\frac{n}{2} \log \sigma^2 + \frac{1}{2\sigma^2} \sum_{i=1}^{n} [(x_i - \mu_1)^2 z_i + (x_i - \mu_2)^2 (1 - z_i)]$$

implying the E and M step as follows:

E Step

$$z_i = E(z_i \mid x_i, \mu_1, \mu_2, \sigma^2)$$
$$= P(\text{patient } i \text{ belongs to the control group} \mid x_i)$$
$$= \frac{\exp[-(x_i - \mu_1)^2/\sigma^2]}{\exp[-(x_i - \hat{\mu}_1)^2/\hat{\sigma}^2] + \exp[-(x_i - \hat{\mu}_2)^2/\hat{\sigma}^2]}$$
$$\text{for } i = 1, \ldots, n, \ k = 1, \ldots, K$$

M Step

$$\hat{\mu}_{1m} = \left[\sum x_i \hat{z}_i\right]\left[\sum \hat{z}_i\right]^{-1}$$
$$\hat{\mu}_{2m} = \left[\sum x_i (1 - \hat{z}_i)\right]\left[n - \sum \hat{z}_i\right]^{-1}$$
$$\hat{\sigma}^2 = \frac{1}{n}\sum_{i=1}^{n}\left[(x_i - \hat{\mu}_1)^2 \hat{z}_i + (x_i - \hat{\mu}_2)^2(1 - \hat{z}_i)\right]$$

We initialize the values $\hat{z}_i$ with random numbers, taking the values 1 and 2, respectively, and alternate the E and M steps (starting with the M step) until the change in parameters in two successive steps is sufficiently small.

The newly acquired information on the variance of the response variable can be used to make decisions whether there is a need to amend the study protocol to increase the sample size. Specifically, assume that N_0 patients were planned with the assumption that the deviation of the response variable is equal to σ_0. With the MLE $\hat{\sigma}$ from the EM algorithm, if $\hat{\sigma} \leq \sigma_0$, no amendment is necessary. On the other hand, if $\hat{\sigma} > \sigma_0$, the study sample size needs to be updated as

$$N_1 = \frac{\hat{\sigma}^2}{\sigma_0^2} N_0$$

Therefore, additional $(N_1 - N_0)/2$ patients should be recruited for each treatment arm.

15.3.3   Generalization to K Samples

The EM algorithm of Gould and Shih (1992) can be generalized to the case of univariate K-sample distributions with different means and unknown common variance as follows:

E Step

$$
\begin{aligned}
z_{ik} &= p(z_{ik} \mid x_i, \mu_1, \ldots, \mu_K, \sigma^2) \\
&= P(\text{patient } i \text{ belongs to treatment group } k \mid x_i) \\
&= \frac{\exp[-(x_i \quad \mu_k)^2/\sigma^2]}{\sum_{k=1}^{K} \exp[-(x_i - \hat{\mu}_k)^2/\hat{\sigma}^2]} \\
&\qquad\qquad \text{for } i = 1, \ldots, n, \ k = 1, \ldots, K
\end{aligned}
$$

M Step

$$
\begin{aligned}
\hat{\mu}_{km} &= \left[\sum x_i \hat{z}_{ik}\right]\left[\sum \hat{z}_{ik}\right]^{-1}, \ k = 1, \ldots, K \\
\hat{\sigma}^2 &= \frac{1}{n}\sum_{i=1}^{n}\sum_{k=1}^{K}\left[(x_i - \hat{\mu}_k)^2 \hat{z}_{ik}\right]
\end{aligned}
$$

The implementation in S-PLUS is straightforward. The code is given in the Appendix.

15.4   Bayesian Methodology

The EM algorithm as described does not yield estimates of the precision of the variance estimates. Bayesian methodology, in particular the Gibbs sampler, a method from the Markov Chain Monte Carlo (MCMC) toolbox, can provide those estimates. Inference is based on samples generated from the posterior distribution of the parameter components. Before going into details of the Gibbs sampler, we set the necessary Bayesian framework.

In a Bayesian context, parameters are treated as random quantities. The point estimates of the "true" parameters in frequentist statistics are replaced by distributions on the parameter space. The posterior distribution of the parameters is derived by updating the prior information with the information contained in the data. For a comprehensive review on Bayesian statistics, see Bernardo and Smith (1994), Draper (1998, Chap. 1), or Berger (1985).

The model is specified as a joint density for parameters and data. The observations are regarded as fixed and inference is based on the posterior distribution of the parameters. The posterior distribution is often high-dimensional and of a nonclosed form, such that derivation of the marginal distribution of one of the parameters can require cumbersome integration.

The following describes the basic Normal–Gamma model. Consider a sample $x = (x_1, \ldots, x_n)$ with independent observations from a Normal distribution $N(\mu, \sigma^2)$. The mean of the Normal distribution comes from a Normal distribution with prior mean μ_* and variance σ^2. The precision

$(1/\sigma^2)$ is distributed according to a Gamma distribution. The prior model is given by

$$
\begin{aligned}
\mu &\sim N(\mu_*, \sigma^2) \\
\sigma^{-2} &\sim \Gamma_2(\sigma_*^2, n_*) \\
x_i &\sim N(\mu, \sigma^2) \text{ iid., } i = 1, \ldots, n
\end{aligned}
$$

The likelihood is given as the product of the $n + 1$ Normal distributions and the Gamma distribution. From that it follows that the posterior distributions are given as follows:

$$
\begin{aligned}
\mu \mid \sigma^2, x &\sim N(\mu_{**}, \sigma^2) \\
\sigma^{-2} \mid \mu, x &\sim \Gamma_2(\sigma_{**}^2, n_{**})
\end{aligned}
$$

with

$$
n_{**}\sigma_{**}^2 = n_*\sigma_*^2 + \sum_{i=1}^{n}(x_i - \mu)^2
$$

$$
n_{**} = n_* + n
$$

$$
n_{**}\mu_{**} = n_*\mu_* + \sum_{i=1}^{n}x_i
$$

One star ($*$) in the index denotes a prior parameter, two stars ($**$) denote a posterior parameter. The term n_* denotes the number of observations assigned to the prior variance σ_*^2. Setting n_* to 3 can be regarded as assigning the weight of three observations to the a priori information.

The posterior mean μ_{**} is a weighted mean of the prior mean and the mean of the data. The notation Γ_2 denotes a special parameterization of the Gamma distribution, such that a random variate with distribution $\Gamma_2(\sigma, \nu)$ has a mean value of $1/\sigma^2$ (see, e.g., Leamer (1978), or Krause (1994)).

15.5   The Sampling Approach

Gibbs sampling has been introduced by Geman and Geman (1984). Gelfand and Smith (1990) have introduced it as a tool to derive marginal distributions in Bayesian statistics. Casella and George (1992) and Chib and Greenberg (1995) illustrate the idea of MCMC sampling in detail.

Markov Chain Monte Carlo (MCMC) methods (including the Gibbs sampler) allow to derive samples of the posterior distribution. Instead of deriving the marginal posterior distributions of a parameter analytically, samples are generated and the distribution can be estimated based on the sample. Samples allow to derive correlations between parameters, distributions of transformations of parameters, and more.

Starting with an initial set of parameters, each parameter is in turn sampled from its conditional distribution given the current values of the other parameters. The stream of random numbers generated converges in distribution (under mild conditions) to the posterior distribution.

Note the difference between point convergence in an algorithm like EM and convergence in distribution in MCMC methods. Point convergence means that a stream of values converges towards a point and will stay constant once convergence is achieved. Convergence in distribution means that the distribution generating the data stream will stay constant once convergence is achieved. However, the data itself will still vary according to the distribution. Convergence in distribution is therefore somewhat difficult to diagnose.

In the context of MCMC methods, many publications discuss convergence diagnosis and whether a single long simulation run or parallel shorter runs are preferable. Raftery and Lewis (1992) developed widely used convergence diagnostics. A review of several criteria is given in Cowles and Carlin (1996). Geyer (1992), Gelman and Rubin (1992b), and Gelman and Rubin (1992a) provide a thorough discussion on single or multiple chains.

15.5.1   The Gibbs Sampling Algorithm

Let the parameter Θ be partitioned into components $(\Theta_1, \ldots, \Theta_r)$. Suppose we cannot sample directly from $p(\Theta \mid x_1, \ldots, x_n) = p(\Theta_1, \ldots, \Theta_r \mid x_1, \ldots, x_n)$. Instead we sample from the lower-dimensional conditional distributions $p(\Theta_1 \mid \Theta_2, \ldots, \Theta_r, x_1, \ldots, x_n)$, $p(\Theta_2 \mid \Theta_1, \Theta_3, \ldots, \Theta_r, x_1, \ldots, x_n)$, $\ldots$, $p(\Theta_r \mid \Theta_1, \ldots, \Theta_{r-1}, x_1, \ldots, x_n)$.

Figure 15.1 illustrates the idea. The data to generate comes from a two-dimensional Normal distribution, shown as contour lines. Instead of generating two-dimensional Normal random numbers $z = (x, y)$ one can generate one-dimensional random numbers from the conditional distributions of $(y \mid x)$ and $(x \mid y)$.

The Gibbs sampling algorithm proceeds as follows. Given the state $\Theta^{(t)} = \theta^{(t)}$ at time t, generate a value for $\Theta^{(t+1)}$ in r steps as follows:

$$\Theta_1^{(t+1)} \sim p(\theta_1 \mid \theta_2^{(t)}, \ldots, \theta_r^{(t)}, x_1, \ldots, x_n)$$

$$\Theta_2^{(t+1)} \sim p(\theta_2 \mid \theta_1^{(t)}, \theta_3^{(t)}, \ldots, \theta_r^{(t)}, x_1, \ldots, x_n)$$

$$\vdots$$

$$\Theta_r^{(t+1)} \sim p(\theta_r \mid \theta_1^{(t)}, \ldots, \theta_{r-1}^{(t)}, x_1, \ldots, x_n)$$

The essential feature of the algorithm is that the limiting distribution of the generated Markov chain converges under mild conditions towards the distribution $p(\Theta)$. The Markov chain must be irreducible and aperiodic and all conditional and marginal distributions must exist. For a sufficiently

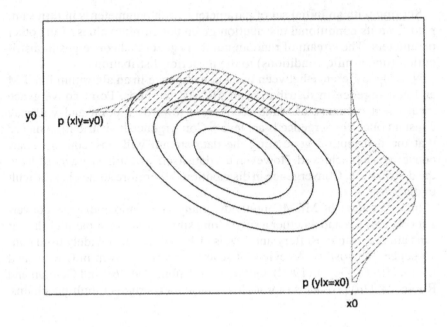

Figure 15.1. Two-dimensional Normal distribution with univariate conditional densities.

large number of iterations t, the generated samples $\theta^{(t)}$ represent a sample from the distribution $p(\Theta)$.

15.5.2 Sampling Using Full Conditional Distributions

The sampling process can be very efficient if the posterior conditional distributions of the parameters are given in closed form ("full conditional"), for example, as Normal and Gamma distributions. The following model is a conjugate model, the prior and posterior distributions belong to the same family. The parameters are updated with the information contained in the data. The prior model is as follows:

$$\mu_k \sim N_{\mu_{k-1}}^{\infty}(\mu_{k*}, \tau_*^{-1}), \; k = 1, \ldots, K$$
$$\tau \sim \Gamma(\alpha_*, \beta_*)$$
$$(\pi_1, \ldots, \pi_K) \sim \text{Dirichlet}(\delta_*, \ldots, \delta_*)$$

where $N(\mu, \tau^{-1})$ denotes the Normal distribution with mean μ and precision τ or variance τ^{-1}, respectively. N_a^b denotes the truncated Normal distribution. A random variate $X \sim N_a^b(\mu, \tau^{-1})$ follows a Normal distribution limited to the interval (a, b). $\Gamma(\alpha, \beta)$ is the usual two parameter

Gamma distribution with mean α/β and variance α/β^2. μ_0 is defined as $\mu_0 = -\infty$.

The prior parameters are set to

$$
\begin{aligned}
\mu_{k*} &= 0, \, k = 1, ..., K \\
\tau_* &= 10^{-6} \\
\alpha_* &= 0.01 \\
\beta_* &= 0.01 \\
\delta_* &= 1
\end{aligned}
$$

Based on the prior model we can derive the likelihood function for the prior model. The common likelihood function of all parameters is given by (up to a constant) the product of the prior parameter distributions and the data.

This allows to derive the full conditional distributions for each parameter component in a similar way to the Normal–Gamma model presented in the previous section (detailed derivations of the full conditional distributions are given in Krause (1994), for example). The full conditional posterior distributions are given as follows:

$$
\begin{aligned}
\mu_k \,|\, \tau^{-1}, x &\sim \mathrm{N}_{\mu_{k-1}}^{\infty}(\mu_{k**}, \tau_{**}^{-1}) \\
\tau \,|\, \mu_1, ..., \mu_K, x &\sim \Gamma(\alpha_{**}, \beta_{**}) \\
(\pi_1, ..., \pi_K) &\sim \text{Dirichlet}(\delta_* + n_1, ..., \delta_* + n_K) \\
z_i \,|\, \tau, \mu_1, ..., \mu_K, x &\sim \text{Multinomial with } p(z_i = k) \propto \pi_k \varphi(x_i; \mu_k, \tau^{-1})
\end{aligned}
$$

with φ the density function of the standard Normal distribution. The posterior parameters are given as

$$
\begin{aligned}
\mu_{k**} &= [\tau \sum_{i:z_i=k} x_i + \mu_* \tau_*]/[n_k \tau + \tau_*] \\
\tau_{**} &= n\tau + \tau_* \\
\alpha_{**} &= \alpha_* + n/2 \\
\beta_{**} &= \beta_* + \frac{1}{2} \sum_{k=1}^{K} \sum_{i:z_i=k} (x_i - \mu_k)^2
\end{aligned}
$$

with

$$
n_k = \sum_{i=1}^{n} I(z_i = k), \, k = 1, ..., K
$$

$I(.)$ denotes the indicator function which is 1 if (.) is true and 0 otherwise. The posterior precision τ_{**} consists of the combined prior and data precision.

15.5.3 Alternatives to Full Conditional Distributions

Working out all full conditional distributions allows to sample efficiently from closed-form distributions. A random number generator is mostly available for standard distributions.

Alternatively, there are algorithms that require almost no analytical work. The samples are generated from the conditional distributions without deriving them explicitly. The adaptive rejection method (Gilks and Wild (1992)) is an efficient tool to generate random numbers from a broad class of functions without the need to have them available in closed form.

The adaptive rejection algorithm forms the basis for a software called BUGS (Bayesian inference Using Gibbs Sampling). BUGS (Spiegelhalter et.al. (1996)) is available via the World Wide Web and allows to run Gibbs sampling by simply specifying the prior model.

15.5.4 Expectation Conditional Maximization (ECM)

The EM algorithm delivers basically a point estimate (the ML estimate) of the posterior distribution obtained via Gibbs sampling. Differences can occur due to the influence of the a priori parameters and the sampling process.

Based on the idea of using conditional distributions to achieve estimates for a high-dimensional distribution, Meng and Rubin (1993) developed the ECM (Expectation Conditional Maximization) algorithm which can be used for complex models when closed-form distributions (and ML estimates) are not available.

15.6 Application

We use a simulated dataset to illustrate the process. In the simulation, the means of two samples are fixed at 0 and 1 (therefore, $\Delta = \mu_1 - \mu_2$ is fixed at 1). The sample size and variance are fixed at 100 patients per treatment arm and $\sigma^2 = 1$ and 5, respectively. As pointed out by Shih (1992), the ratio Δ/σ is usually in the range of 0.2 and 0.5 for most clinical trials. The parameter selection of the simulation reflects this observation.

15.6.1 EM (Expectation Maximization)

The EM algorithm is implemented in a straightforward manner in S-PLUS. The function EM.GouldShih for $K \geq 2$ is given in the Appendix. Figure 15.2 illustrates the convergence of the parameter values over iterations. In a "well-behaved" model like ours, it takes not more than 20 iterations to achieve convergence (one further iteration step would change each param-

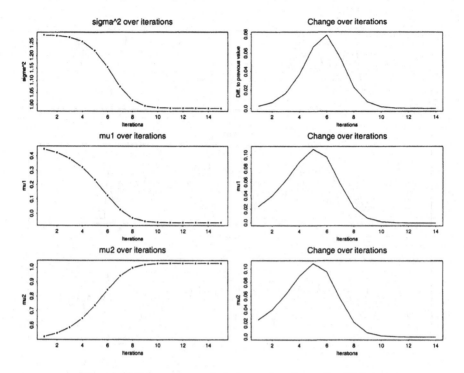

Figure 15.2. Parameter estimates over iterations of an EM run.

eter by less than 10^{-5}. The runtime on a standard PC is just a few seconds for a dataset with 100 observations in each of two groups.

The data underlying Figure 15.2 was obtained by generating data using the function EM.GenerateData with default settings. It generates 100 random numbers with mean 0 and another 100 with mean 1, both with standard deviation 1, from the Normal distribution:

```
> x <- EM.GenerateData()
```

The EM algorithm is run by entering

```
> x.em <- EM.GouldShih(x)
```

and the graph was generated by

```
> EM.plot(x.em)
```

All functions are provided in the Appendix or on the supplementing Web page.

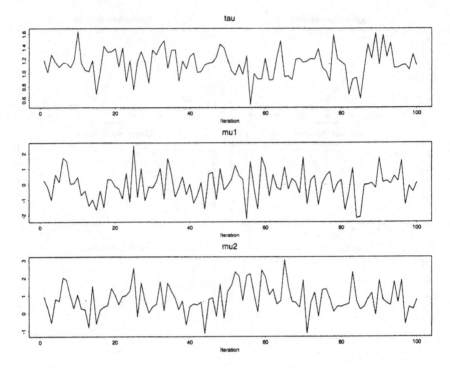

Figure 15.3. Parameter values generated by a single run of the Gibbs sampler.

15.6.2   Gibbs Sampling

We have implemented the Gibbs sampler as an S-PLUS routine. However, the nature of the Gibbs sampler is somewhat incompatible with S-PLUS. Each iteration step depends on the previous step, and each iteration cycle depends on the previous cycle. Double or triple loops are therefore unavoidable. As this is one of S-PLUS's major weaknesses, it is not advisable to use S-PLUS for Gibbs sampling unless only a few and short runs are needed.

The model with 100 observations in $K = 2$ groups is fairly simple however, such that a single chain with 100 iterations takes no more than one minute to complete. Running 100 chains of length 100 will already take about one hour on an average (Intel-based) PC.

The data generated in a single chain of 100 iterations is shown in Figure 15.3. It illustrates that there is only a very short burn-in period and moderate autocorrelation of the chains generated.

We have run 50 chains of length 100 each. We have discarded the first 50 values of each chain to take the burn-in period into account. Every tenth value of each chain entered the sample (to reduce autocorrelation). A density estimate of the posterior distribution of each parameter is shown in Figure 15.4.

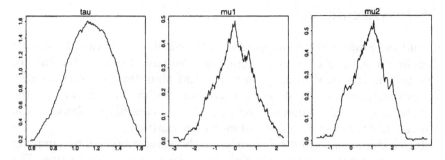

Figure 15.4. Density estimates for posterior samples.

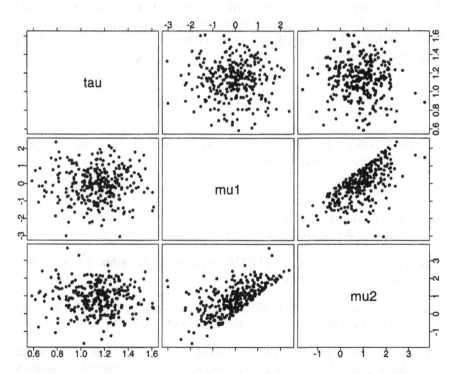

Figure 15.5. Pairwise scatter plots of the samples generated from the posterior distribution with $K = 2$ groups.

Finally, Figure 15.5 shows a scatter plot matrix for the posterior samples. The μ parameters do not exhibit an obvious correlation to the variance τ, and the limitation $\mu_2 > \mu_1$ is clearly visible.

15.7   Discussion

Gould and Shih (1992) have provided a means of reestimating the sample size in clinical trials without unblinding. We have extended the case of two treatment groups to K groups. The EM algorithm provides an easy to implement means for deriving maximum likelihood estimates for the parameters. The algorithm for the case $K = 2$ is easily implemented in S-PLUS, and the case $K > 2$ is not much more difficult.

If one is interested in the dispersion of the parameters, Gibbs sampling provides a means to obtain a sample from the posterior distribution. The sample allows to use a more conservative bound for the variance estimate, for example, the (estimated) upper quartile of the posterior distribution.

For simple problems it is possible to use S-PLUS for Gibbs sampling. However, nested for loops are a well-known weakness, such that it might be advisable to switch to more efficient programs like BUGS (Spiegelhalter et al. (1996)). S-PLUS and BUGS code written by the authors is available from the supplementary web page.

15.8   References

Berger, J. (1985). *Statistical decision theory and Bayesian analysis*. 2nd ed. Springer-Verlag, New York.

Bernardo, J., and Smith, A. (1994). *Bayesian Theory*. Wiley, New York.

Bernardo, J., Berger, J., Dawid, A., and Smith, A., eds. (1992). *Bayesian Statistics 4*. Oxford University Press.

Casella, G., and George, E. (1992). Explaining the gibbs sampler. *The American Statistician* **46**, 167–174.

Chib, S., and Greenberg, E. (1995). Understanding the metropolis-hastings algorithm. *The American Statistician* **49**(4), 327–35.

Cowles, M., and Carlin, B. (1996). Markov chain monte carlo convergence diagnostics: A comparative review. *Journal of the American Statistical Association* **91**(434), 883–904.

Dempster, A., Laird, N., and Rubin, D. (1977). Maximum likelihood from incomplete data via the em algorithm (with discussion). *Journal of the Royal Statistical Society, Series A* **132**, 234–244.

Diebolt, J., and Robert, C. (1994). Estimation of finite mixture distributions through bayesian sampling. *Journal of the Royal Statistical Society, Series B* **56**, 363–375.

Draper, D. (1998). Bayesian hierarchical modeling. Manuscript pre-print, 2nd version, available from: http://www.bath.ac.uk/~masdd/home.html.

Escobar, M., and West, M. (1995). Bayesian density estimation and inference using mixtures. *Journal of the American Statistical Association* **90**(2), 577–588.

Gelfand, A., and Smith, A. (1990). Sampling–based approaches to calculating marginal densities. *Journal of the American Statistical Association* **85**, 398–409.

Gelman, A., and Rubin, D. (1992a). Inference from iterative simulation using multiple sequences (with discussion). *Statistical Sciences* **7**(4), 457–511.

Gelman, A., and Rubin, D. (1992b). A Single Series From the Gibbs Sampler Provides a False Sense of Security. In: Bernardo et.al. (1992).

Gelman, A., Carlin, J., Stern, H., and Rubin, D. (1995). *Bayesian Data Analysis*. Chapman & Hall, London.

Geman, S., and Geman, D. (1984). Stochastic relaxation, gibbs distributions and the bayesian restoration of images. *IEEE Transactions on Pattern Analysis and Machine Intelligence* **6**, 721–741.

Geyer, C. (1992). Practical markov chain monte carlo. *Statistical Sciences* **7**(4), 473–483.

Gilks, W., and Wild, P. (1992). Adaptive rejection sampling for gibbs sampling. *Journal of the Royal Statistical Society, Series C* **41**, 337–348.

Gilks, W., Richardson, S., and Spiegelhalter, D., eds. (1996). *Markov Chain Monte Carlo in practice*. Chapman & Hall, London.

Gould, A., and Shih, W. (1992). Sample size re-estimation without un-blinding for normally distributed outcomes with unknown variance. *Communications in Statisctics* **21**(10), 2833–2853.

Krause, A. (1994). *Computerintensive statistische Methoden– Gibbs Sampling in Regressionsmodellen*. Fischer, Stuttgart.

Leamer, E. (1978). *Specification Searches*. Wiley, New York.

Meng, X., and Rubin, D. (1993). Maximum likelihood estimation via the ecm algorithm: A general framework. *Biometrika* **80**(2), 267–78.

Peace, E., ed. (1992). *Biopharmaceutical Sequential Statistical Applications*. Marcel Dekker, New York.

Raftery, A., and Lewis, S. (1992). How many iterations in the Gibbs Sampler?. In: Bernardo et al. (1992). pp. 763–773.

Richardson, S., and Green, P. (1997). On bayesian analysis of mixtures with unknown number of components. *Journal of the Royal Statistical Society, Series B* **59**, 731–792.

Robert, C. (1996). Mixture of distribution: inference and estimation. In: Gilks et.al. (1996).

Rubin, D. (1976). Inference and missing data. *Biometrika* **63**, 581–592.

Shih, W. (1992). Sample size reestimation in clinical trials. In: Peace (1992). pp. 285–301.

Shih, W. (1993). Sample size reestimation for triple blind clinical trials. *Drug Information Journal* **27**, 761–764.

Shih, W., and Gould, A. (1995). Re-evaluating design specifications of longitudinal clinical trials without unblinding when the key response is rate of change. *Statistcs in Medicine* **14**, 2239–2248.

Spiegelhalter, D., Thomas, A., Best, N., and Wilks, W. (1996). *The BUGS 0.5 Manual.* Available from: http://www.mrc-bsu.cam. ac.uk/bugs/welcome.shtml.

Stephens, M. (1997). Bayesian methods for mixtures of normal distributions. Unpublished Ph.D. thesis.

Wittes, J., and Brittan, E. (1990). The role of internal pilot studies in increasing the efficiency of clinical trials. *Statistics in Medicine* **9**, 65–72.

15.A Appendix

The central S-PLUS functions of this chapter are given in the following. Auxiliary functions, for example, to generate random numbers for the use of the EM algorithm and for graphing the EM output, are provided on the Web page accompanying this book.

Note that all functions using random numbers set the random number generator's seed, such that they are exactly reproducible using any version of S-PLUS on any system.

15.A.1   Gould–Shih EM Algorithm Implementation

```
# Gould-Shih EM algorithm with extension to K groups
#
# Parameters:
# x            - data.
#                Can be generated using EM.GenerateData(n).
# K            - number of groups
# epsilon      - abortion criterion (max. change in parameters
#                between two successive iteration steps)
# seed         - random number generator seed for random data.
# quiet        - if T, all intermediate output is suppressed
#
# Returns:
# z            - matrix (n x K) of probabilities that patient
#                i belongs to group k
# sigma2       - all sigma2 values
# mu           - all mu values (matrix with K columns)
# max.change   - max. change of parameters in last iteration
#
# Notes:
# Use EM.plot to graph the iteration results.
# ---
EM.GouldShih <- function(x, K = 2, epsilon =
  1E-5, seed = 27, quiet = F)
{
  # error checks
  if(missing(x)) stop("Data must be supplied.")
  # set random number generator seed
  set.seed(seed)
  # auxiliary variables
  n <- length(x)
  # if quiet then nothing is shown on the screen.
  # use a dummy cat function for this purpose.
  ccat <- if(quiet) function(...)
    {
    }
  else cat
  # store results of each iteration step
  sigma2.store <- NULL
  mu.store <- NULL
  # we need starting values.
  # z[i, j] contains P(patient i in treatment group j)
  z <- matrix(runif(n * K), n, K)
  # normalize probabilities to have row sum 1
```

```
z <- z/apply(z, 1, sum)
# the abortion criterion
max.change <- epsilon + 1
i <- 0
# The E-M cycle
while(max.change > epsilon) {
  # iteration counter
  i <- i + 1
  # M step
  nominator <- apply(x * z, 2, sum)
  # vector (K)
  denominator <- apply(z, 2, sum)
  # vector (K)
  mu <- nominator/denominator
  # vector (n)
  X <- matrix(x, n, K)
  squared.diffs <- t((t(X) - mu) * (t(X) - mu))
  # matrix (nxK)
  sigma2 <- mean(apply(squared.diffs * z, 1,
    sum))
  # E step
  term <- exp( - squared.diffs/2 * sigma2)
  z <- term/apply(term, 1, sum)
  # matrix (nxK)
  sigma2.store <- c(sigma2.store, sigma2)
  mu.store <- rbind(mu.store, mu)
  # update abortion criterion
  if(i > 1) max.change <- max(abs(sigma2.store[
      i - 1] - sigma2.store[i]),
      abs(mu.store[i, ] - mu.store[
      i - 1, ]))
  # show max change on the screen
  ccat(i, ":\tmax. change:", max.change, "\n")
}
# return the results
return(list(z = z, s2 = sigma2.store, mu = mu.store,
  max.change = max.change))
}
```

15.A.2   Generation of Generalized Gamma Random Numbers

Older versions of S-PLUS provide only the standard Gamma function to generate random numbers from. If this is the case, we can use the fact that if $x \sim \Gamma(a, 1)$ then $x/b \sim \Gamma(a, b)$ to generate generalized Gamma random numbers with mean a/b and variance a/b^2.

```
rgammageneral <- function(n, a, b)
{
  return(rgamma(n, a)/b)
}
```

15.A.3   Generation of Dirichlet Random Numbers

We use the following relation between the Dirichlet and the Gamma distribution: If independent $Z_i \sim \Gamma(a_i)$, $i = 1, \ldots, n$, then $\{X_i = Z_i / \sum Z_i\} \sim$ Dirichlet$(a_1, \ldots, a_n)$ (Gelman et.al. (1995)).

```
rdirichlet <- function(a)
{
  x <- rgamma(length(a), a)
  return(x/sum(x))
}
```

15.A.4   Generation of Truncated Normal Random Numbers

A random number following a truncated Normal distribution can be generated from

$$x = \Phi^{-1}[\Phi(a) + U(\Phi(b) - \Phi(a))] \sim N_a^b(\mu, \sigma^2)$$

with $U \sim \mathcal{U}(0, 1)$ and Φ the cumulative distribution function of the Normal distribution (see Krause (1994)). Note that the dispersion parameter of the Normal distribution in S-PLUS is the standard deviation, the square root of the variance. We take over this convention and specify the standard deviation of the truncated Normal distribution:

```
# ---
# Generation of truncated Normal random numbers (univariate).
# Random numbers have a Normal distribution limited to
# the interval (a, b).
```

```
# See Devroye 1986 or Krause 1994 for the algorithm.
# ---
rtruncnorm <- function(n, m = 0, s = 1, a =  - Inf, b = Inf)
{
  qnorm(pnorm(a, m, s) + runif(n) * (pnorm(b, m, s) -
    pnorm(a, m, s)), m, s)
}
```

15.A.5   Gibbs Sampling Routine for the Extended Gould–Shih Model

```
# ---
# Gibbs sampler for the extended Gould-Shih (1992) model
# with K groups
#
# Parameters
# x          - data (patient data, univariate)
# mu.x       - prior mean (smallest mean, scalar, typically 0)
# tau.x      - prior precision (1/variance)
# alpha.x    - prior parameter for the variance distribution
# beta.x     - prior parameter for the variance distribution
# delta.x    - prior for the Dirichlet distrib. of the P[k]
# K          - number of (treatment) groups
#
# iterations - number of iteration cycles
# seed       - random seed
#
# WARNING:
# If the seed is not set differently for multiple runs,
# the result will always be exactly the same.
# ---
gibbs.GouldShih <- function(x = c(rnorm(50, 0, 1), rnorm(
  50, 1, 1)), mu.x = 0, tau.x = 1E-6,
  alpha.x = 0.01, beta.x = 0.01, delta.x = 1, K = 2,
  iterations = 10, ..., seed = 27)
{
  # random number seed
  set.seed(seed)
  # assign each patient to a group randomly
  n <- length(x)
  z <- sample(1:K, n, replace = T)
  # Initialization
  tau.store <- mu.store <- list()
```

```
mu <- rep(NA, K)
tau <- tau.x
# iteration loop
for(i in 1:iterations) {
  cat(".")
  # show we're doing something
  xsplit <- split(x, z)
  # split x by (treatment) group
  # ---
  # posterior parameters
  # ---
  n.k <- sapply(xsplit, length)
  # group sizes
  group.means <- sapply(xsplit, mean)
  # group-wise means
  # catch the case of an empty group
  if(length(group.means) < K) {
    group.means <- n.k <- rep(NA, K)
    for(k in 1:K) {
      group.means[k] <- mean(x[
        z == k])
      n.k[k] <- sum(z == k)
    }
    group.means[is.na(group.means)] <-
      0
  }
  mu.xx <- (tau * group.means + mu.x * tau.x)/
    (tau * n.k + tau.x)
  tau.xx <- n * tau + tau.x
  alpha.xx <- alpha.x + n/2
  squared.diffs <- sapply(xsplit, function(x)
  sum((x - mean(x))^2))
  beta.xx <- beta.x + sum(squared.diffs)/2
  # random numbers from full conditional posteriors
  tau <- 1/rgammageneral(n = 1, alpha.xx,
    beta.xx)
  mu[1] <- rnorm(n = 1, mu.xx[1], sqrt(1/tau))
  for(j in 2:K)
    mu[j] <- rtruncnorm(1, mu.xx[j],
      sqrt(1/tau), mu[j - 1])
  P <- rdirichlet(delta.x + n.k)
  # probs[i, j] = prob(z[i]=k, i=1,...,n; k=1, ... K)
  probs <- matrix(P, n, K, byrow = T) * dnorm(
    x, matrix(mu, n, K, byrow = T), sqrt(
    tau))
```

```
    probs <- probs/apply(probs, 1, sum)
    # norm to 1
    for(i in 1:n)
      z[i] <- sample(1:K, 1, prob = probs[
        i, ])
    tau.store[[i]] <- tau
    mu.store[[i]] <- mu
  }
  # end iterations
  return(list(tau = unlist(tau.store), mu = mu.store))
}
```

Figure 15.3 is created by running a single chain of length 100 on the data generated before, x, and graphing the data as follows:

```
> x <- EM.GenerateData()
> gibbs.result <- gibbs.GouldShih(x, iterations=100,
  seed=1)
> gibbs.singlerun.plot(gibbs.result)
```

Figures 15.4 and 15.5 are created by using

```
> gibbs.result <- list()
> For(i=1:50, gibbs.result[[i]] <- gibbs.GouldShih(x,
  iterations=100, seed=i))
> gibbs.posterior.sample(gibbs.result)
```

16

Meta-Analysis of Clinical Trials

Keith O'Rourke
Ottawa Hospital / University of Ottawa, Ottawa, Ontario, Canada

Beverley Shea
Ottawa Hospital / University of Ottawa, Ottawa, Ontario, Canada

George A. Wells
University of Ottawa, Ottawa, Ontario, Canada

16.1   Introduction

Meta-analysis (MA) may loosely be defined as the explicit analysis and summary of multiple investigations or studies. Being such, it offers the potential to reach the goal of keeping up to date in the drug research and development process without sacrificing thoroughness. Theoretically, meta-analysis can effectively summarize the accumulated research on a topic, promote new questions on the matter, as well as channel the stream of clinical research toward relevant horizons. Consequently, meta-analyses can be very important in the drug research and development process—from initial consideration of a new drug through the conduct of Phase III studies to possible removal of a drug.

Initial consideration of a drug may involve deliberation of what the current evidence is for any sort of treatment effect or for any proposed standard of care that the placebo regimen would need to include. As for the conduct of later phases of research, in Australia, the government is starting to regulate the use of MAs (or systematic reviews) in drug registration submissions (see http://www.health.gov.au/tga/pubs/pubs.htm). As for removal of a drug, an MA may establish that there is little benefit or even some possible harm—especially in regard to side effects that may not show up in individual trials.

As an application of biostatistics, MA may seem modern and novel but it has a long and fascinating history. What is perhaps most fascinating about the history is that it reveals that new or special techniques are not really required for MAs, but rather old and familiar techniques need only be better understood and applied. Essentially nothing new is required, just a slightly different perspective and perhaps some more careful thinking. Flexible modern software like S-PLUS is also helpful. In order to convey this we have chosen to briefly review a number of different MAs (rather than analyze one in depth), as well as provide a historical motivation in an Appendix (Section 16.A.1). We also provide an overview of the necessary but largely nonstatistical preparatory work that is required for MAs.

The data for the MAs are usually publicly available and very simple to extract and enter into S-PLUS. In fact, as we will point out later, a good written report of an MA will provide the necessary individual study summaries so that others can reanalyze it. All of the MAs we present here were actually published by methodologists reanalyzing an original MA to illustrate various suggested statistical approaches and methods for MA. This provides not only an opportunity for us to note how the results varied between their analyses and the analyses given here, but also a source of second opinions about suggested statistical approaches and methods for MA. The downside of using this nonrandom selected set of MAs is that they are not representative. It is our opinion that they may represent fairly extreme cases that exaggerate the sensitivity of results to the methods of analysis used. This may be what made them attractive to us and to other methodologists.

Drawing on a quote from W.G. Cochran given in the History Appendix, MAs are much more like observational studies—being much less straightforward than randomized clinical trials (RCTs) and requiring large amounts of input from subject matter experts. The "half-empty or half-full glass of water" perspective on this is that the statistical techniques are unlikely to be the "critical" part of an MA. On the other hand, in our opinion, the completely "empty glass of water" perspective arises when biostatisticians refuse to be involved in MAs because of a dislike for the admittedly large and nonquantifiable uncertainties that can only be qualitatively addressed using input from subject matter experts. There really are no alternatives to doing an MA. The inappropriate avoidance of it in the past by biostatisticians is now being coined as "the statistical myth of a single study" by Chatfield (1995) and "the cult of a single study" by Longford and Nelder (1999).

In the Appendix (Section 16.A.1) we briefly comment on the history of MA, starting with astronomers in the 1800s, and point out an early application in medicine by Pearson (1904). We then review the adaptation of these techniques to agricultural trials by Fisher and Cochran in the 1930s. Readers in a hurry may skip this Appendix if they wish but—in hindsight—this is what we wish we had known when we had first started to get involved in MA.

In the chapter itself, we further comment on what we think the primary motivation for an MA should be—the quantitative and rigorous investigation of the replication of study results. Some authors have coined this the analytic task (i.e., quantitatively investigating and modeling how individual study results differed) as opposed to the synthetic task (i.e., quantitatively summarizing the individual studys' common results by combining or pooling them) (Rothman and Greenland, 1998). (The choice of the terms analytic/synthetic perhaps follow from the dictionary meaning of analysis as being the separation of a whole into its parts and synthesis as being the assembling of separate parts into a whole.) Arguably, the analytic task is first and foremost, and given the way RCTs are conducted and reported on, one should expect some limitations in this investigation and sometimes very severe limitations. Perhaps the largest temptation that needs to be guarded against in an MA is that of trying "somehow" to combine or pool studies or "some subset of the studies" when any combining is clearly questionable (O'Rourke and Detsky, 1989).

Meta-analysis means different things in different circles and has numerous names, so we will define what it is we mean by MA and other terms such as systematic reviews in Section 16.2. Apparently, MA is a largely North American term. We will restrict ourselves to the MA of RCTs as we feel that the issues involved in the MA of non-RCTs either alone or in combination with RCTs are quite different and as of yet still very controversial. Some clarity can perhaps be found in Rosenbaum et al. (1999) and Egger et al. (1998).

In Section 16.3 we outline the preparatory work that needs to be done so that the investigation of replication (again the primary task in an MA) can be quantitatively and rigorously carried out. A much larger amount of preparatory work is required than with a single RCT and most of it is of a nonstatistical nature (specification of inclusion/exclusion criteria, literature searches, data abstraction and adjudication, methodological quality assessment of individual studies, etc). We will then offer some guidelines for the reporting of MAs. As it is our purpose here to identify and provide some sense of what needs to be done in MA, rather than provide a self-contained "cookbook" approach to the non-statistical aspects, this section may need to be supplemented by reading references such as Normand (1999) or following guidelines given by MA methods research groups given on web sites that will be given below.

Given that some reasonable degree of replication is found, the synthetic task (summarization or pooling) becomes of interest and methods for this will be described in section 16.4. Here it is useful to divide the tasks into within-study and between-study estimates and tests. The within-study part is usually a large sample situation and virtually all reasonable estimates and tests will agree whereas the between-study part is usually a very small sample situation and different approaches can give very different answers.

In Section 16.5, we reanalyze the example MAs of RCTs using S-PLUS procedures glm, coxph, and gls. The intent in using these procedures is to provide a modeling framework in which to carry out the analysis. This provides facilities for model checking and graphing, as well as easy and immediate extensions of the basic analysis. The code is, we feel, exceedingly straightforward and we have been using this approach since 1986. It uses a very simplified allowance for random effects that has both theoretical support and the very practical advantage of always being conservative (i.e., will always provide a confidence interval that includes the null value if the fixed effects confidence interval does). This is not the case with most other random effects methods (Poole and Greenland, 1999).

One of the MA examples we redo is from a paper on Bayesian methods (Smith et al., 1996) using Markov Chain Monte Carlo (MCMC) via BUGS software that provides a pointer for both those who wish to take a Bayesian approach or alternatively those just interested in conveniently investigating more complicated forms of allowance for random effects. Current challenges and limitations are then briefly addressed. Some readers familiar with MA may wish—at this point—to directly proceed to the statistical analysis section (Section 16.4).

16.2   Motivations for Meta-Analysis: The Insurmountable Opportunity of Meta-Analysis

That most researchers realize the need for stating their results in the context of previous studies is clear from the inclusion of a literature review section in almost all articles published in scientific journals. MA is a further development and refinement of this approach, offering a more rigorous and coherent treatment of past research work. It is tempting to propose that no individual RCT results should be published without the inclusion of an appropriate MA in place of the ad hoc literature review section (O'Rourke and Detsky, 1989).

Now, perhaps the first thing that needs to be said is that MA is an inevitable but certainly less than ideal enterprise and that the selection and evaluation of statistical methods needs to reflect this "reality." An individual study analysis is simply an MA were all study weights are "0" except one. Problems do not arise just because studies are grouped together—the problems just become more apparent. For instance, the bias from selective completion and publication of studies (perhaps based on p-values) needs to be addressed even when interpreting an individual study on its own. How could you know or prove that an individual study was exempt from such selection pressures?

We need to view unplanned or uncoordinated replication of study results as an opportunity—both for confirmation and contrast of results. The driving question should be: Should another trial should be undertaken? And if so, how? (i.e., sample size, blocking, eligibility criteria, etc.). Given that the replication of studies is largely unplanned and uncoordinated, problems and deficiencies should be expected. On occasion one is tempted to just use the best study (or studies) or even declare that there are no "interpretable" studies at all. In extreme situations this may be justified.

A number of terms are used concurrently to describe the process of systematically reviewing and integrating research evidence, including "systematic review," "meta-analysis," "quantitative review," "overview," and "pooling." Some writers prefer the term systematic review, because in some circles the term MA has come to mean just the statistical analysis that combines studies (i.e., just the synthetic task), whereas a systematic review may, or may not, actually combine the results of several studies. Since it is always appropriate and desirable to systematically review a body of data, but often is inappropriate, or even misleading to statistically combine results of studies, this preference may be justified.

Indeed, it is our impression that reviewers often find it hard to resist the temptation of combining studies even when such combining is questionable or clearly inappropriate. As pointed out by Egger et al. (1998) some meta-analysts combine non-RCTs without fully considering the motivations and justification. Some meta-analysts have even suggested combining RCTs with non-RCTs without fully considering the problematic variance/bias trade-off involved (Droitcour et al., 1993). Since it is crucial to understand the limitations of combining and the importance of exploring between-study heterogeneity and sources of potential bias when conducting MAs, the use of the term systematic review rather than MA may indeed be better (Egger et al., 2000).

16.3 Preparatory Work

There are several steps that have been identified in initiating and conducting an MA. One of the key messages is to carefully consider the methodological steps that have been outlined by researchers in MA. One possible useful source for these is from the Cochrane Collaboration—an international organization that actively promotes the methodology and conduct of systematic reviews. Their web site is at http://www.cochrane.org.

As an MA is first and foremost a scientific endeavor, it must be done in a manner that is explicit and fully replicable by others. Very briefly (L'Abbe et al., 1987), there should be adequate compliance with the following requirements:

1. A clearly defined set of questions.
2. An explicit and detailed working protocol.
3. A literature search strategy that can be replicated.
4. Inclusion and exclusion criteria for research reports and a list of exclusions with reasons given.
5. Verification of independence of published studies (use of separate patients in different studies).
6. A careful exploration of differences in treatment effect estimates, with the aim of explaining them on the basis of relevant clinical differences (biological variation), differences in quality of research (methodological variation), or simply sampling variability, with appropriate combination of treatment effect estimates if and where indicated.
7. A listing of individual study results or inputs, along with a listing of what is believed to be the most relevant clinical and methodological differences between studies, so that the analyses can be easily replicated by others.
8. A set of conclusions which includes a summary of what was believed to be done adequately and what was done inadequately along with how this is reflected in estimates of uncertainty.
9. A set of suggestions and directions for future research—either for the particular questions or areas of similar research.

When determining exactly what the MA will entail it is important to clearly define the main components (requirement 1 above). Poorly focused questions can lead to unclear decisions about what research to include and how to summarize it. There may be several key components to a well-formulated question (Counsell, 1996; Richardson et al., 1995). A clearly defined question should specify the types of people (participants), the types of interventions or exposures, the types of outcomes of interest, and as well as the types of studies that are relevant to answering the questions.

Once a well-formulated question has been developed, it is very important to decide on how to search the literature (requirement (3) above). A comprehensive and ambitious search of the literature may be very worthwhile. It has been documented that although Medline is one of the most useful electronic databases for searching for published studies, only searching Medline could result in missing anywhere from 20–70% of the studies (Dickersin et al., 1994). To be reassured that all relevant data are included in a review, it is important to use multiple sources to identify studies. Some well-known electronic databases include, Medline, Embase, the Cochrane Controlled Trials Register, Scisearch, registers of clinical trials, and other bibliographical databases such as those developed by review groups within the Cochrane Collaboration. Other forms of identifying additional studies can be very crucial and include checking reference lists of retrieved articles, personal communication, and hand searching specific journals or proceedings such as conference proceedings.

It is important to note that when developing a search strategy specific to a topic, it is highly recommended that a librarian be consulted. Developing an adequate search strategy for an electronic database often requires either investing a fair amount of time in learning the vagaries of the search terms or soliciting the assistance of an expert in searching the database. In addition, a reference management system is useful for tracking articles identified in the search.

Once the literature has been retrieved, the articles must be assessed for eligibility using the predefined inclusion/exclusion criteria as stated in the a priori protocol (requirement (4) above). Some suggestions include having two or more reviewers to assess the studies, masking the study reports to hide information such as names of authors, institutions, journal of publication and results, and deciding in advance how disagreements will be handled.

Careful thought should be given to the data that will be extracted from each of the included studies. Reviewers can design either paper or electronic data collection forms to help with the process. Mulrow and Oxman (1997) in their recommendations to Cochrane Collaboration reviewers pointed out that many reviewers use a double extraction process whereby two independent appraisals of each study can be compared and differences reconciled.

Some authors feel that assessing the quality of the trials included in the MA is essential. There have been several international conferences on this topic and the contributors have supported the notion that the assessment of quality of RCTs is important to the conduct of a high-quality meta-analyses (Anonymous, 1997; Cook et al., 1995). There continues to be controversy both in how to appraise the individual study quality and in ways of incorporating these appraisals into the formal analyses (Detsky et al., 1992).

Guidelines for reporting meta-analyses have been suggested and a recent consensus conference has put together some possibly useful guidelines. Following on a past initiative to improve the quality of reporting of individual RCTs (the CONSORT statement (Begg et al., 1996)) a conference on the Quality of Reporting of Meta-analyses [QUOROM] was conducted. This conference resulted in a checklist that addressed issues regarding the abstract, introduction, methods, and results section of a report of an MA of RCTs. Further details are provided at http://www.thelancet.com/newlancet/eprint/2/ and in Moher et al. (1999) and Shea et al. (2000). The inclusion of the individual study results used in the analyses so that they can be reanalyzed has long been encouraged in MA (L'Abbe et al., 1987).

16.4   Statistics Methods of Analysis

Conceptually, the results from separate studies can be taken as independent observations "in their own right." These then can be both graphically and quantitatively analyzed for consistency (replication, homogeneity) and, if consistent, combined together in an overall summary (possibly stratified by subgroups). Simple summaries of the individual study's data, usually maximum likelihood estimates or approximations (often just "means" and "standard errors") are taken as the study level results. (With binary outcomes the number of successes and failures determines both the mean and variance.)

Consistency is usually quantitatively analyzed by comparing the observed "variance" between the individual study summaries with that predicted from the within-study "standard errors." The combined overall summary is usually taken as some "weighted average" of the study summaries. And an allowance for some reasonable lack of homogeneity is usually made via a random effects model using the excess observed "variance" of the summaries as an estimate of the allowance required for the random effects (Cox, 1982).

Modern statistical methods and software (and history given in Section 16.A.1) would suggest the use of the study log-likelihoods as the most appropriate summaries of the study results and the summing of log-likelihoods as the most appropriate way to get the combined overall summary (O'Rourke, 2001). An immediate measure of consistency would then be the difference in log-likelihoods given a separate maximum likelihood estimate (MLE) for each study versus a single common MLE (from the combined log-likelihood): A simple rescaling of this quantity is called the scaled deviance and is essentially a variance on the log-likelihood scale (McCullagh and Nelder, 1989). (See also Venables and Ripley (1994), for a discussion about the term "scaled" in the definition of scaled deviance.) Starting with the definition of log-likelihood, the scaled deviance D is defined as follows:

$$l(\mathbf{u};\mathbf{y}) = \sum_i \log\left[f_i\left(y_i;\theta_i\right)\right] \tag{16.1}$$

$$D(\mathbf{u};\mathbf{y}) = 2l(\mathbf{y};\mathbf{y}) - 2l(\mathbf{u};\mathbf{y}) \tag{16.2}$$

The excess in this can be taken as an estimate of the allowance required for the random effects. Fortunately with modern software, this is quite straightforward and (perhaps sometimes unknowingly) done all the time by biostatisticians when they use generalized linear models (GLMs) such as logistic regression.

The technical details are covered in O'Rourke (2001) and we will not dwell too much on them here. Essentially GLMs use profile log-likelihoods to generate their estimates and tests. And the combined profile log-likelihood for a common treatment effect—given the inclusion of an indicator term for study—is simply the sum of the individual study profile log-likelihoods. That is, when there is no estimation of a scale parameter. This will be the case, for instance, with binary, count, and survival outcomes (analyzed by proportional hazards methods) as well as with continuous outcomes when the within-group estimate of variance is assumed to be known. Fortunately, almost all standard MA analysis techniques can be carried out using these models. Fairly straightforward programs can be written in S-PLUS to demonstrate these claims and are available from the authors.

Perhaps the simplest case is that of binary outcomes in a two-group RCT. Here the likelihood methods are implemented in standard logistic regression. We will describe this, but first give the generic S-PLUS code using the `glm` function, as it may provide some hints:

```
> # provides individual study treatment effect estimates
> glm.individual <- glm(cbind(sucess, failure) ~
    study * treatgroup, family = binomial)
> # provides a common (pooled) treatment effect estimate
> glm.pooled <- glm(cbind(sucess, failure) ~
    study + treatgroup, family = binomial)
> # provides a test for homogeneity - warning low power
> anova(glm.individual, glm.pooled, test = "Chisq")
> # provides fixed effect pooled estimates & tests
> summary(glm.pooled)
> # provides random effects pooled estimates & tests
> summary(glm.pooled, dispersion =
    glm.pooled$deviance/(number of studies - 1))
```

First the logistic regression is set up to include an indicator term for study, a term for treatment group, and an interaction term (treatment by study). The indicator term for study allows a separate baseline estimate for each study so that each study's treatment effect estimate contribution is relative to its own control group. The treatment group term allows for a common treatment effect estimate and the interaction term allows for a separate treatment effect estimate for each individual study (the same as one would get using each study's data alone).

The consistency of study results is then quantitatively analyzed by investigating the variation in the estimated individual study treatment effects and less preferably the statistical significance of omitting the interaction term in the logistic regression. With the omission of the interaction term, a common "pooled" treatment effect is constructed along with estimates and tests.

Of course numerous sensitivity analyses should be considered. The use of a modeling framework facilitates this in that one can investigate the effects of other covariates (sometimes referred to as doing meta-regression—watch out for aggregation bias), investigate subgroups (beware of multiplicity problems), and utilize the full array of diagnostic model checking and plotting facilities that is provided in glm. We feel that this advantage over the use of formulas and special software should not be under-estimated and especially not under-used. Because we are covering a number of MA examples we have chosen not to actually do this here. The diagnostic and plotting techniques would be very similar to those in standard statistical modeling approaches as, for instance, discussed in Venables and Ripley (1994).

Random effects are easily allowed for by simply increasing the standard errors. This is done in order to widen—but not shift—the confidence intervals for the combined treatment effect. This can be accomplished by multiplying the standard errors by the square root of some measure of heterogeneity or excess variation (Cox, 1982). Because random effects are motivated to make allowance for extra uncertainty in the combined estimate, either due to flaws in the individual studies or true underlying differences in treatment effects in the individual studies, this multiplication is done only when this measure is greater than one.

In order to remain within a modeling framework, instead of multiplying the standard errors "by hand" we modify the log-likelihood for combined treatment effect by using the "dispersion = measure of heterogeneity" statement in the summary command. When such an option does not exist the log-likelihood quantities are divided by the "measure of heterogeneity." The effect is the same—standard errors of estimates and tests are multiplied by the square root of a "measure of heterogeneity." There are many other allowances and modeling of random effects used in MA, usually much more complicated.[1] Most usually increase the standard errors by about the same amount but in addition shift the confidence intervals. As this shift may be away from the null hypothesis these ways of "allowing for extra uncertainty" can puzzlingly be nonconservative.

Again our preferred choice for the measure of heterogeneity is scaled deviance as defined in (16.2) above. The default choice of measure of heterogeneity in S-PLUS is the Pearson chi-squared statistic. There is seldom much difference and no clear-cut reasons in the literature to prefer one above the other (Cox and Snell, 1989).

There are some differing approaches suggested for MA in the literature (Bailey, 1987; Berkey et al., 1995; Bucher et al., 1997; DerSimonian and Laird, 1986; Eddy et al., 1990; Efron, 1996; Mosteller and Chalmers, 1992; Normand, 1999; Rubin, 1992; Smith et al., 1996), and given the small "sample size" for

---

[1] In choosing simply to increase the standard errors rather than formally modeling the random effects, we are in a sense not following our own advice of using a modeling approach. For instance, with a lack of a model for the random effects, diagnostic model checking of random effects cannot be carried out. One of the motivations for formal modeling of random effects is to facilitate model checking (Longford and Nelder, 1999; Lee and Nelder, 1996). However, given the small sample sizes and lack of motivation for any particular distribution form for the random effects in MA, we believe this will seldom be productive. Influence diagnostics and plots of individual study contributions to analyses do provide some informal checks on how random the random effects appear without formal random effects modeling.

investigating study consistency and estimating required allowances for random effects, there may well be differences between different approaches. Because these allowances for random effects actually determine the individual study weights in most random effects approaches, these differences may be quite important. In theory most of the methods can be related to each other as being different approximations to different likelihood-based methods which converge to each other with increasing sample sizes. Because of this, if differing approaches do in fact lead to the similar individual study weights, then given the "large sample size" over all studies, there will be almost no differences in the combined summaries. Again, the challenging statistical aspect of MA arises in the small sample size for between-study estimates and tests.

16.5 Examples

We provide four examples of MAs, chosen to cover a number of issues and challenges that may arise. The first example (Section 16.5.1) involves binary outcomes from RCTs that were of varying appraised quality and differed in the direction of treatment effect estimates. The second (Section 16.5.2) also involves binary outcomes but here the direction of treatment effect estimates was quite consistent. On the other hand, the estimates differed with regard to the size of the effect and issues about the particular choice of random effects models arose. The third example (Section 16.5.3) again involves binary outcomes and this time illustrates that the "between-study" estimates and tests can be quite sensitive to the method of analysis chosen. The forth example (Section 16.5.4) is really a repeat of the first example, but this time using survival outcomes from the same studies. The fifth example (Section 16.5.5) involves continuous outcomes, which in MA are perhaps harder to deal with than binary or survival outcomes.

16.5.1 Parenteral Nutrition in Cancer Patients

Our first example is taken from an MA of total parenteral nutrition (TPN) in cancer patients undergoing chemotherapy (McGeer et al., 1990). This MA summarized the evidence on parenteral nutrition in cancer patients undergoing chemotherapy. The authors reported the results of pooled analyses of the effect of TPN support on survival, tumor response, and toxicity in cancer patients undergoing chemotherapy. The authors suggested that routine use of TPN in patients undergoing chemotherapy should be strongly discouraged, and that trials involving specific groups of patients or modifications in TPN should be undertaken with caution. In conclusion, in this paper the authors were unable to detect a beneficial effect of TPN.

The primary outcome in this MA was 3-month mortality and we reanalyze this outcome here. Logistic regression modeling was carried out as described above, the essential code being:

```
> glm.individual <- glm(cbind(success,failure) ~
    study * treatgroup, family = binomial)
> glm.pooled<-glm(cbind(success,failure) ~
    study + treatgroup, family = binomial)
> anova(glm.pooled, glm.individual, test = "Chisq")
> summary.fixed <- summary(glm.pooled)
> summary.random <- summary(glm.pooled,
    dispersion = glm.pooled$deviance/7)
```

The individual study odds ratios (from glm.individual) varied from 0.5 to 5.25, with five of the eight studies having odds ratios greater than or equal to 1. The test for heterogeneity had a chi-squared value of 10.36 on seven degrees of freedom and p-value = 0.17 (which as argued above should not be taken as any assurance of homogeneity.) For an explanation of anova and its output, see Venables and Ripley (1994), especially pages 215 and 216. The fixed effect pooled odds ratio (from summary.fixed) was 0.74 with an approximate 95% Wald-based confidence interval of 0.42 to 1.31. The random effects pooled odds ratio (from glm.pooled using summary.random) was 0.74 with an approximate 95% Wald-based confidence interval of 0.36 to 1.48. Complete S-PLUS code for all calculations is given in the Appendix (Section 16.A.2).

This particular MA turned out to be fairly controversial as there were both estimates of benefit and harm, and the estimates of little benefit or harm came from studies that were arguably better conducted (at least given information in the published reports). Note that the test for heterogeneity was not statistically significant. Additionally, it was very difficult for some involved in the debate to entertain that nutritional supplementation could be anything but good for malnourished patients. Others suggested that tumors were better able to access the nutrients supplied by the treatment than the patients. This example led to the publication of a methods paper to incorporate quality appraisals of the individual studies (Detsky et al., 1992)—but such methods are still controversial.

16.5.2 Prevention of Respiratory Tract Infections

Our second example is taken from a paper on how to do Bayesian MA using Markov Chain Monte Carlo (MCMC) and involved studies on the prevention of respiratory tract infections by selectively decontaminating the digestive tract using antibiotics (Smith et al., 1996). As an example it provides both an opportunity to argue for a simple "always conservative" method of random effects adjustment as well as to provide pointers to conducting MA using Bayesian methods via MCMC methods and BUGS software.

In this MA, the authors described an MA conducted by an international collaborative group investigating the clinical benefits of selective decontamination of the digestive tract. This was an MA of 22 randomized controlled trials. In each trial, patients in intensive care units were randomized to either a treatment or control group, where treatment consisted of different combinations of oral nonabsorbable antibiotics, with some studies in addition including a systematic component of the treatment. Patients in the control groups were given no treatment.

In all of the 22 trials there was less infection in the treatment group—highly suggestive of a treatment effect using a sign test to rule out the null of a 50% chance advantage of the treatment group over the control group:

```
# Two sided p-value from a sign test
> 2 * 0.5^22
[1] 4.768372e-007
```

Estimates of the size of the treatment effect though seemed to vary, suggesting the use of a random effects model. The authors of the Bayesian methods paper implemented a standard random effects model (i.e., exchangeable estimates normally distributed on the logit scale) with the additional twist of Bayesian priors on the parameters of the random effects model. They used MCMC to carry out the analysis. They noticed (as have others before) that the standard random effects model down-weights the larger studies. In the MA at hand, this resulted in a larger pooled treatment effect estimate and a less conservative confidence interval (i.e., a lower limit further away from the null value). This is an example of the standard random effects models not being conservative. In fact, if it is believed that the larger studies were better conducted (as happened to be the case in this MA and may be true more often than not) this down-weighting of larger studies is clearly wrong (i.e., the assumption of exchangeability is not appropriate).

To avoid this possibility, as explained earlier we choose to use a simpler, perhaps less well-known random effects model that is always conservative. The essence of the model is to increase the standard error by the multiple of the square root of some measure of heterogeneity but only if greater than one. We chose to use the mean deviance while others (including the default in S-PLUS) use the mean Pearson chi-square for heterogeneity. As mentioned earlier, there is seldom a large difference between the two. Again, rather than actually multiplying the standard errors by hand, we use the dispersion option in the summary command to make the allowance for various tests and estimates of interest more automatic. The essential code is:

```
glm.pooled <- glm(cbind(success, failure) ~
  study + treatgroup, family = binomial)
summary.random <- summary(glm.pooled,
  dispersion = glm.pooled$deviance/21)
```

The odds ratio estimate from glm.pooled was 0.35 with a 95% approximate Wald-based confidence interval from summary.random of 0.24 to 0.49. Using MCMC the hierarchical Bayes estimate of the odds ratio was 0.25 with a 95% probability interval or 0.16 to 0.36. Again less conservative than the confidence interval from the analysis given here.

16.5.3   *Pneumocystis carinii* Pneumonia in HIV Infection

Our third example presents a slight twist, in that a comparison of two drugs was investigated using studies that directly compared the drugs as well as studies that just compared the two drugs to placebo (Bucher et al., 1997). This MA was comprised of RCTs that compared two experimental drugs (trimetho-prim-sulphamethoxazole and dapsone/pyrimethamine) to one standard "placebo" regimen (aeosolized pentamidine), or directly compared (trimetho-prim-sulphamethoxazole to either dapsone or dapsone/pyrimethamine) for the primary and secondary prevention of *Pneumocystis carinii* pneumonia in HIV infection. In the first set of studies, what is investigated is whether there is an "interaction" between drug type (i.e., is the drug A effect larger compared to placebo than the drug B effect compared to placebo). It should be noted, as was pointed out in the paper, patients were not randomized in this comparison—patients in the drug B studies may differ in treatment responsiveness from those in the drug A studies—and therefore methods for nonrandomized comparisons are actually required. As mentioned earlier, these are too controversial for inclusion in this chapter, but the closed-form formulas based on normality assumptions derived by the authors for indirect comparisons were also applied to direct comparison studies. It is only with these direct comparison studies that we contrast their methods with logistic regression:

```
# Direct comparison studies
glm.individual <- glm(cbind(x, n - x) ~
   factor(study) * treatgroup, family = binomial)
glm.pooled <- glm(cbind(x, n - x) ~
   factor(study) + treatgroup, family = binomial)
anova(glm.pooled, glm.individual, test = "Chisq")
summary.fixed <- summary(glm.pooled)
```

The "direct comparison" odds ratio effect estimate from `glm.pooled` was 0.616 with a 95% approximate Wald-based confidence interval from `summary.fixed` of 0.43 to 0.89. The test for heterogeneity yielded $p = 0.003$. From the analyses presented in reference, the "direct comparison" odds ratio fixed effect estimates were very similar, 0.64 with a 95% confidence interval of 0.45 to 0.90, but their test for heterogeneity yielded $p = 0.41$ which is very different. Our results were confirmed by exact logistic regression (program available by request from the authors).

16.5.4   Survival Analysis, Parenteral Nutrition in Cancer Patients Revisited

Of methodological note with the example presented in Section 16.5.1, the complete survival data (in addition to 3-month mortality) was largely extractable from the Kaplan–Meier plots in the individual studies as they had censoring marks. At the time, an analysis using Cox proportional hazards regression

showed similar results. The actual, by hand, extracted data from the Kaplan–
Meier curves are no longer at hand, but simulated survival data for three
fictitious studies allows us to outline the required S-PLUS coding for such an
MA. It is important to highlight that the correct analysis is to stratify on
baseline hazard (i.e., allow a separate baseline hazard function for each study)
rather than use an indicator term for study (i.e., force study baseline hazard
functions to be a multiple of some combined baseline hazard function):

```
# Cox proportional hazards regression
ph.individual <- coxph(Surv(time, censor) ~
    treatgroup * strata(study))
ph.pooled <- coxph(Surv(time, censor) ~
    treatgroup + strata(study))
# random effects calculations and estimates
mean.dev<-(2 * (ph.individual$loglik[2] -
    ph.pooled$loglik[2]))/2
ph.pooled$loglik <- ph.pooled$loglik/mean.dev
ph.pooled$wald.test <- ph.pooled$wald.test/mean.dev
ph.pooled$score <- ph.pooled$score/mean.dev
summary(ph.pooled)
```

16.5.5   Specialist Stroke Care and Length of Hospital Stay

Our last example involves continuous outcomes: comparisons of length of hospital stay for stroke patients under two different management protocols (Normand, 1999). The objective in the individual studies was to determine whether
specialist stroke care (managed by specialty units) results in a shorter length of
hospitalization compared to routine management on a general ward. This is not
an example from the pharmaceutical industry—but could well be of interest to a
pharmaceutical firm considering the development and/or evaluation of a drug for
the same condition (specifically questions about what should the placebo regimen consist of). It is also possibly an example of the temptation to combine
studies when such combining is questionable, and was for some reason irresisti-
ble. There is extreme observed heterogeneity!

Unfortunately in MA, continuous outcomes, even when assumed to be normally distributed, raise some obscure methodological questions mainly because
of the need to estimate within-study standard errors. With binary or count data,
the standard errors are not estimated but calculated from the observed proportions or counts. The same is true with Cox proportional hazards regression for
survival data. (In fact, Cox proportional hazards regression is actually an "MA"
of single event tables within a given study, and an S-PLUS program to demonstrate this is available from the authors.)

But with continuous data, the full maximum likelihood (ML) based estimates
use both within-study and between-study information when estimating the
within-study variance and this is believed by some to be "inappropriate." The
issue is not specific to MA, it is often discussed under the name reduced maxi-

mum likelihood estimation (REML) or partial likelihood, and some comments on the issues are available in Ripley (1999). From a practical point of view, it would seem sensible to take just the within-study estimates of the within-study standard errors as known (or estimated without error) given the usually fairly large sample size within studies. This was the approach adopted by Normand (1999).

The analysis can be done with the gls procedure (which allows different variances in different strata) by setting the values of the within-study variances to the observed within-study group or within-study pooled variances using the strata error term option. In keeping with our strategy of using "conservative" random effects, we will not use the random effects lme procedure but will use gls and divide the log-likelihoods by the mean deviance. Using lme (or nlme) would implement an normal theory exchangeable random effects model that again in some cases may inappropriately down-weight larger studies.

Additionally, gls is a fairly new procedure and we needed to both download a new release (NLME 3.3 can be obtained at http://nlme.stat.wisc.edu) and make a manual adjustment to get it to work correctly. One of the developers—Jose Pinheiro, from Bell Laboratories—has indicated that newer releases will correct the problem we encountered. We also were unable to fit a saturated model to calculate the observed deviance directly. In the Appendix (Section 16.A.6) we also reanalyze the MA using hand formulas from Normand (1999) and direct likelihood calculations. These turn out to be fairly simple, but do not provide the modeling framework that gls does:

```
# gls with varIdent options set to known variances
gls.pooled <- summary(gls(difflos ~ 1,
  weights = varIdent(form = ~ 1 | stdy, fixed = 11),
  method = "ML",
  control = glsControl(sigma = sqrt(vardifflos[1])))))
# for the correct standard error estimate,
# multiply ML estimate by sqrt(8/9)^2.
gls.pooled$tTable[2] <- gls.pooled$tTable[2] * sqrt(8/9)
```

The fixed effect estimates from all three methods were identical, as they should be, but the random effects estimates and standard errors were quite different. Our pooled estimate was −3.5 with a standard error of 4.3 while their estimates ranged from −10 to −15 and with standard errors from 5 to 9.5 depending on the particular random effects model they used.

---

[2] The problem results from the fact that gls currently applies a "correction" to the standard errors on the assumption that they are being estimated even though the options were set to "no estimation"—the incorrect correction being to multiply the standard errors by $\sqrt{N/(N-1)}$ where N = number of studies.

16.6 Further Challenges

Theory would suggest the use of conditional or exact methods for certain kinds of outcomes such as with binary outcomes. We have been unable to construct an example MA that has resulted in a practical difference in the pooled estimates and tests, but there can be some differences in estimates and tests of study homogeneity. With reasonably sized individual studies there is little or no discernable difference between the profile versus conditional likelihoods (consistent with what the asymptotic theory suggests), and all estimates and tests will be approximately the same. The exact methods are actually fairly straightforward for MA and example programs in S-PLUS are available from the authors. (On the other hand, the theory of random effects for both conditional and unconditional methods is not quite fully developed. For instance, although it is standard practice in MA to form confidence intervals and tests for pooled effects using critical levels from the chi-squared distribution, some have suggested the use instead of the F-distribution. The issue is reviewed in an MA context by O'Rourke (2001) and commented on in general by Venables and Ripley (1994, p. 215).)

Empirical Bayes or borrowing strength is a topic that often arises in MA. Given background knowledge about how RCTs are conducted and reported leads us to believe this will seldom be of important interest in most MAs. (Also see the discussion in the History Appendix (Section 16.A.1.) The apparent differences in treatment effect estimates will likely be due to the differing quality of the studies and not an intrinsically different treatment effect for a given study's "population." In such a case, a study-specific estimate (either borrowing strength from other estimates or not) would seem to be of little value. Exceptions of course will occur.

Quality appraisals of individual studies and their quantitative inclusion into MA remains controversial. The actual appraisal can be of great value for the discussion section of an MA, and not knowing how to quantitatively incorporate them should not lead to their abandonment. Sensitivity analysis may seem quite reasonable and a cumulative quality-based method was suggested in Detsky et al. (1992). Sensitivity analyses, however, are not always guaranteed to be more likely do more good than harm in MA (O'Rourke, 2001) or in other areas of applied statistics (Senn, 1996).

16.7 Conclusion

We had hoped to show that essentially nothing new or very different is required in MA other than the application of the usual principles and methods of statistics, and that S-PLUS software provides standard, useful, and flexible techniques for carrying out both the analytic and synthetic tasks of MA. We have tried to do this using a modeling framework (i.e., glm, coxph, gls) that facilitates model checking and graphing as well as allows an easy extension of the basic analysis (i.e., investigation of covariates and treatment interactions). We have provided only the most cursory sketches of the analysis required—almost all MAs require extensive sensitivity analyses. Hopefully we have done something to dispute the myth of a single study and distracted some membership away from the cult of a single study.

16.8 References

Anonymous (1997). Meta-analyses under scrutiny. *Lancet* **350**, 675.

Bailey, K.R. (1987). Inter-study differences: How should they influence the interpretation and analysis of results? *Statistics In Medicine* **6**, 351–358.

Begg, C., Cho, M., Eastwood, S., Horton, R., Moher, D., Olkin, I., Pitkin, R., Rennie, D., Schultz, K.F., Simel, D., and Stroup, D.F. (1996). Improving the quality of reporting of randomized controlled trials: The consort statement. *Journal of the American Medical Association* **276**(8), 637–639.

Berkey, C.S., Hoaglin, D.C., Mosteller, F., and Colditz, G.A. (1995). A random-effects regression model for meta-analysis. *Statistics in Medicine* **14**, 395–411.

Bucher, H.C., Guyatt, G.H., Griffith, L.E., and Walter, S.D. (1997). The results of direct and indirect treatment comparisons in meta-analysis of randomized controlled trials. *Journal of Clinical Epidemiology* **50**(6), 683–691.

Chatfield, C. (1995). Model uncertainty, data mining and statistical inference. *Journal of the Royal Statistical Society Association* **Series A 158**, 418–466.

Cochran, W.G. (1937). Problems arising in the analysis of a series of similar experiments. *Journal of the Royal Statistical Society* **4**(1), 102–118.

Cochran, W.G. (1980). Summarizing the results of a series of experiments. *Proceedings of the 25th Conference on the Design of Experiments in Army Research Development and Testing,* Durham, NC. ARO Report. U.S. Army Research Office.

Cook D.J., Sackett, D.L., and Sptizer, W.O. (1995). Methodologic guidelines for systematic reviews of randomized controlled trials in health care from the Potsdam consultation of meta-analysis. *Journal of Clinical Epidemiology* **48**, 167–171.

Counsell C. (1996). Formulating the questions and locating the studies for inclusion in systematic reviews. *Annals of Internal Medicine* **127**, 380–387.

Cox, D.R. (1982). Combination of data. In: Kotz, S. and Johnson, N.L., eds., *Encyclopedia of Statistical Science*, Vol. 2. Wiley, New York.

Cox, D.R. and Snell, E.J. (1989). *The Analysis of Binary Data*. Second edition. Chapman Hall, London.

DerSimonian, R. and Laird, N.M. (1986). Meta-analysis in clinical trials. *Controlled Clinical Trials* **7**, 177–188.

Detsky, A.S., Naylor, C.D., O'Rourke, K., McGeer, A.J., and L'Abbe, K.A. (1992). Incorporating variations in the quality of individual randomized trials into meta-analysis. *Journal of Clinical Epidemiology* **45**(3), 255–265.

Dickersin K., Scherer R., and Lefebvre, C. (1994). Identifying relevant studies for systematic reviews. *British Medical Journal* **309**(6964), 1286–1291.

Droitcour, J., Silberman, G., and Chelimsky, E. (1993). Cross-design synthesis—A new form of meta-analysis for combining results from randomized clinical trials and medical-practice databases. *International Journal of Technology Assessment in Health Care* **9**(3), 440–449.

Eddy, D.M., Hasselblad, V., and Shachter, R. (1990). An introduction to a Bayesian method for meta-analysis: The confidence profile method. *Medical Decision Making* **10**, 15–23.

Efron, B. (1996). Empirical Bayes methods for combining likelihoods. *Journal of the American Statistical Association* **91**(434), 538–565.

Egger, M., Smith, G.D., and O'Rourke, K. (2000). Rationale, potentials and promise. In: Egger, M., Smith, D.G., and Altman, eds., *Systematic Reviews in Health Care: Meta-Analysis in Context*. BMJ Books, London.

Egger, M., Schneider, M., and Smith, G.D. (1998). Spurious precision? Meta-analysis of observational studies. *British Medical Journal* **316**, 140–144.

Fisher, R.A. (1935). *The Design of Experiments*. Oliver and Boyd, Edinburgh, Scotland.

L'Abbe, K.A., Detsky, A.S., and O'Rourke, K. (1987). Meta-analysis in clinical research. *Annals of Internal Medicine* **107**, 224–233.

Lee, Y. and Nelder, J.A. (1996). Hierarchical generalized linear models. *Journal of the Royal Statistical Society* **58**(4), 619–678.

Longford, N.T. and Nelder, J.A. (1999). Statistics versus statistical science in the regulatory process. *Statistics In Medicine,* **18**(17–18), 2311–2320.

McCullagh, P. and Nelder, J.A. (1989). *Generalized Linear Models*, 2nd ed. Chapman Hall, New York.

McGeer, A.J., Detsky, A.S., and O'Rourke K. (1990). Parenteral nutrition in cancer patients undergoing chemotherapy: A meta-analysis. *Nutrition* **8**(3), 233–240.

Moher, D., Cook, D.J., Eastwood, S., Olkin, I., Rennie, D., and Stroup, D.F. (1999). Improving the quality of reports of meta-analyses of randomised controlled trials: The QUOROM statement. Quality of Reporting of Meta-analyses. *Lancet* **354**(9193), 1896-1900.

Mosteller, F. and Chalmers, T.C. (1992). Some progress and problems in meta-analysis of clinical trials. *Statistical Science* **7**(2), 227–236.

Mulrow, C.D. and Oxman, A., eds. (1997). How to conduct a Cochrane systematic review. *Cochrane Collaboration Handbook.* The Cochrane Library. The Cochrane Collaboration, Issue 4. Update Software, Oxford.

Normand, S.-L.T. (1995). Meta-analysis software: A comparative review. *The American Statistician* **49**(3), 298–309.

Normand, S.-L.T. (1999). Tutorial in biostatistics meta-analysis: Formulating, evaluating, combining, and reporting. *Statistics in Medicine* **18**, 321–359.

O'Rourke, K. (2001). Meta-analysis: Conceptual issues of addressing apparent failure of individual study replication or "inexplicable" heterogeneity. In: Ahmed, S.E. and Reid, N., eds., Lecture notes in statistics: *Empirical Bayes and Likelihood Inference.* Springer-Verlag, New York.

O'Rourke, K. and Detsky, A.S. (1989). Meta-analysis in medical research: Strong encouragement for higher quality in individual research efforts. *Journal of Clinical Epidemiology* **42**(10), 1021–1024.

Pearson, K. (1904). Report on certain enteric fever inoculation statistics. *British Medical Journal* **3**, 1243–1246.

Poole, C. and Greenland, S. (1999). Random-effects meta-analyses are not always conservative. *American Journal of Epidemiology* **150**(5), 469–475.

Richardson, W.S., Wilson, M.S., Mishikawa, J., and Hayward, R.S.A. (1995). The well-built clinical question: A key to evidence-based decisions. *American College of Physicians Journal Club* **123**(3), A12–A13.

Ripley, B. (1999). Modern data analysis in S-PLUS. *Proceedings of the 1999 International S-PLUS User Conference Workshop.* MathSoft, Seattle, WA.

Rosenbaum, P.R. (1999). Choice as an alternative to control in observational studies. *Statistical Science* **14**(3), 259–304.

Rothman, K.J. and Greenland, S. (1998). *Modern Epidemiology.* Lippincott–Raven, Philadelphia, PA.

Rubin, D.B. (1992). Meta-analysis: Literature synthesis or effect-size surface estimation? *Journal of Educational Statistics* **17**(4), 363–374.

Senn, S.J. (1996). The AB/BA cross-over: How to perform the two-stage analysis if you can't be persuaded that you shouldn't. In: Hansen, B. and De Ridder, M., eds. *Liber Amicorum Roel van Strik.* Erasmus University, Rotterdam, 93-100.

Shea, B., Dube, C., and Moher, D. (2000). Assessing the quality of reports of meta-analyses: a systematic review of scales and checklists. In: Egger, M., Smith, G.D., and Altman, D., eds., *Systematic Reviews in Health Care: Meta-Analysis in Context.* BMJ Books, London.

Smith, T.C., Spiegelhalter, D.J., and Parmar, M.H.K. (1996). Bayesian meta-analysis of randomized trials using graphical models and BUGS. In: Berry, D.A. and Stangl, D.K., *Bayesian Biostatistics*. Marcel Dekker, New York.

Stigler, S.M. (1986). *The History of Statistics: The Measurement of Uncertainty before 1900*. Belknap Press of Harvard University Press, Cambridge, MA.

Venables, W.N. and Ripley, B.D. (1994). *Modern Applied Statistics with S-PLUS*. Springer-Verlag, New York.

Yates, F. and Cochran, W.G. (1938). The analysis of groups of experiments. *Journal of Agricultural Science* **28**, 556–580.

16.A. Appendix

16.A.1   A Brief History of Meta-Analysis

As was mentioned earlier, MA is first and foremost an investigation of replication. Now the apparent lack of replication is a very old problem and figured prominently in astronomy and geodesy in the 1700s and 1800s. For instance, in spite of a firm belief that the diameter of Venus was essentially fixed, different observers, under different conditions, kept getting different answers. (Even the same observers under similar conditions, got different answers but that would not be considered an MA.)

If there is replication, then intuitively some combination of the estimates should be better—but WHICH combination and exactly HOW much better will it be? These two key questions attracted the attention of many early "meta-analysts." Two particularly well-known ones, Laplace and Gauss, essentially solved these questions. (That early "meta-analysts" developed the field of statistics may seem a bit of a stretch but it is a thesis advanced by Stigler in his book *The History of Statistics: The Measurement of Uncertainty Before 1900* (Stigler, 1986).

The challenge of combining the observations was taken up by Laplace, Gauss, Legendre, and others, and resulted in the development of the techniques and the probabilistic justification of least squares around 1810. We will start with Laplace in 1772. Now, probability models were being used to represent the uncertainty of observations caused by measurement error by the late 1700s. Laplace decided to rewrite these probability models not as the probability that an observation equalled "the true value plus some error" but simply as the truth plus the "probability of some error." This focused attention on the mean error rather than on the mean observation. Stigler (1986) suggests that this was important in order to recognize that for independent observations the probability of the joint error was simply a multiple of the probability of individual errors (i.e., Prob(error1, error2) = Prob(error1) × Prob(error2)).

Here, both the problem of how to combine estimates—and determining the amount of uncertainty in the combination—has a readily apparent solution. Since the likely error is proportional to Prob(error1) × Prob(error2), a combination is suggested by taking the most likely error as the error that actually occurred. From a pseudo-Bayesian perspective, with the additional specification of a prior probability of error, the probable error is Prior × Prob(error1) × Prob(error2), and a combination is suggested by taking the most probable error as the error that actually occurred. (Laplace used a uniform prior.)

The first is what we now call "likelihood" and the combination suggested by taking the most likely error is called the "maximum likelihood estimate." The second one is a kind of Bayesian approach resulting in a posterior distribution for the error in the combination, and the combination suggested by taking the most probable error could be called the maximum posterior estimate. (Note that it is probable error rather than probable value that this posterior represents.)

The only important difference between the two is the use of a prior to convert "most likely error" to "most probable error"—the combinations are both obtained via a multiplication of likelihoods. Hence, the WHICH combination question is simply answered by noting that one should multiply the likelihoods for independent observations. (And with some modification, dependent observations can also be dealt with.)

But Laplace's method of combining and quantifying uncertainty in the combination of observations was "only in principle," as it required an explicit probability distribution for errors in the individual observations and no acceptable ones existed. Later, Gauss drew on empirical experience that the mean of observations (prone to equal errors) was the best apparent combination of the observations. Reasoning that it might in fact be the ideal combination, Gauss worked out what the probability distribution of individual errors needed to be so that Laplace's method of maximum likelihood would result in the mean being the required combination. The probability distribution turned out to be what is today referred to as the Normal or Guassian distribution.

Laplace later showed that Gauss's method was more than just a good guess—but justified by the central limit theorem—at least for large sample sizes. Gauss however worried about the small sample sizes that often occurred in astronomy and geodesy, and returned to the problem in 1821. In this, he justified the least squares technique for small sample sizes by noting that the least squares combination would be unbiased (given that all the individual estimates were unbiased) and would have minimum variance. Most statistical techniques used today in MA follow from either this minimum variance unbiased justification or Laplace's maximum likelihood approach.

In 1904, Karl Pearson was perhaps the first to combine estimates not from astronomy or geodesy, but from medicine (Pearson, 1904). It is clear that Pearson was familiar with Laplace's and Gauss's work in astronomy and geodesy, having published a critique on part of a 1861 book by Airy that had reviewed their work for astronomers. Fisher, who was also aware of Gauss's work, addressed the combination of estimates in agriculture as early as 1935 (Fisher, 1935). Fisher's discussion was cast in the context of a particular example in a chapter entitled "Comparisons with interactions—cases in which we compare primary effects with interaction." In this, Fisher pointed out that treatments (often fertilizers) might react differently to different types of soil, but that often we want to ascertain that a treatment was not merely good on the aggregate of fields used but fields suitable for treatment. Where, as in astronomy, random effects were used to make allowances for varying amounts of measurement error, Fisher realized that in agricultural trials it was more likely that the treatment effect varied. This is quite different, in that if you were merely interested in the aggregate of fields used then the variation need not be allowed for, where as with varying amounts of measurement error (i.e., "constant day error" or "constant study error") it always needs to be allowed for.

Fisher's implementation of the random effects model is also interesting—he simply replaced the error in the analysis of variance table that is usually estimated from the within-study variances with the variance observed in treatment effects across the studies (hence his curious title above). Unfortunately, in the MA, he presented that the study sample sizes were equal (and the within-study variability was assumed to be equal), so it is difficult to tell how Fisher would have treated the more common situation of unequal sample sizes and variances.

In 1937, using likelihood methods, Cochran (1937) extended the problem Fisher looked at by allowing unequal variances and sample sizes and explicitly assuming that the biological variation was normally distributed. With these extensions and assumptions he derived the semiweighted mean as the maximum likelihood estimate, where the semiweights were determined both by the within-study estimate of error and the between-study variation of the observed effect. This is the modern random effects model (also known as the hierarchical linear model, mixed linear model, or multilevel model). Cochran did appear to be uncomfortable with some of the implications of his assumed model (especially the semiweights which down-weighted larger studies and perhaps differed from what Fisher would do) and commented over 40 years later that he was still working on it (Cochran, 1980). In 1938, Yates and Cochran (1938) presented extensive worked examples of the analysis of groups of experiments. Here, they pointed out that if the error variance is associated with the effect under-estimation, precision weighting would produce a biased estimate. They also recognized if there were random effects (due to variation in treatment effects) there would be "endless complications" in drawing conclusions.

16.A.2   S-PLUS Code for MA of Parenteral Nutrition in Cancer Patients (Section 16.5.1)

```
# data
study <- rep(1:8, 2)
treatgroup <- c(rep(0, 8), rep(1, 8))
success <- c(16, 19, 21, 13, 21, 13, 16, 57, 17, 19, 13,
    13, 10, 15, 20, 52)
```

```
n <- c(19, 21, 24, 18, 25, 15, 20, 62, 19, 20, 19, 14, 20,
   18, 25, 57)
failure <- n - success
# Generalized Linear Model
glm.individual <- glm(cbind(success, failure) ~
   factor(study)* treatgroup, family = binomial)
glm.pooled <- glm(cbind(success, failure) ~
   factor(study) + treatgroup, family = binomial)
glm.null <- glm(cbind(success, failure) ~
   factor(study), family = binomial)
# heterogeneity test
anova(glm.pooled, glm.individual, test = "Chi")
# pooled effect test(s) and estimates
anova(glm.null, glm.pooled, test = "Chi")
anova(glm.null, glm.pooled, test = "F")
summary.fixed <- summary(glm.pooled)
exp(summary.fixed$coef[9, 1])
exp(c(
   summary.fixed$coef[9, 1] + 2 * summary.fixed$coef[9, 2],
   summary.fixed$coef[9, 1] - 2 * summary.fixed$coef[9, 2]))
summary.random <- summary(glm.pooled,
   dispersion = glm.pooled$deviance/7)
exp(summary.random$coef[9, 1])
exp(c(summary.random$coef[9, 1] +
      2 * summary.random$coef[9, 2],
    summary.random$coef[9, 1] -
      2*summary.random$coef[9, 2]))
```

16.A.3 S-PLUS Code for MA of Prevention of Respiratory Tract Infections (Section 16.5.2)

```
# data input
success <- c(25, 24, 37, 11, 26, 13, 38, 29, 9, 44, 30, 40,
   10, 40, 4, 60, 12, 42, 26, 17, 23, 6, 7, 4, 20, 1, 10, 2,
   12, 1, 1, 22, 25, 31, 9, 22, 0, 31, 4, 31, 7, 3, 14, 3)
failure <-c(29, 17, 58, 6, 23, 71, 132, 31, 11, 3, 130,
   145, 31, 145, 42, 80, 63, 183, 31, 75, 0, 62, 40, 34, 76,
   13, 38, 99, 149, 27, 18, 27, 137, 169, 30, 171, 45, 100,
   71, 189, 48, 88, 11, 62)
treatgroup <- c(rep(1, 22), rep(0, 22))
study <- paste("s", rep(1:22, 2), sep = "")
# Generalized Linear Model
glm.pooled <- glm(cbind(success, failure) ~
   study + treatgroup, family = binomial)
```

```
summary.random <- summary(glm.pooled,
  dispersion = glm.pooled$deviance/21)
# Pooled random effects estimates.
# To match parameterization in original publication
# we take minus the coefficient.
exp(-summary.random$coef[23, 1])
exp(c(-(summary.random$coef[23, 1] +
      2 * summary.random$coef[23,2]),
    -(summary.random$coef[23,1] -
      2*summary.random$coef[23,2])))
# pooled random effects estimates,
# using default dispersion estimate
summary.random <- summary(glm.pooled, dispersion = 1)
exp(-summary.random$coef[23, 1])
exp(c(-(summary.random$coef[23, 1] +
      2 * summary.random$coef[23, 2]),
    -(summary.random$coef[23, 1] -
      2 * summary.random$coef[23, 2])))
```

16.A.4 S-PLUS Code for MA of *Pneumocystis Carinii* Pneumonia in HIV Infection (Section 16.5.3)

```
# data input
study <- rep(letters[15:22], each = 2)
# so that we can use a character label for treatment groups
# and still get odds ratios
options(contrasts = c("contr.treatment"))
treatgroup <- rep(c("t", "d"), 8)
success <- c(1, 9, 3, 8, 0, 1, 42, 41, 1, 1, 3, 13, 0, 6,
  6, 9)
failure <- c(65, 54, 104, 108, 15, 14, 234, 247, 38, 46,
  78, 72, 104, 90, 109, 96)
# Generalized Linear Model
glm.individual <- glm(cbind(success, failure) ~
  study * treatgroup, family = binomial, maxit = 25)
glm.pooled <- glm(cbind(success, failure) ~
  study + treatgroup, family = binomial)
# heterogeneity test
anova(glm.pooled, glm.individual, test = "Chi")
summary.fixed <- summary(glm.pooled)
exp(summary.fixed$coef[9, 1])
```

```
exp(c((summary.fixed$coef[9, 1] +
    2 * summary.fixed$coef[9, 2]),
  (summary.fixed$coef[9, 1] -
    2 * summary.fixed$coef[9, 2])))
summary.random <- summary(glm.pooled,
  dispersion = glm.pooled$deviance/7)
exp(summary.random$coef[9, 1])
exp(c((summary.random$coef[9, 1] +
    2 * summary.random$coef[9, 2]),
  (summary.random$coef[9, 1] -
    2 * summary.random$coef[9, 2])))
```

16.A.5 S-PLUS Code for MA of Survival Analysis, Parenteral Nutrition in Cancer Patients Revisited (Section 16.5.4)

```
.Random.seed <- c(33, 45, 4, 43, 55, 2, 62, 51, 30, 39, 5,
  3)
n <- 50
s1 <- rbind(cbind(1, 1, T.exp <- rexp(n, rate = .1),
  d.exp <- runif(n, 0, quantile(T.exp, .95)) > T.exp),
  cbind(1, 0, T.exp <- rexp(n, rate = .3),
  d.exp <- runif(n, 0, quantile(T.exp, .95)) > T.exp))
n <- 100
s2 <- rbind(cbind(2, 1, T.exp <- rexp(n, rate = .2),
  d.exp <- runif(n, 0, quantile(T.exp, .95)) > T.exp),
  cbind(2, 0, T.exp <- rexp(n, rate = .3),
  d.exp <- runif(n, 0, quantile(T.exp, .95)) > T.exp))
n <- 50
s3 <- rbind(cbind(3, 1, T.weibull <- rweibull(n, 0.5, 5.1),
  runif(n, 0, quantile(T.weibull, .9)) > T.weibull),
  cbind(3, 0, T.exp <- rweibull(n, 0.5, 5.1),
  runif(n, 0, quantile(T.weibull, .9)) > T.weibull))
#simmulated data
xx <- data.frame(rbind(s1, s2, s3))
dimnames(xx)[[2]] <- c("study", "treatgroup", "time",
  "censor")
#Cox proportional hazards regression
ph.individual <- coxph(Surv(time, censor) ~
  treatgroup * strata(study), data = xx)
ph.pooled<-coxph(Surv(time, censor) ~
  treatgroup + strata(study), data = xx)
```

```
# test for heterogeneity
1 - pchisq(
  2 * (ph.individual$loglik[2] - ph.pooled$loglik[2]), 2)
# fixed effect estimates
summary(ph.pooled)
# random effects calculations and estimates
mean.dev <- (2 * (ph.individual$loglik[2] -
  ph.pooled$loglik[2]))/2
ph.pooled$loglik <- ph.pooled$loglik/mean.dev
ph.pooled$wald.test <- ph.pooled$wald.test/mean.dev
ph.pooled$score <- ph.pooled$score/mean.dev
summary(ph.pooled)
```

16.A.6 S-PLUS Code for MA of Specialist Stroke Care and Length of Hospital Stay (Section 16.5.5)

```
# This code is based on using the NLME 3.3 library
# downloaded from http://nlme.stat.wisc.edu
library(nlme, first = T)
xx <- matrix(c(1, 1, 155, 55, 47, 1, 0, 156, 75,
  64, 2, 1, 31, 27, 7, 2, 0, 32, 29, 4, 3, 1, 75,
  64, 17, 3, 0, 71, 119, 29, 4, 1, 18, 66, 20, 4,
  0, 18, 137, 48, 5, 1, 8, 14, 8, 5, 0, 13, 18,
  11, 6, 1, 57, 19, 7, 6, 0, 52, 18, 4, 7, 1, 34,
  52, 45, 7, 0, 33, 41, 34, 8, 1, 110, 21, 16, 8,
  0, 183, 31, 27, 9, 1, 60, 30, 27, 9, 0, 52, 23,
  20), ncol = 5, byrow = T)
# using weighted least squares hand formulaes
ntreated <- xx[xx[ , 2] == 1, 3]
meantreated <- xx[xx[ , 2] == 1, 4]
sdtreated <- xx[xx[ , 2] == 1, 5]
ncontrol <- xx[xx[ , 2] == 0, 3]
meancontrol <- xx[xx[ , 2] == 0, 4]
sdcontrol<-xx[xx[ , 2] == 0, 5]
difflos <- meantreated - meancontrol
pooleds <- ((ntreated - 1) * sdtreated^2 +
    (ncontrol - 1) * sdcontrol^2)/(ntreated + ncontrol - 2)
vardifflos <- pooleds * (1/ntreated + 1/ncontrol)
losweights <- 1/vardifflos
losbarweighted <- sum(losweights * difflos)/sum(losweights)
varlosbarweighted <- 1/sum(losweights)
c(losbarweighted, sqrt(varlosbarweighted))
```

```
# random effects allow
overdisp <- sum(losweights *
  (difflos - losbarweighted)^2)/(9 - 1)
c(losbarweighted, sqrt(varlosbarweighted) * sqrt(overdisp))
# direct use of likelihoods
llik <- function(x, m, s)
  sum(log(dnorm(x, m, s)))
llik.m <- function(m)
  return(-llik(difflos, m, sqrt(vardifflos)))
mle <- optimize(llik.m, c(-5, 0))$min
dev.mle <- 2 * (llik(difflos, difflos, sqrt(vardifflos)) -
  llik(difflos, mle, sqrt(vardifflos)))
dev.null <- 2 * (llik(difflos, difflos, sqrt(vardifflos)) -
  llik(difflos, 0, sqrt(vardifflos)))
se <- abs(mle/sqrt(dev.null - dev.mle))
c(mle, se)
# random effects
re.llik <- function(x, m, s)
  sum(log(dnorm(x, m, s)))/(dev.mle/(9 - 1))
re.llik.m <- function(m)
  -re.llik(difflos, m, sqrt(vardifflos))
re.mle <- optimize(re.llik.m, c(-5, 0))$min
re.dev.mle  <- 2 *
  (re.llik(difflos, difflos, sqrt(vardifflos)) -
  re.llik(difflos, re.mle, sqrt(vardifflos)))
re.dev.null <- 2 *
  (re.llik(difflos, difflos, sqrt(vardifflos)) -
  re.llik(difflos,0,sqrt(vardifflos)))
re.se <- abs(re.mle/sqrt(re.dev.null - re.dev.mle))
c(re.mle, re.se)
# using gls
stdy <- LETTERS[1:9]
nn <-c("B" = 1,"C" = 1, "D" = 1, "E" = 1, "F" = 1,
  "G" = 1, "H" = 1, "I" = 2)
ll <- sqrt((vardifflos/vardifflos[1])[-1])
names(ll) <- names(nn)
gls.pooled <- summary(gls(difflos ~ 1,
  weights = varIdent(form= ~ 1 | stdy, fixed = ll),
  method = "ML",
  control = glsControl(sigma = sqrt(vardifflos[1]))))
gls.pooled$tTable[2] <- gls.pooled$tTable[2] * sqrt(8/9)
gls.pooled$tTable[1:2]
# summary fixed effects
# direct likelihood
c(mle, se)
```

```
# gls mle
gls.pooled$tTable[1:2]
# summary random effects
# direct likelihood
c(re.mle, re.se)
# gls mle with hand formula adjustment
c(gls.pooled$tTable[1],
  gls.pooled$tTable[2] * sqrt(overdisp))
```

Part 6:

Phase IV Studies

17

Analysis of Health Economic Data

John R. Cook, George W. Carides, and Erik J. Dasbach
Merck Research Laboratories, Blue Bell, PA, USA

17.1 Introduction

The purpose of this chapter is to demonstrate the application of S-PLUS in the analysis of health economic data. We have divided the chapter into two major sections. The first section is intended to provide the reader with a background on the field of pharmacoeconomic evaluation. In this section, we review the significance and relevancy of this type of analysis for health policy decisions as well as discuss the features which make this type of analysis unique with respect to other types of evaluation problems. In the remainder of the chapter, we demonstrate, via a case study, how we used S-PLUS in a pharmacoeconomic analysis. This analysis is based on a clinical study comparing the safety and efficacy of two drugs in patients with heart failure.

17.2 Background

Safety and efficacy have been the primary focus for the approval and marketing of pharmaceutical products. However, because of the growing concern over the cost of health care, decision-makers are increasingly utilizing additional information concerning patient outcomes, resource utilization, cost, and cost-effectiveness. Many hospital and managed care pharmacies have a strict process for budgeting pharmaceutical products. To ensure compliance with this process, lists, called *formularies*, are often developed to specify pharmaceuticals and the dosages that are deemed most appropriate or cost-advantageous for patient care. On a larger scale, some countries maintain national drug formularies to determine the level of reimbursement for drugs, while other countries use reference pricing to control the cost of pharmaceuticals at the national level. Because of these pressures to contain the cost of medical care, economic evaluations of pharmaceutical products have become increasingly important.

The emergence of economic analyses has brought about the development of guidelines for the conduct of these evaluations (Commonwealth of Australia, 1995; Ontario Ministry of Health, 1994; FDA Division of Drug Marketing, Advertising and Communications, 1995; Gold et al., 1996; Siegel et al., 1997). Moreover, the *New England Journal of Medicine* (Kassirer and Angell, 1994) and the *British Medical Journal* (Drummond and Jefferson, 1996) have issued guidelines for authors and reviewers of economic evaluations to be published in these journals. Additionally, several textbooks are now available that discuss methods for conducting economic evaluations of health care programs (Gold et al., 1996; Haddix et al., 1996; Drummond et al., 1997). These initiatives have increased interest among pharmaceutical companies to make comparative claims on the basis of economic endpoints, such as resource use, cost, and cost-effectiveness. As a result, many clinical development programs in pharmaceutical companies now include pharmacoeconomic objectives.

17.2.1   Modeling Studies

To meet these objectives, several research methods may be used. Modeling studies simulate patient outcomes and costs by making assumptions about the effects of therapy. Decision-analytic models list the possible consequences, or pathways, of treatment in a decision tree with the likelihood and expected cost of each pathway specified. These designs are useful for treatments for acute diseases that are characterized by recurrent disease states. Alternatively, epidemiological models have been used to assess the impact of therapy on diseases that may take years to develop. In these models, clinical and economic outcomes are typically related to a set of risk factors, one or more of which is affected by treatment.

17.2.2   Clinical–Economic Trials

A second general research method is the use of clinical–economic trials, where the scope of the trial is extended from the usual clinical evaluation to include economic endpoints. While these studies have the advantage of being randomized, double-blind, controlled trials that allow a great deal of internal validity, they have important drawbacks for economic evaluations (Drummond and Davies, 1991). The studies may have insufficient power to detect significant differences in both resource use and cost. The studies may also lack generalizability and external validity because of the use of a comparator (i.e., placebo) in a restrictive patient population that may be inappropriate to the decision-maker. In addition, the clinical trial period may be too short to assess the full impact of therapy on cost and outcomes. Finally, given these clinical studies typically are not designed to answer an economic question, an assessment of economic effects may be difficult because important covariates that affect resource use and costs may not be captured at baseline.

Strategies for undertaking economic assessments within the clinical trial have been discussed (Glick, 1995; Schulman et al., 1996; Mauskopf et al., 1996; Heyse et al., 2000). Because of the lack of power, recommendations have been made to use confidence intervals, rather than formal tests of hypotheses, to permit publication of results while allowing the readers to assess the magnitude of the significance (Drummond and O'Brien, 1993; O'Brien and Drummond, 1994). The limited time horizon can also be overcome by using data external to the trial in conjunction with prediction models based on the data collected during the trial.

17.2.3   Resource Utilization and Cost Analyses

When data on actual resources utilized (i.e., hospitalizations, procedures, outpatient visits) are collected in a clinical trial, one should first evaluate whether there are treatment differences in the rate of utilization. While this analysis may seem straightforward, the data will often present some challenges. First, utilization data will tend to include a few individuals who consume a large number of resources, while the majority of the patients utilize little or no resources. Thus, methods to test for treatment differences that rely upon the normality assumption (such as the t-test or ANOVA model) may be inappropriate. In addition, methods that ignore or dampen the influence of the high-resource patients (such as the median test or the Wilcoxon rank-sum test to compare the location shift in the distributions) may yield misleading results. The same problem exists with the analysis of cost data. For the payer, high-cost individuals are very important, so statistical methods that dampen their effect or ignore these extreme cost individuals may yield biased estimates of the expected costs associated with a particular treatment. Thus, the relevant statistic to the payer for decision-making with respect to costs is the cost per patient. Resampling methods, such as the bootstrap, are often used in this setting because they do not require strong assumptions about the form of the underlying distribution (Desgagne et al., 1998).

Another challenge arises when resource utilization and cost data are subject to censoring. Patient data may be incomplete due to study design considerations (e.g., study with a long enrollment period with a fixed study termination date) or because a patient is lost to follow-up. While censoring has been recognized as a real problem in the analysis of cost data (Fenn et al., 1995), traditional methods for incorporating censored information do not work well with cost data (Carides and Heyse, 1996). These methods assume that survival and censoring times are independent. Even if this assumption holds, the cumulative cost at the time of censoring is typically not independent of the total (uncensored) cumulative cost. Recently, several statistical methods have been proposed to deal specifically with censored cost data (Lin et al., 1997; Carides, et al., 2000; Bang and Tsiatis, 2000).

Finally, there is controversy regarding which costs should be included (Gold et al., 1996). Generally, health care costs associated with the intervention and the disease(s) related to the intervention are included. There is less consensus as to whether to include costs for diseases unrelated to the intervention and incurred during added years of life, and whether gains in productivity associated with decreased morbidity and mortality should be included.

17.2.4   Cost-Effectiveness Analysis

While an analysis of resource utilization and cost should be conducted first, one ultimately wants to answer the question of whether the benefits of a new treatment "outweigh" its cost. It is recommended that an incremental cost-effectiveness analysis be conducted to answer this question (Gold et al., 1996; Drummond et al., 1997). Such an analysis balances the difference in the average cost of care against the difference in the average health outcomes achieved with the new treatment relative to the comparator or standard treatment. This is typically evaluated using the incremental cost-effectiveness ratio, defined as the ratio of the additional (incremental) cost per additional (incremental) unit of health benefit for one therapy relative to another. Health benefits are typically measured by the gain in life years or quality-adjusted life years.

The duration of patient follow-up in the clinical–economic trial may be too short to evaluate gains in life years. In this case, models are typically used to project health benefits and health care costs beyond the observed trial period to the appropriate time horizon. These projections may be based directly on the limited mortality data if observed within the trial and/or on prediction models that link clinical events observed during the trial with reductions in life expectancy.

In cost-effectiveness analyses, both costs and benefits are typically discounted to present value. (With discounting, costs and benefits are assigned a lower value the farther into the future they are incurred or received to account for the time value of money; see Gold et al., 1996; Drummond et al., 1997). The practice of discounting costs has not been controversial. It is well recognized that a dollar is worth more today than a dollar in the future. However, there is debate as to whether health benefits should also be discounted. The issue rests on whether a year of life saved today should be valued more than a year of life saved in the future. Drummond et al. (1997, pp. 107–108) present arguments for both sides of the issue, and they, and others (e.g., Gold et al., 1996), conclude that both costs and benefits should be discounted at the same rate.

A great deal of attention has recently been placed on statistical approaches to assess uncertainty in the cost-effectiveness ratio (Mullahy and Manning, 1994; Manning et al., 1996; Drummond et al., 1997). Various methods have been proposed to construct 95% confidence intervals for the ratio (O'Brien et al., 1994; Wakker and Klaassen, 1995; Chaudhary and Stearns, 1996; Cook and Heyse, 2000). Polsky et al. (1997) compared several of these approaches in a simulation study, recommending either a bootstrap resampling technique or the use of Fieller's theorem (Fieller, 1954). Because of the often modest incremental health benefit and inadequate sample size in clinical trials for economic endpoints, analyses often uncover a great deal of uncertainty associated with the incremental cost-effectiveness estimate.

Drawing conclusions from cost-effectiveness analyses can be a complex task for a health care decision-maker because the decisions usually involve tradeoffs between costs and health outcomes. The key question for a decision-maker is "How high a cost am I willing to incur in order to achieve the additional health benefits associated with the new therapy?" To assist the decision-maker, the incremental cost-effectiveness ratio for the therapy can be ranked relative to the cost-effectiveness ratios for other well-accepted health care interventions. Presumably, if the intervention falls within the broad range of well-accepted interventions, then it should also be considered an acceptable therapy based on cost-effectiveness.

In some cases, a fixed threshold may be established by the decision-maker for how high a cost he or she is willing to incur in order to achieve the additional health benefits. If the cost effectiveness ratio is below this threshold, then the therapy would be deemed cost-effective (not withstanding the uncertainty associated with the point estimate). This acceptance region for a given threshold is represented in Figure 17.1 by Sectors IA, IIIB, and IV on the incremental cost-effectiveness plane. If the incremental cost-effectiveness ratio falls within Quadrant IV (where the new therapy is more effective and less costly), the new treatment is said to dominate the comparator and would be deemed acceptable regardless of the threshold. However, if the incremental cost-effectiveness ratio falls within Quadrants I or III, the acceptability of the therapy will depend on the value of the threshold.

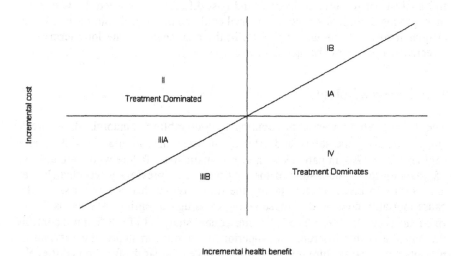

Figure 17.1. Cost-effectiveness quadrants with region of acceptance (IA, IIIB, IV) for a given threshold (shown as a diagonal line bisecting Quadrants I and III).

Because this threshold is likely to vary from one decision-maker to another it may not be possible to reach a definitive conclusion as to whether the therapy is cost-effective. In addition, even for a given threshold there is uncertainty associated with the estimated incremental cost-effectiveness ratio and, hence, uncertainty associated with the decision of whether the therapy is cost-effective relative to the threshold. One useful graphical display for the decision-makers is the cost-effectiveness acceptability curve proposed by vanHout et al. (1994). For any given threshold, the probability that the cost-effectiveness ratio falls below the threshold (i.e., in Sectors IA, IIIB, and IV where the therapy is acceptable) is estimated. The acceptability curve is generated by plotting the probability of acceptance (on the y-axis) over a range of threshold values (on the x-axis). It should be noted that there are no general guidelines for the decision-maker as to how high the acceptance probability should be in order to declare a therapy cost-effective. Instead, it should be used as a guide to the decision-maker as to the relative certainty associated with the decision.

17.3   Case Study: Evaluation of Losartan

This section presents an analysis of pharmacoeconomic data from a clinical trial, the losartan heart failure study, ELITE (Evaluation of Losartan in the Elderly Study), in which projections beyond the trial are required due to the limited time horizon of the clinical trial. It is assumed that unit cost data are available from external sources to estimate the cost for the resources incurred during the trial (see Copley-Merriman and Lair (1994) for a discussion of this issue). While a full analysis of resource utilization and cost differences between the two treatment groups during the course of the trial could be performed using S-PLUS, this chapter focuses on the use of S-PLUS in the assessment of the long-term cost-effectiveness of a new therapy.

17.3.1   Background

The ELITE Study was a multinational, double-blind, randomized, 48-week study comparing the safety and efficacy of two drugs, losartan ($n = 352$) and captopril ($n = 370$), in patients with symptomatic heart failure who were at least 65 years of age. The results of the trial have been previously reported (Pitt et al., 1997). In brief, patients taking the drug losartan had a lower risk of all-cause mortality compared to those taking the drug captopril (4.8% versus 8.7%; risk reduction: 46%, $p = 0.035$). A subsequent study, ELITE II, did not confirm the hypothesis that losartan was superior to captopril in improving survival in patients with heart failure due to systolic left ventricular dysfunction (Pitt et al., 2000). Prior to the completion of ELITE II, a pharmacoeconomic evaluation was conducted with the objective of estimating the incremental cost per life year gained for patients randomized to losartan compared to captopril (Dasbach et al., 1999). We reproduce much of that evaluation here using S-PLUS. In addition, we present a new survival projection to highlight S-PLUS features associated with evaluating time series data.

   In Section 17.3.2, the method used to estimate the life years gained with losartan relative to captopril is presented. The estimate is obtained by merging data from the clinical trial with United States age–sex specific mortality data. Because United States mortality rates are currently available only up through age 85, mortality rates beyond age 85 were predicted based on a model from the available data. The approach to estimate lifetime costs is presented in Section 17.3.3. The estimates are based on the observed utilization of resources (hospitalizations, ER visits, and study drug) during the trial and projected utilization rates beyond the trial for the survivors. External sources are used to obtain unit cost estimates to apply to the estimated utilization rates. Finally, in Section 17.3.4, results from Sections 17.3.2 and 17.3.3 are combined to estimate the incremental cost per life year gained. The method for constructing the cost-effectiveness acceptability curve based on the nonparametric bootstrap method is also provided.

17.3.2   Estimation of Life Years Gained

The first step in the cost-effectiveness analysis of the ELITE study is to estimate the gain in life years associated with losartan relative to captopril. In the ELITE study, all patients (including all who discontinued study medication) were followed to the minimum of time of death or 48 weeks. Patients surviving to the end of the trial period, therefore, are treated as censored with respect to lifetime survival. In this section we describe the method used to project the remaining lifetime for the censored patients and compute the life years gained.

Because only a small number of deaths occurred during the relatively short study duration (48 weeks), we use external data to project survival beyond the trial period rather than use the within-trial survival experience. Although survival was greater among patients randomized to losartan, we conservatively assume a common mortality hazard for the two treatment groups beyond the trial period. The starting point is the United States life table information by gender for the general population up to age 85 (National Center for Health Statistics, 1996). These data are read directly into S-PLUS to create the data frame life.table (see Appendix 17.A.1):

	Age	Male	Female
[1,]	60	81943	90105
[2,]	61	80710	89324
[3,]	62	79383	88478
....			
[25,]	84	26546	45736
[26,]	85	23532	42282

The values are the numbers of persons still alive at each age out of 100,000 live births for males and females. It is important to note that the life table information is not adequate to estimate remaining survival time as some patients have already reached age 85 by the end of the trial while others have a nonzero probability of surviving beyond age 85. We therefore estimate the gender-specific survival probabilities beyond age 85. To this end, we first compute discrete approximations to the hazard functions for males and females separately utilizing the life table data:

$$\hat{h}(t) = \frac{S(t-1) - S(t)}{S(t-1)} \tag{17.1}$$

where $t = 61, \ldots, 85$. We compute the log of the hazard (for males) as

```
> male <- rts(life.table$Male, start = life.table$Age[1])
> hazard.male <- (lag(male, -1) - male)/lag(male, -1)
> log.hazard.male <- log(hazard.male)
```

where `male` is a regular time series object (`rts`), the reason for which will become clear in what follows. Similar commands are used to create the log of the hazard for females (see Appendix 17.A.1).

The top-left plot in Figure 17.2 displays the natural logarithm of $\hat{h}$ versus age for males, along with the fitted least-squares line. The top-right plot displays residuals from this fit versus age. The bottom two plots display the same thing for females (the commands to produce these plots are given in Appendix 17.A.1). The apparent near linearity of these relationships shown in the left-hand plots suggests that the Gompertz distribution $h(x) = \exp(\alpha + \beta x)$ (Lawless, 1982) might be appropriate for projecting survival beyond age 85. However, the residual versus age plots reveal a curious wave pattern indicative of either the need to include higher-order terms or the presence of autocorrelated errors. Inclusion of second- and third-degree terms did not remove the pattern in the residuals, so we consider methods appropriate for time-dependent structures.

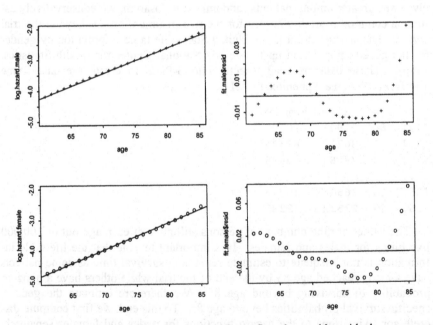

Figure 17.2. Log hazards versus age with linear fits, along with residuals versus age. The symbol "o" denotes female, and "+'" denotes male.

Because we wish to "forecast" survival beyond age 85 and it appears that the series log.hazard.male and log.hazard.female exhibit autoregressive structures, we applied the ARIMA ("autoregressive integrated moving average") methodology of Box and Jenkins (1976). Under this framework an autoregressive process of order p (AR(p)) is defined by

$$X_t = \sum_{i=1}^{p} \phi_i X_{t-i} + \varepsilon_t \tag{17.2}$$

where the ϕ_i's are a series of coefficients for the prior observations, and the ε_i's are a series of uncorrelated random variables with mean zero and variance σ^2. A moving average process of order q (MA(q)) is defined by

$$X_t = \sum_{j=1}^{q} \beta_j \varepsilon_{t-j} + \varepsilon_t \tag{17.3}$$

and an ARMA(p, q) process is defined by

$$X_t = \sum_{i=1}^{p} \phi_i X_{t-i} + \sum_{j=1}^{q} \beta_j \varepsilon_{t-j} + \varepsilon_t \tag{17.4}$$

An ARIMA(p, d, q) process is a process whose d^{th} difference is an ARMA(p, q) process.

Much has been written on identifying appropriate ARIMA models (Venables and Ripley, 1999). In this section we rely primarily on S-PLUS graphical tools for identifying an appropriate model. The usual starting point is inspection of the autocorrelation function (ACF) and partial autocorrelation function (PACF). The ACF gives the correlations between the series and successive lagged values, and decays in an exponential, sinusoidal, or geometric manner, approximately, for an AR(p) process. The PACF gives the correlations between X_s and X_{s+t} after regression on $X_{s+1}, \ldots, X_{s+t-1}$, and is zero for $t > p$ for an AR(p) process. We can easily obtain plots of both of these functions, as shown in Figure 17.3, using the acf function:

```
> par(mfrow = c(2, 2))
> acf(log.hazard.male)
> acf(log.hazard.male, type = "partial")
> acf(log.hazard.female)
> acf(log.hazard.female, type = "partial")
> par(mfrow = c(1, 1))
```

The PACFs strongly suggest an AR(1) structure for both series, but the ACFs appear to dampen too slowly. Thus, the series may not be stationary, i.e., ϕ may not be less than 1. The usual remedy to nonstationarity is to take differences until we obtain a stationary series.

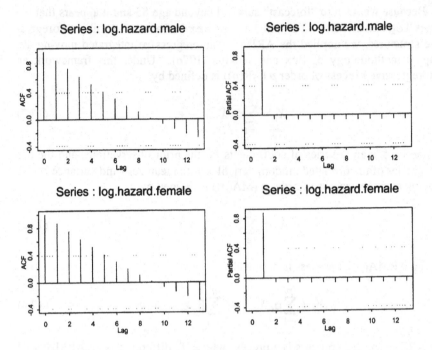

Figure 17.3. ACFs and PACFs for log hazards.

Because it is difficult to determine the degree of differencing required to obtain a stationary series through visual inspection of the ACF alone, we consider other graphical means. One useful device is the cumulative periodogram of the residuals. An S-PLUS function for this, cpgram, is available in the MASS library that comes with S-PLUS. (The latest version of the MASS library can be downloaded from http://www.stats.ox.ac.uk/pub/MASS3/sites.html.) The details of the cumulative periodogram, are beyond the scope of this chapter and can be found in Venables and Ripley (1999). The spirit of this plot is similar to that of a quantile–quantile plot; i.e., a roughly straight line indicates that the model assumptions, in our case stationarity of the process, are met. We use the function arima.mle to fit once, twice, and thrice difference AR(1) models for males and females, and the function arima.diag to extract the residuals:

```
> ar110.male <- arima.mle(log.hazard.male,
    model = list(order = c(1, 1, 0)))
> diag110.male <- arima.diag(ar110.male, resid = T,
    gof.lag = 10, plot = F)
> library(MASS)
> cpgram(diag110.male$resid,
    main = "ARIMA(1,1,0) Fit Males")
```

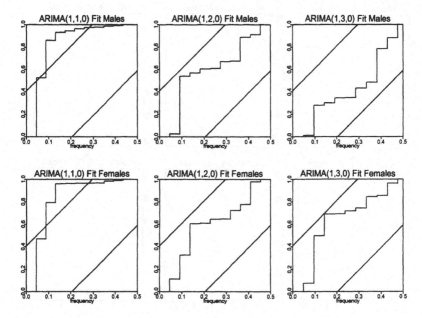

Figure 17.4. Cumulative periodograms of residuals for various models.

Figure 17.4 displays the cumulative periodograms for the three models for males and females (see Appendix 17.A.2 for the code). The ARIMA(1, 1, 0) models are clearly inappropriate as the curves are markedly concave and break through the upper 95% confidence band. An ARIMA(1, 3, 0) model seems like the best fit for males, although some curvature remains. An ARIMA(1, 2, 0) model seems most appropriate for females.

In order to obtain greater confidence about our model choices, we apply the Ljung-Box-Pierce portmanteau test (Ljung and Box, 1978; Box and Pierce, 1970)

$$Q_m = n(n+2)\sum_{k=1}^{m} \frac{r_k^2}{(n-k)} \qquad (17.5)$$

where r_k is the sample autocorrelation function of the residuals and m is the number of lags. Large values of Q_m (and small p-values) are indicative of auto-correlated errors and hence, an inappropriate model. Figure 17.5 shows the p-values associated with the portmanteau tests for up to 10 lags, where a horizontal line is drawn for the 0.05 significance level. These plots are generated by modifying the function `arima.diag.plot` so that only the portmanteau test p-values are produced. The results lead us to the same conclusions as did the cumulative periodograms; i.e., the ARIMA(1, 3, 0) and ARIMA(1, 2, 0) models are reasonable for males and females, respectively.

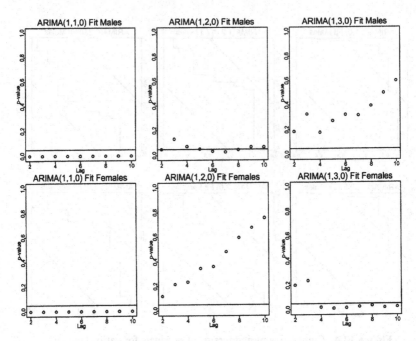

Figure 17.5. Ljung-Box-Pierce portmanteau goodness-of-fit tests for various models.

Next, we use `arima.forecast` to obtain predicted values for the hazard rates beyond age 85 using the above fitted models:

```
male.forecast <- arima.forecast(log.hazard.male, n = 18,
  model = ar130.male$model)
hazard2.male <- rts(c(hazard.male,
  exp(male.forecast$mean) * exp(ar130.male$sigma2/2)),
    start = 61)
hazard2.male[hazard2.male > 1] <- 1
```

where the second term in the product for `hazard2.male` is required to obtain the estimated mean hazard in original (as opposed to log) units. The actual (to age 85) and predicted (beyond age 85) hazards are shown in Figure 17.6 (see Appendix 17.A.3 for the code). Although the predicted hazards are greater than 1 for men beyond age 98 and women beyond age 102, we force these hazards to be 1.

Recursive application of (17.1) yields the lifetime survival function

$$S(t) = S(85) \prod_{i=0}^{t-86} \left[1 - h(t-i)\right], \quad t \geq 86 \qquad (17.6)$$

which we then use to estimate the mean residual lifetime as follows:

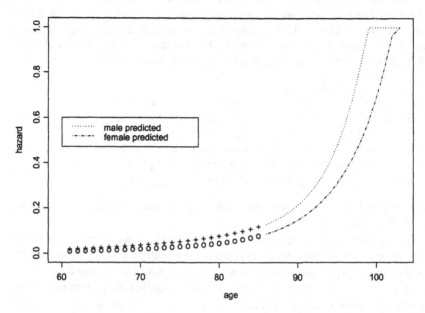

Figure 17.6. Actual ("o" denotes female, "+" denotes male) and predicted hazards for death.

$$m(x) = \frac{\int_{x}^{\infty} S(t)\, dt}{S(x)} \qquad (17.7)$$

where x is the age of the surviving patient at the end of the trial.

```
> age <- time(male)
> surv.male <- male/100000
> surv.male.85 <- surv.male[age == 85]
> age <- time(hazard2.male)
> predicted.surv.male <- surv.male.85 *
    cumprod(1 - hazard2.male[age >= 86])
> surv2.male <- rts(c(surv.male, predicted.surv.male),
    start = 60)
> rlife.male <- rev(cumsum(rev(surv2.male)))/surv2.male
> rlife.male[is.na(rlife.male)] <- 0
```

In this model, we assume that patients in the ELITE heart failure cohort will most likely experience greater mortality than those from the general population matched by age and gender. For the reference case, we assume that the expected remaining lifetime is one-half that of the general population based on the ratio of the observed mortality rate in the captopril arm to the mortality rate of an age–gender matched cohort from the general population. This assumption implies that a multiplier of $k = 0.5$ be applied to the estimated remaining lifetime of patients surviving to the end of the trial. We also include other choices of k (0.2, 0.8, 1.0) to see how sensitive the results are to this assumption:

```
> rlife2.male[16:20,]
 age    rlife.5    rlife.2    rlife.8    rlife
  75   4.887849   1.955140   7.820559   9.775698
  76   4.625212   1.850085   7.400340   9.250425
  77   4.368405   1.747362   6.989448   8.736809
  78   4.117118   1.646847   6.587389   8.234237
  79   3.871047   1.548419   6.193676   7.742095
```

(See Appendix 17.A.4 for the code to produce `rlife2.male` and `rlife2.female`.)

We are now ready to access the ELITE data (which we stored as a SAS dataset) in order to estimate the average remaining lifetime for the two treatments. For this, patient-level information is needed on gender (SEX), age at the end of the trial (AGE.END), follow-up time (RELDAY), whether the patient died during the trial (CENSOR), and an indicator variable for treatment group (TRTDUM). For the UNIX version of S-PLUS we would use the `importData` function. For the Windows version, we click on **File > Import Data > From File**:

```
> Elite[c(1, 15, 16, 18), ]
  ID   SEX   RELDAY   CENSOR   TRTDUM   AGE.END
0001    M      336       0        0        70
0002    M      336       0        1        67
0003    M       50       1        0        66
0004    F      336       0        1        75
```

After splitting the dataset into separate files for males and females, we merge the ELITE data with the residual lifetime calculated above:

```
> surv4.male <- merge(Eltmale, rlife2.male,
    by.x = "AGE.END", by.y = "age")
```

Note that, unlike in SAS, it is not necessary to sort these files by age prior to merging. We assume that patients surviving to the end of the 11-month trial period will survive to at least 1 year, and calculate the remaining lifetime in years from the start of the trial as follows:

```
> elife.5 <- ifelse(surv4.male[,'CENSOR'] == 0,
    surv4.male[ , 'rlife.5'] + 1,
    surv4.male[ , 'RELDAY']/365.25)
```

The variable CENSOR is the indicator for death (1) or survival to the end of the trial period (0), and RELDAY is the follow-up time in days. We compute remaining lifetime for females in the same way and then combine the two to form the dataset surv.dat using the code:

```
> surv.dat <- rbind(surv5.male, surv5.female)
```

Part of the dataset appears below:

AGE.END	ID	SEX	RELDAY	CENSOR	TRTDUM	elife.5
9	66	0003	M	50	1	0	0.1368925
40	67	0002	M	336	0	1	8.2059043
126	70	0001	M	336	0	0	7.2922358
602	75	0004	F	336	0	1	7.1015151

The difference in the average remaining lifetimes is 6.236 years (losartan) − 6.033 years (captopril) = 0.203. For the cost-effectiveness analysis, the esti-mated lifetime for each patient needs to be discounted. A simple way to accom-plish this is to view survival time as a continuous annuity and discount mean re-sidual lifetime, $m(x)$, with rate r using the well-known formula $[1-e^{-r\ m(x)}]/r$ (Haeussler and Paul, 1993). This is implemented in S-PLUS with a discount rate of 3% ($r = 0.03$). The difference in the average discounted remaining survival time (DT) is 5.787 (losartan) − 5.595 (captopril) = 0.192 for the reference case of $k = 0.5$. The other choices for k (0.2, 0.8, 1.0) provide estimates of average discounted life-years gained of 0.094, 0.282, and 0.338, respectively.

17.3.3 Estimation of Lifetime Cost

Next, we estimate the mean cost per patient in the two treatment groups to ob-tain the incremental cost of the new therapy. Just as with survival time, lifetime costs are censored for those who survived to the end of the trial. The total cost for patients over their remaining lifetime is based on their cost during the trial and an estimated cost over their projected remaining lifetime beyond the trial. As was the case for patient-level survival and demographics, we read a SAS da-taset of resource utilization and costs (1997 dollars) into S-PLUS:

```
> Eltcost[c(1, 15, 16, 18), ]
```

	ID	HDAYS	CUM.ER	IN.MED	FUT.MED
1	0001	0	0	564.435	635.535
15	0002	3	0	371.030	0.000
16	0003	0	0	53.385	0.000
18	0004	0	0	402.750	445.605

where HDAYS is the cumulative number of days hospitalized, CUM.ER is the cumulative number of emergency room (ER) visits, IN.MED is the study medi-cation cost within-trial, and FUT.MED is the estimated future annual cost of study medication (based on the randomized study treatment dose at the end of the trial). Patients who discontinued study medication during the trial are as-signed zero future study medication costs (see patients 0002 and 0003 above). (By assigning zero future medication costs following discontinuation, the total cost is underestimated. But, due to the greater discontinuation rate in the captopril arm, the analysis will be conservative for losartan.) The within-trial total cost, within.cost, is computed as

```
> within.cost <- 1051 * Eltcost[ , 'HDAYS'] +
    521 * Eltcost[ , 'CUM.ER'] + Eltcost[ , 'IN.MED']
```

where $1051 is the average per day Medicare hospital payment, and $521 is the average Medicare payment for an emergency room visit (U.S. Department of Health and Human Services, 1997).

For patients dying during the trial, the discounted lifetime cost (*DLC*) is given by their within-trial cost (within.cost). Patients surviving to the end of the trial are expected to incur additional costs for study medication, hospitalizations, and ER visits during their estimated remaining lifetime. This additional cost is estimated for each patient by multiplying their estimated residual lifetime with their expected annual cost for study medication (FUT.MED), hospitalizations, and ER visits. Prior analysis revealed a similar rate of hospitalizations and ER visits during the trial for the two treatments. Consequently, we assume future costs associated with hospitalizations and ER visits will be the same for surviving patients regardless of their randomized treatment. For hospitalizations and ER visits, the expected annual cost was estimated by the annualized average cost during the trial among all survivors ($3839 per year). Based on the higher cost during the trial among patients who died, it is expected that the costs will also be higher near the time of death for patients who survive beyond the trial. The "exit" cost for hospitalizations and ER visits among patients dying during the trial is estimated based on the excess cost of patients dying within-trial as compared with the survivors ($12,020). We merge these cost data with the expected remaining lifetime data to obtain the discounted lifetime cost, *DLC*, for patients surviving beyond-trial:

$$DLC = \text{within.cost} + \frac{1}{12}(\$3839 + \text{FUT.MED}) +$$

(17.8)

$$\frac{(\$3839 + \text{FUT.MED})\left[1 - e^{-0.03\,m(x)}\right]}{0.03} + \$12,020\,e^{-0.03[1+m(x)]}$$

where $3839 is the annual emergency room plus hospitalization cost of patients surviving to the end of the trial period (11 months), and $12,020 is the estimated cost associated with patient death.

For the reference case ($k = 0.5$), the average discounted lifetime cost per patient randomized to losartan is $34,981 and to captopril is $34,585. Thus, the estimated incremental cost per patient is $396. For choices for k of 0.2, 0.8, and 1.0, the estimated incremental cost per patient was $182, $599, and $727, respectively.

Next, with S-PLUS it is easy to obtain a confidence interval for the difference between mean lifetime cost for the two treatment groups. We simply use the function `bootstrap` to obtain a 95% BCa confidence interval (see Efron and Tibshirani, 1993) for the reference case of $k = 0.5$:

```
> bts.cost <- bootstrap(Elite2.dat,
    mean(discount.life.cost.5[TRTDUM == 1]) -
    mean(discount.life.cost.5[TRTDUM == 0]),
    group = TRTDUM)
> limits.bca(bts.cost, probs = c(25, 975)/1000)
          2.5%     97.5%
Param -1378.036 2535.778
```

Thus, the 95% BCa percentile confidence limits are (−$1378 to $2536) with the value of $k = 0.5$. More discussion of the bootstrap method of inference is provided in Section 17.3.4.

17.3.4   Cost-Effectiveness Analysis

As in many studies comparing both costs and benefits of new therapies, the ELITE study indicates that losartan numerically increases life expectancy over captopril while also numerically increasing cost. The purpose of the cost-effectiveness analysis is to provide a measure to assess whether the added benefit of a therapy justifies its added cost and to provide some guidance regarding its interpretation. To this end, we estimate the incremental cost-effectiveness ratio, defined as

$$ICER = \frac{E(DLC_L) - E(DLC_C)}{E(DT_L) - E(DT_C)} \qquad (17.9)$$

where DT is the discounted remaining survival time and the subscripts L and C denote losartan and captopril, respectively. The estimated $ICER$ for ELITE is $2065 per year of life gained ($396/0.1918) for the reference case of $k = 0.5$. The other choices for k (0.2, 0.8, 1,0) provide $ICER$ estimates of $1965, $2124, and $2151, respectively.

The key question to the decision maker is whether they are willing to incur a cost of roughly $2000 to $3000 to gain an extra year of life with the new therapy. These ratios (for the various choices for k) are all well within the range generally accepted as cost-effective (Goldman et al., 1992). By comparison, the cost-effectiveness ratio for hemodialysis in end-stage renal disease has been estimated to range from $20,000 to $79,000 per life year gained (Tengs et al., 1995). Thus, losartan should be accepted on the basis of cost-effectiveness if hemodialysis is considered a cost-effective therapy.

While the $ICER$ estimate falls below most common thresholds for declaring a therapy as cost-effective ($20,000 to $100,000), there is uncertainty associated with this decision. One way to characterize the uncertainty of the decision is to estimate the probability that the $ICER$ falls below a given threshold. Because some decision-makers may use a different threshold, the "acceptance" probabil-

444   J.R. Cook, G.W. Carides, and E.J. Dasbach

ity is typically estimated across a range of possible threshold values. The cost-effectiveness acceptability curve (VanHout et al., 1994) is used to display the results of this analysis. For a given threshold, we use a nonparametric bootstrap method (Efron and Tibshirani, 1993) to compute the probability that the cost-effectiveness ratio is acceptable (i.e., compute the probability that the cost-effectiveness ratio is in Sectors IA, IIIB, or IV of Figure 17.1). The acceptance probability is simply the percentage of bootstrap replicates that fall within the acceptance region for each candidate threshold. The bootstrap method involves sampling with replacement from the empirical joint distribution of lifetime costs and life expectancy. Specifically, we take a random sample of patients with replacement and compute:

1. $[\,E(DLC_L) - E(DLC_C)\,]^*$ (incremental discounted cost),
2. $[\,E(DT_L) - E(DT_C)\,]^*$ (incremental discounted remaining survival time),
3. $ICER^*$ (incremental cost-effectiveness ratio),

where the asterisks refer to statistics computed from a random sample with replacement. We repeat this process $B = 1000$ times thus creating bootstrap distributions of each of these three statistics.

Two S-PLUS functions are written to carry easily out this analysis (see Appendix 17.A.5). The first function, theta.fct, calculates the mean cost difference, the mean lifetime difference, and the $ICER$ for a single bootstrap replication. The second function, elite.bt, calls this function B times, plots the bootstrap replicates of incremental costs versus incremental life years gained, and computes the probabilities of acceptance for a range of potential cost-effectiveness thresholds (e.g., $10,000, $20,000, ..., $100,000).

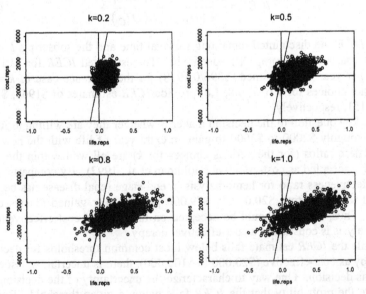

Figure 17.7. Bootstrap pairs of incremental costs and survival for values of k.

The function is run with $B = 1000$ for each of the life expectancy assumptions. Figure 17.7 shows the plots of the bootstrap replicates of the estimated incremental cost versus incremental life years gained for $k = 0.2$, 0.5, 0.8, and 1.0. In addition to the scatterplot, a reference cost-effectiveness threshold of $50,000 (diagonal line with slope of 50,000) is drawn in each of the four plots. Based on this threshold, the probability that losartan is cost-effective relative to captopril is 0.91 for the reference value of $k = 0.5$; i.e., 91% of the bootstrap replicates lie to the southeast of the threshold line (in Sectors IA, IIIB, and IV in Figure 17.1).

The other choices of k (0.2, 0.8, 1.0) provide probabilities of acceptance of 0.93, 0.91, and 0.92, respectively. Figure 17.8 shows the acceptance curves for the chosen values of k.

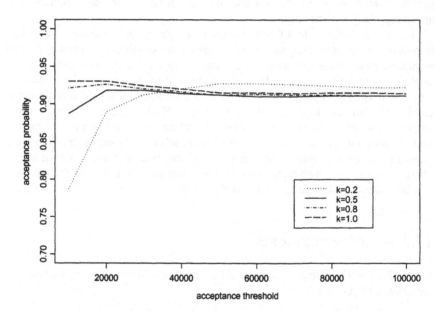

Figure 17.8. Cost-effectiveness acceptability curves for losartan compared to captopril based on the ELITE Study for alternative assumptions concerning survival relative to the general population (k).

17.4   Summary and Conclusions

In this chapter, we demonstrate how S-PLUS can handle the analytic challenges posed in a pharmacoeconomic evaluation. While S-PLUS can be used to examine treatment differences in resource utilization and cost during the course of a clinical trial, we focused on how it can be used to conduct a cost-effectiveness evaluation based on a lifetime projection of survival benefit and cost.

One challenge is to obtain an appropriate model for projecting survival beyond the trial. In our example, a time series model was used to forecast the hazard rates beyond age 85, the age limit for the United States life tables at that time. The built-in functions and graphics capabilities of S-PLUS allow one to thoroughly conduct an exploratory analysis to suggest an appropriate class of survival models. With the estimated hazard rates, it was easy in S-PLUS to estimate the annual survival probabilities and then the mean residual lifetime.

Another challenge with pharmacoeconomic evaluations is the need to use resampling techniques to assess the uncertainty of not only simple statistics (such as the discounted incremental cost) but also more complex statistics such as the acceptance probability. With S-PLUS, BCa confidence intervals can easily be obtained with the function bootstrap. In addition, user-defined functions can be developed (such as theta.fct) to bootstrap more complex statistics. The graphical interface with S-PLUS allows one to display the bootstrap replicates to aid interpretation of the results.

Economic evaluations of new treatment programs will become even more important as concern over the cost of health care continues to grow. Formal economic evaluations can assist health care decision-makers to decide whether new treatment programs provide sufficient benefit to the patient relative to their cost. Increasingly, these evaluations for new pharmaceutical products are being conducted alongside randomized clinical trials. Such evaluations often require external cost data to be merged with resource utilization data captured within the trial. In addition, projections of health benefit and cost beyond the trial are often required due to the limited time horizon of the clinical trial. Consequently, pharmacoeconomic evaluations can be fairly complex and require the use of a flexible statistical software package such as S-PLUS.

17.5 Acknowledgments

The authors wish to thank Shannon Allen for her collaboration in the analysis presented in Section 17.3.2.

17.6 References

Bang, H., and Tsiatis, A.A. (2000). Estimating medical costs with censored data. *Biometrika* **87**, 329–343.

Box, G.E.P. and Jenkins, G.M. (1976). *Time Series Analysis: Forecasting and Control*. Revised edition. Holden-Day, San Francisco, CA.

Box G.E.P. and Pierce, D.A. (1970). Distribution of residual autocorrelation in autoregressive-integrated moving average time series models. *Journal of the American Statistical Association* **80**, 580–619.

Carides, G.W. and Heyse, J.F. (1996). Nonparametric estimation of the parameters of cost distributions in the presence of right-censoring. *Proceedings of the Biopharmaceutical Section.* American Statistical Association Annual Meetings 1996, pp. 186–191.

Carides G.W., Heyse, J.F., and Iglewicz, B. (2000). A regression-based method for estimating mean treatment cost in the presence of right-censoring. *Biostatistics* **1**, 299–313.

Chaudhary, M.A. and Stearns, S.C. (1996). Estimating confidence intervals for cost effectiveness ratios: An example from a randomized trial. *Statistics in Medicine* **15**, 1447–1458.

Commonwealth of Australia. (1995). *Guidelines for the pharmaceutical industry on preparation of submissions to the Pharmaceutical Benefits Advisory Committee: Including economic analyses.* Department of Health and Community Services, Canberra, Australia.

Cook, J.R. and Heyse, J.F. (2000). Use of an angular transformation for ratio estimation in cost-effectiveness analysis. *Statistics in Medicine* **19**, 2989–3003.

Copley-Merriman, C. and Lair, T.J. (1994). Valuation of medical resource units collected in health economic studies. *Clinical Therapeutics* **16**, 553–568.

Dasbach, E.J., Rich, M.W., Segal, R., Gerth, W.C., Carides, G.W., Cook, J.R., Murray, J.F., Snavely, D.B., and Pitt, B. (1999). The cost-effectiveness of losartan versus captopril in patients with symptomatic heart failure. *Cardiology* **91**, 189–194.

Desgagne, A., Castilloux, A.M., Angers, J.F., and LeLorier, J. (1998). The use of the bootstrap statistical method for the pharmacoeconomic cost analysis of skewed data. *PharmacoEconomics* **13**, 487–497.

Drummond, M.F. and Davies, L.M. (1991). Economic analysis alongside clinical trials: Revisiting the methodological principles. *International Journal of Technology Assessment in Health Care* **7**, 561–573.

Drummond, M.F. and Jefferson, T.O., on behalf of the BMJ Economic Evaluation Working Party (1996). Guidelines for authors and peer reviewers of economic submissions to the BMJ. *British Medical Journal* **313**, 275–283.

Drummond, M.F. and O'Brien, B.J. (1993). Clinical importance, statistical significance and the assessment of economic and quality of life outcomes. *Health Economics* **2**, 205–212.

Drummond, M.F., O'Brien, B.J., Stoddart, G.L., and Torrance, G.W. (1997). *Methods for Economic Evaluation of Health Care Programmes,* 2nd ed., Oxford University Press, Oxford, UK.

Efron, B. and Tibshirani, R.J. (1993). *An Introduction to the Bootstrap.* Chapman & Hall, New York.

448   J.R. Cook, G.W. Carides, and E.J. Dasbach

Fenn, P., McGuire, A., Phillips, V., Backhouse, M., and Jones, D. (1995). The analysis of censored treatment cost data in economic evaluation. *Medical Care* 33, 851–863.

Fieller, E.C. (1954). Some problems in interval estimation. *Journal of the Royal Statistical Society Series B* 16, 175–185.

Food and Drug Administration. (1995). *Comparing treatments: Safety, effectiveness and cost-effectiveness.* Masur Auditorium, National Institutes of Health, Bethesda, MD, March 23–24.

Glick, H. (1995). Strategies for economic assessment during the development of new drugs. *Drug Information Journal* 29, 1391–1403.

Gold, M.R., Siegel, J.E., Russell, L.B., and Weinstein, M.C., eds. (1996). *Cost-Effectiveness in Health and Medicine.* Oxford University Press, New York.

Goldman, L., Gordon, D.J., Rifkind, B.M., Hulley, S.B., Detsky, A.S., Goodman, D.W., Kinosian, B., and Weinstein, M.C. (1992). Cost and health implications of cholesterol lowering. *Circulation* 85(5), 1960–1968.

Haddix, A.C., Teutsch, S.M., Shaffer, P.A., and Dunet, D.O., eds. (1996). *Prevention Effectiveness: A Guide to Decision Analysis and Economic Evaluation.* Oxford University Press, New York.

Haeussler, E.F. and Paul, R. (1993). *Introductory Mathematical Analysis for Business, Economics, and the Life and Social Sciences,* 7th ed. Prentice Hall, Englewood Cliffs, NJ, pp. 581–582.

Heyse, J.F., Cook, J.R., and Drummond, M.F. (2000). Statistical considerations in pharmacoeconomic evaluations. In: Chow, S.C., ed., *Encyclopedia of Biopharmaceutical Statistics.* Marcel Dekker, New York.

Kassirer, J.P. and Angell, M. (1994). Editorial: The Journal's policy on cost-effectiveness analysis. *New England Journal of Medicine* 331, 669–670.

Lawless, J.F. (1982). *Statistical Models and Methods for Lifetime Data.* Wiley, New York, p. 26.

Lin, D.Y., Fewer, E.J., Etzioni, R., and Wax, Y. (1997). Estimating medical costs from incomplete follow-up data. *Biometrics* 53, 113–128.

Ljung, G.M. and Box, G.E.P. (1978). On a measure of lack of fit in time series models. *Biometrika* 65, 297-303.

Manning, W.G., Fryback, D.G., and Weinstein, M.C. (1996). Reflecting uncertainty in cost-effectiveness analysis. In: Gold, M.R., Siegel, J.E., Russell, L.B., and Weinstein, M.C., eds., *Cost-Effectiveness in Health and Medicine.* Oxford University Press, New York, Chap. 8, pp. 247–275.

Mauskopf, J., Schulman, K., Bell, L., and Glick, H. (1996). A strategy for collecting pharmacoeconomic data during Phase II/III clinical trials. *PharmacoEconomics* 3, 264–277.

Mullahy, J. and Manning, W. (1994). Statistical issues in cost-effectiveness analyses. In: Sloan, F., ed., *Valuing Healthcare Costs, Benefits, and Effectiveness of Pharmaceuticals and Other Medical Technologies.* Cambridge University Press, New York, Chap. 8, pp. 149–241.

National Center for Health Statistics. (1996). *Vital Statistics of the United States*, 1992, Vol. II, Sec. 6, life tables. Public Health Service, Washington, DC.

O'Brien, B.J. and Drummond, M.F. (1994). Statistical versus quantitative significance in the socioeconomic evaluation of medicines. *PharmacoEconomics* **5**, 389–398.

O'Brien, B.J., Drummond, M.F., LaBelle, R.J., and William, A. (1994). In search of power and significance: Issues in the design and analysis of stochastic cost-effectiveness studies in health care. *Medical Care* **32**, 150–163.

Ontario Ministry of Health. (1994). *Ontario guidelines for economic analysis of pharmaceutical products.* Ministry of Health, Toronto, Canada.

Pitt, B., Poole-Wilson, P.A., Segal, R., et al., on behalf of the ELITE II Investigators. (2000). Effect of losartan compared with captopril on mortality in patients with symptomatic heart failure: randomised trial — the Losartan Heart Failure Survival Study – ELITE II. *Lancet* **355**, 582–1587.

Pitt, B., Segal, R., and Martinez, F.A. (1997). Randomised trial of losartan versus captopril in patients over 65 with heart failure (Evaluation of Losartan in the Elderly Study, ELITE). *Lancet* **349**, 747–752.

Polsky, D., Glick, H.A., Willke, R., and Schulman, R. (1997). Confidence intervals for cost-effectiveness ratios: A comparison of four methods. *Health Economics* **6**, 243–252.

Schulman, K.A., Llana, T., and Yabroff, K.R. (1996). Economic assessment within the clinical development program. *Medical Care* **34**, DS89–DS95.

Siegel, J.E., Torrance, G.W., Russell, L.B., Luce, B.R., Weinstein, M.C., Gold, M.R., and the members of the Panel on Cost-Effectiveness in Health and Medicine (1997). Guidelines for pharmacoeconomic studies: Recommendations from the panel on cost effectiveness in health and medicine. *PharmacoEconomics* **11**, 159–168.

Tengs, T.O., Adams, M.E., and Pliskin, J.S. (1995). Five-hundred life-saving interventions and their cost-effectiveness. *Risk Analysis* **15**, 369–390.

U.S. Department of Health and Human Services. (1997). *Health Care Financing Review Statistical Supplement.* U.S. Department of Health and Human Services Health Care Financing Administration. Office of Research and Demonstrations, Baltimore, MD, p. 93, Table 33.

Van Hout, B.A., Maiwenn, J.A., Gordon, G.S., and Rutten, F.F.H. (1994). Costs, effects and C/E-ratios alongside a clinical trial. *Health Economics* **3**, 309–319.

Venables, W.N. and Ripley, B.D. (1999). *Modern Applied Statistics with S-PLUS*. Third edition. Springer-Verlag, New York.

Wakker, P. and Klaassen, M.P. (1995). Confidence intervals for cost/effectiveness ratios. *Health Economics* **4**, 373–381.

17.A. Appendix

17.A.1 S-PLUS Code to Create Figure 17.2

```
# Read in the Life Table data
life.table <- read.table("lifetable.txt", header = T,
  skip = 7)
# Compute hazard and log-hazard for males
male <- rts(life.table$Male, start = life.table$Age[1])
hazard.male <- (lag(male, -1) - male)/lag(male, -1)
log.hazard.male <- log(hazard.male)
# Compute hazard and log-hazard for females
female <- rts(life.table$Female, start = life.table$Age[1])
hazard.female <- (lag(female, -1) - female)/lag(female, -1)
log.hazard.female <- log(hazard.female)
# Fit linear model of log.hazard.male vs. age
age <- time(log.hazard.male)
fit.male <- lm(log.hazard.male ~ age)
# Fit linear model of log.hazard.female vs. Age
fit.female <- lm(log.hazard.female ~ age)
# Plot log.hazard.male vs. age with fitted line and
#       Residuals from fit vs. age,
#       then do the same thing for Females
orig.mfrow <- par(mfrow = c(2, 2))
plot(age, log.hazard.male, pch = "+", ylim = c(-5, -2))
abline(fit.male)
plot(age, fit.male$resid, pch = "+")
abline(h = 0)
plot(age, log.hazard.female, pch = "o", ylim = c(-5, -2))
abline(fit.female)
plot(age, fit.female$resid, pch = "o")
abline(h = 0)
par(orig.mfrow)
```

17.A.2 S-PLUS Code to Create Figure 17.4

```
library(MASS)
par(mfrow = c(2, 3))
# Create plots for males
ar110.male <- arima.mle(log.hazard.male,
  model = list(order = c(1, 1, 0)))
diag110.male <- arima.diag(ar110.male, resid = T,
  gof.lag = 10, plot = F)
cpgram(diag110.male$resid, main = "ARIMA(1,1,0) Fit Males")
ar120.male <- arima.mle(log.hazard.male,
  model = list(order = c(1, 2, 0)))
diag120.male <- arima.diag(ar120.male, resid = T,
  gof.lag = 10, plot = F)
cpgram(diag120.male$resid, main = "ARIMA(1,2,0) Fit Males")
ar130.male <- arima.mle(log.hazard.male,
  model = list(order = c(1, 3, 0)))
diag130.male <- arima.diag(ar130.male, resid = T,
  gof.lag = 10, plot = F)
cpgram(diag130.male$resid, main = "ARIMA(1,3,0) Fit Males")
# Create plots for females
ar110.female <- arima.mle(log.hazard.female,
  model = list(order = c(1, 1, 0)))
diag110.female <- arima.diag(ar110.female, resid = T,
  gof.lag = 10, plot = F)
cpgram(diag110.female$resid,
  main = "ARIMA(1,1,0) Fit Females")
ar120.female <- arima.mle(log.hazard.female,
  model = list(order = c(1, 2, 0)))
diag120.female <- arima.diag(ar120.female, resid = T,
  gof.lag = 10, plot = F)
cpgram(diag120.female$resid,
  main = "ARIMA(1,2,0) Fit Females")
ar130.female <- arima.mle(log.hazard.female,
  model = list(order = c(1, 3, 0)))
diag130.female <- arima.diag(ar130.female, resid = T,
  gof.lag = 10, plot = F)
cpgram(diag130.female$resid,
  main = "ARIMA(1,3,0) Fit Females")
par(mfrow = c(1, 1))
```

17.A.3 S-PLUS Code to Create Figure 17.6

```
# Create forecasts of log.hazard beyond age 85
# for Males and Females
male.forecast <- arima.forecast(log.hazard.male, n = 18,
  model = ar130.male$model)
hazard2.male <- rts(c(hazard.male,
  exp(male.forecast$mean) * exp(ar130.male$sigma2/2)),
    start = 61)
hazard2.male[hazard2.male > 1] <- 1
female.forecast <- arima.forecast(log.hazard.female,
  n = 18, model = ar120.female$model)
hazard2.female <- rts(c(hazard.female,
  exp(female.forecast$mean) * exp(ar120.female$sigma2/2)),
    start = 61)
hazard2.female[hazard2.female > 1] <- 1 #
# Create Figure 17.6
age <- time(hazard2.male)
plot(age, hazard2.male, type = "n", ylab = "hazard",
  ylim = c(0, 1))
points(age[age <= 85], hazard2.male[age <= 85], pch = "+")
lines(age[age > 85], hazard2.male[age > 85], lty = 2)
points(age[age <= 85], hazard2.female[age <= 85],
  pch = "o")
lines(age[age > 85], hazard2.female[age > 85], lty = 3)
legend(61, 0.6,
  legend = c("male predicted", "female predicted"),
  lty = c(2, 3))
```

17.A.4 S-PLUS Code to Estimate Lifetime Survival Function and Mean Residual Lifetime

```
# Create estimates for Males
age <- time(male)
surv.male <- male/100000
surv.male.85 <- surv.male[age == 85]
age <- time(hazard2.male)
predicted.surv.male <- surv.male.85 *
  cumprod(1 - hazard2.male[age >= 86])
surv2.male <- rts(c(surv.male, predicted.surv.male),
  start = 60)
rlife.male <- rev(cumsum(rev(surv2.male)))/surv2.male
```

```
rlife.male[is.na(rlife.male)] <- 0
rlife2.male <- cbind(age = time(rlife.male),
  rlife.5 = 0.5 * rlife.male, rlife.2 = 0.2 * rlife.male,
  rlife.8 = 0.8 * rlife.male, rlife = rlife.male) #
# Create estimates for Females
age <- time(female)
surv.female <- female/100000
surv.female.85 <- surv.female[age == 85]
age <- time(hazard2.female)
predicted.surv.female <- surv.female.85 *
  cumprod(1 - hazard2.female[age >= 86])
surv2.female <- rts(c(surv.female, predicted.surv.female),
  start = 60)
rlife.female <- rev(cumsum(rev(surv2.female)))/surv2.female
rlife.female[is.na(rlife.female)] <- 0
rlife2.female <- cbind(age = time(rlife.female),
  rlife.5 = 0.5 * rlife.female,
  rlife.2 = 0.2 * rlife.female,
  rlife.8 = 0.8 * rlife.female, rlife = rlife.female)
```

17.A.5 S-Plus Code to Perform Cost-Effectiveness Analysis

This section contains the functions used to conduct the cost-effectiveness analysis and to plot the results. The core of the outer function elite.bt is the built-in S-Plus function bootstrap. This function can bootstrap virtually any statistic in order to provide bias and variance estimates, and nonparametric confidence intervals. The inner function theta.fct computes the difference in mean lifetime cost (losartan–captopril), the difference in mean lifetime (losartan–captopril), and the cost-effectiveness ratio for a single set of bootstrap resampled patients. The code appears below.

```
elite.bt <- function(num.reps) {
  results <- bootstrap(Elite2.dat, theta.fct,
  B = num.reps, group = TRTDUM)$replicates #
##############################################
# Function to calculate CE ratio and
# probability of acceptance -- ELITE.
# This function is the theta (statistic)
# for which the bootstrap is applied.
#
  theta.fct <- function(x) {
    delta.cost <-
      mean(x$discount.life.cost.5[x$TRTDUM == 1]) -
      mean(x$discount.life.cost.5[x$TRTDUM == 0])
```

```
  delta.life <-
    mean(x$elife.5d[x$TRTDUM == 1]) -
    mean(x$elife.5d[x$TRTDUM == 0])
  ce.ratio <- delta.cost/delta.life
  c(delta.cost,delta.life,ce.ratio)
}
###################################################
  cost.reps <- results[ , 1]
  life.reps <- results[ , 2]
  ce.reps <- results[ , 3]
  plot(life.reps, cost.reps, xlim = c(-1, 1),
    ylim = c(-6000, 6000))
  abline(0, 0)
  abline(0, 1000000000)
  abline(0, 50000)
  title('k=0.5') #
#
# Compute probability of acceptance.
#
  prob.accept <- single(10)
  for(i in 1:10) {
    ind.1A <- ifelse(cost.reps > 0 & life.reps > 0 &
      ce.reps < 10000 * i, 1, 0)
    ind.3B <- ifelse(cost.reps < 0 & life.reps < 0 &
      ce.reps > 10000 * i, 1, 0)
    ind.4 <- ifelse(cost.reps < 0 & life.reps > 0, 1, 0)
    num.1A <- sum(ind.1A)
    num.3B <- sum(ind.3B)
    num.4 <- sum(ind.4)
    prob.accept[i] <- (num.1A + num.3B + num.4)/num.reps
  }
#
# Write acceptance probabilities to external file
# to access later for plotting acceptability curves.
  write.table(cbind(
    c(10, 20, 30, 40, 50, 60, 70, 80, 90, 100),
      prob.accept), file = 'c:\\bootreps\\reps5.txt',
    dimnames.write = F)
}
```

Part 7:

Manufacturing and Production

18

Evaluation of the Decimal Reduction Time of a Sterilization Process in Pharmaceutical Production

Jeffrey Eisele
Department of Biostatistics, Novartis Pharma AG, Basel, Switzerland

Mauro Gasparini
Dipartimento di Matematica, Politecnico di Torino, Torino, Italy

Amy Racine
Department of Biostatistics, Novartis Pharma AG, Basel, Switzerland

18.1   Introduction

Consider measuring the effectiveness of sterilization during the production of a pharmaceutical substance. The decimal reduction time, D-value, of a sterilization process is defined to be the time required to reduce the number of microorganisms present by a factor of 10. Because the issue of concern is the rate of bacterial death this process is often referred to as "Death Kinetics." Different methods for sterilization are possible depending on the nature of the pharmaceutical product, for example, heat treatment. Since the D-value depends to a large extent on environmental conditions, it is necessary to determine experimentally the substance specific D-value for each substance that is sterilized by means of moist heat. If the D-value of the substance is the same as that of a reference substance, typically water, the sterilization process is deemed successful.

Various international guidelines (e.g., *American National Standard*, 1991; *International Standard*, 1994) describe and define the parameters within which the experiments for estimating the D-value should be carried out. These include experimental design considerations such as time points for sampling, minimum number of samples, number of replicates, and methods for analysis and modeling.

The survivor curve is defined to be the mean of the logarithm (base 10) of the bacterial counts versus time from onset of sterilization. Although there is widespread evidence that the survivor curve is not linear over the full range of inactivation, the *American National Standard* (1991) and *International Standard* (1994) guidelines prescribe the use of a linear model. The D-value is then the negative of the inverse of the slope of the linear regression.

Often the lack of log-linearity is caused by a "shoulder" that occurs during the initial phase of the heat treatment. To get around this the *International Standard* (1994) guidelines state that "Survivor data points within 0.5 log of the initial population shall not be included in the regression analysis." The guidelines also state that "The value obtained for the correlation coefficient for the linearity of the survivor curve shall be not less than 0.80."

Working within the restrictions set by the guidelines, the task at hand is to provide the laboratory scientists responsible for quality assurance with an automated statistical analysis of their experiments. This requires development of a practical tool for every-day use by the laboratory scientists and not a one-time analysis of a dataset.

At the time the problem arose in our company SAS was the only statistical programming language available to the laboratory scientists. Thus, the original implementation was in SAS. We found it easier to first solve the problem in S-PLUS and then to rewrite our interactive S-PLUS code in SAS. We present here our original solution to the problem using S-PLUS. For illustrative purposes, the most important S-PLUS commands are introduced and commented on step-by-step in this chapter. In reality, the organization of the work in the laboratory is such that batch programming is preferred (rather than writing all the S-PLUS operations described in this chapter as one or more functions). This can be achieved by putting all commands in a source file and dumping the results to output files using the `sink` command. A similar mechanism is required for outputting graphics; for example, in this chapter the `postscript` command is used.

The remainder of this chapter is organized as follows. In Section 18.2 we describe the experimental design, including some regulatory restrictions. This is followed in Section 18.3 with a description of the data and the data manipulation necessary to put it into a useable form. In Section 18.4 the statistical methodology is introduced and an application is presented in Section 18.5. A discussion of the statistical methods is presented in Section 18.6. The Appendix contains graphical presentation of the data, the complete S-PLUS code in batch format, and a listing of the data used for the application.

18.2   Experimental Design

For each assay, the microbial count of the test substance and water are determined at time 0 and at equally spaced time intervals thereafter. At each time point there are replicate (usually three) determinations of the microbial count. Several assays are then conducted under similar conditions. The *D*-value of the test substance is compared to that of water by pooling the replicate assays.

This is a regulated process and as such conditions may be imposed on the experimental design by regulatory authorities. The *International Standard* (1994) guidelines include the following conditions:

1.   A minimum of five exposures (measurements).

2. The bacterial count must be measured at equally spaced time intervals.
3. An equal number of replicates must be taken for each exposure.

The *American National Standard* (1991) suggests a minimum of 18 samples, three replicate determinations at each of six exposure times.

18.3 Data Processing

The data is originally stored in an ASCII file in a matrix format with columns indicating assay, time, microbial counts for the test substance, and microbial counts for water. In the dataset used for illustration purposes here, there are three assays with three replications at each time point. In the first assay the sampling times are 0, 5, 10, 15, and 20 min, in the second assay 0, 4, 8, 12, 16, and 20 min, and in the third assay 0, 5, 8, 11, 14, 17, and 20 min. A sample of the data in the form of the ASCII file is presented in Table 18.1.

Table 18.1. Data in ASCII format.

Assay	Time	Test substance	Water
1	0	135000	660000
1	0	875000	585000
...	...	...	...
3	20	119	4
3	20	77.5	10.5

Data manipulation within S-PLUS is needed to stack the last two columns and to introduce a new variable (column) labeling the substance (test or water). This can be achieved using `read.table` to read the ASCII data into a temporary array in S-PLUS. The substance and water counts can then be stacked using the `cbind` command in combination with adding a column indicating whether the counts are for water or the test substance (labeled "abc123"). A sample of the resulting data array, stored in the data frame `data`, is shown below. The complete dataset can be found in Appendix 18.A.5. A graphical display of the data can be found in Section 18.5 when one considers the dots without the interpolating lines.

```
> data
      treatment  assay  time    counts
  1     abc123      1      0  135000.0
  2     abc123      1      0  875000.0
...
107     water       3     20       4.0
108     water       3     20      10.5
```

18.4 Statistical Methodology

The standard approach to estimation of the D-value of a sterilization process is to fit a linear regression of the $\log_{10}$ of the microbial count, N, on the time in process, T (see Heintz et al., 1976). That is

$$\log_{10}(N) = \alpha + \beta \cdot T + \varepsilon \tag{18.1}$$

where α and β are the intercept and slope and depend on the substance under consideration (test substance or water), and ε is a random error term. The parameters α and β may vary between assays. The D-value is then $-1/\beta$.

A problem with this approach is that in linear regression observations at the low and high ends of the time axis will have the greatest influence on the estimation of α and β. This is because the leverage of a point in linear regression is a function of the distance of its x-value from the mean of all of the x-values (Cook and Weisberg, 1982). At later time points, the microbial count may fall below the detection limit and thus be assigned an arbitrary value (e.g., the detection limit). At the beginning of the sterilization process there may be a lag in time to response (death) due to low temperatures. This means that values which cannot be determined and/or lie in the region of little influence of the sterilization process may have the greatest weight on the estimation of the D-value. This can result in an unstable estimation of the regression line and hence an unreliable estimate of the D-value. The inclusion or exclusion of these early and late values from the estimation process could make a large difference in the estimate of the D-value. See Section 18.6 for further discussion of this issue.

Thus, to avoid the lag time problem we follow the *International Standard* (1994) guidelines and leave to the laboratory scientist the responsibility for excluding, when necessary, any time point for which the $\log_{10}(N)$ is within 0.5 of $\log_{10}(N)$ at time zero from the model fitting. In the application presented in Section 18.5, no data points were excluded from the model fitting.

18.4.1 Test for Linearity

The first step in the estimation of the D-value is to determine if a simple linear regression model is appropriate. This is done by fitting a quadratic regression model to each assay, for both the test substance and water, and testing for the significance of the quadratic term with a t-test. A nonsignificant test along with observation of the graphical output (plots of the data and fitted linear regression curves) will indicate the plausibility of the linearity assumption. Time points may then be deleted manually by the laboratory scientist to assure compliance with the requirement of linearity on the part of the guidelines. A minimum of three equally spaced time points is required.

18.4.2 Modeling Individual Assays

Because there are often replicates of assays, separate estimates of the D-value are made for each assay and a pooled estimate of the D-value is made by combining the results from all assays. For the individual assays a linear regression of the form given in (18.1) is fitted to each assay (separately for the test substance and water). This allows for each substance within an assay to have a different error variance. The least squares estimates of β, b, and its standard error, s, are then used to estimate the D-value, $-1/\beta$, and to calculate its 100 $(1-\alpha)\%$ confidence interval for each assay. The confidence interval for the D-value is obtained by taking the negative reciprocal of the endpoints of the confidence interval for β:

$$b \pm t_{1-\alpha/2, n-2} \cdot s \qquad (18.2)$$

where $n =$ (number of time points) $\times$ (number of replicates) within each assay.

For each assay, the ratio of the D-values is estimated using water as the reference. The confidence interval for this ratio is calculated using Fieller's theorem, a clever way to obtain an exact confidence interval for the ratio of the means of two jointly normal variates obtained by Fieller (1954). See Kotz et al. (1983) and references therein for a discussion. Let (X, Y) be a bivariate normal random vector with means μ and ν, variances σ^2 and τ^2, and covariance γ, and let $\wedge$ indicate the corresponding sample quantities, to be viewed as estimates. Then a $100(1-\alpha)\%$ confidence interval for μ/ν is given by

$$\frac{\hat{\mu} \cdot \hat{\nu} - z^2 \cdot \hat{\gamma} \pm \left[\left(\hat{\mu} \cdot \hat{\nu} - z^2 \cdot \hat{\gamma} \right)^2 - \left(\hat{\mu}^2 - z^2 \cdot \hat{\sigma}^2 \right) \cdot \left(\hat{\nu}^2 - z^2 \cdot \hat{\tau}^2 \right) \right]^{0.5}}{\hat{\nu}^2 - z^2 \cdot \hat{\tau}^2} \qquad (18.3)$$

where z is the $100(1-\alpha/2)$ percentile of the standard normal distribution. In the application presented here, γ is 0, since the slopes of the test substance and water in the same assay are estimated separately.

A ratio of 1 indicates that the test substance and water have the same D-value. Thus, if this interval contains the value 1, then there is no significant evidence that the D-value of the test substance differs from that of water. If the lower (upper) end of the interval is greater (less) than 1, then there is evidence that the D-value for the test substance is greater (less) than that of water.

18.4.3 Pooling of Assays

Obtaining a global estimate of the ratio of the D-value of the test substance to that of water requires pooling the results from the assays. The approach taken here is to fit to the assays a nonlinear model including a common parameter representing this ratio. Estimation of this parameter then provides a global estimate of the ratio of the D-values. The nonlinear model has the following form:

$$\log_{10}\left(N_{ijk}\right) = \alpha_{ij} + \theta_i \cdot \beta_j \cdot T_{ijk} + \varepsilon_{ijk}, \quad \text{where} \quad \theta_i = \begin{cases} \theta & \text{if } i = 1 \\ 1 & \text{if } i = 2 \end{cases} \quad (18.4)$$

where $i = 1$ (water), 2 (substance), j runs from 1 to the number of assays, and k runs from 1 to the number of time points within each assay. The nonlinear model described by (18.4) is linear within each assay but it allows for different slopes for water and substance. The ratio of such slopes, θ, is assumed constant across assays. Finally, it is also assumed that the within-assay error is the same for each assay.

In (18.4), the parameter θ represents the ratio of the D-values since

$$\frac{\text{D-value substance}}{\text{D-value water}} = \frac{-1/\beta}{-1/\theta \cdot \beta} = \theta \quad (18.5)$$

The maximum likelihood estimate of θ is then used to estimate the ratio of the D-value of the test substance to that of water.

A standard $100(1-\alpha)\%$ Wald confidence interval

$$\hat{\theta} \pm z_{1-\alpha/2} \cdot \hat{\sigma}_\theta, \quad (18.6)$$

is calculated for the ratio.

18.5   Application

The data introduced in Section 18.3 are used here to illustrate the methodologies described in Section 18.4. The outline of this section follows that of Section 18.4. Only the most important S-PLUS commands are noted. The complete S-PLUS code for the analysis can be found in the Appendix.

18.5.1   Test for Linearity

To obtain a visual inspection of the data, a first step in determining whether the linearity assumption is reasonable, the data are plotted using the plot command and least squares lines added with abline. The resulting plot for assay 1 is displayed in Figure 18.1. (The plots for assays 2 and 3 can be found in Appendix 18.A.1, and the code to produce the plots for all three assays is in Appendix 18.A.2.) The assumption of linearity does not appear to be completely unreasonable, although a "shoulder" can be clearly seen in all three assays.

The quadratic model described in Section 18.4.1 is fitted using the lm command. To test for linearity, a second model is fitted without the quadratic term and using the update command. The results of the test for linearity at level $\alpha = 0.05$ are given separately for each assay, and for the test substance and water in Table 18.2. Since none of the tests is significant, we can assume that the log-linear model is reasonable.

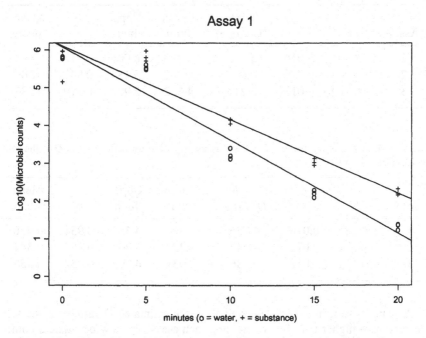

Figure 18.1. Plot of data and least squares fits for test substance and water for assay 1.

Table 18.2. Test for linearity of the individual assays.

Substance	Assay	p-Value	Linearity
abc123	1	0.201	OK
abc123	2	0.962	OK
abc123	3	0.153	OK
Water	1	0.320	OK
Water	2	0.121	OK
Water	3	0.796	OK

18.5.2   Modeling Individual Assays

Estimates of the D-value and 90% confidence intervals for it are obtained for each assay from the fit of (18.1) in Section 18.5.1 using the summary command to extract the results of the 1m fit. The estimates and confidence intervals, calculated using (18.2), are given in Table 18.3 for the test substance and in Table 18.4 for water. The code to produce these results is given in Appendix 18.A.3.

Table 18.3. Results for test substance. For each assay, separate estimates of the D-value and 90% confidence intervals.

Assay	b	s	$-1/b$ (D-value)	Lower limit	Upper limit	r^2	Mean square
1	−0.195	0.016	5.125	4.470	6.006	0.918	0.443
2	−0.203	0.012	4.918	4.457	5.484	0.947	0.349
3	−0.192	0.013	5.211	4.662	5.907	0.919	0.387

Table 18.4. Results for water. For each assay, separate estimates of the D-value and 90% confidence intervals.

Assay	b	s	$-1/b$ (D-Value)	Lower limit	Upper limit	r^2	Mean square
1	−0.247	0.016	4.056	3.648	4.566	0.951	0.426
2	−0.211	0.012	4.747	4.322	5.264	0.952	0.344
3	−0.233	0.012	4.286	3.935	4.705	0.952	0.355

As can be seen in Tables 18.3 and 18.4, the estimated D-values for the test substance are higher than for water. For each assay, we now calculate a confidence interval for the ratio of the D-value of the test substance to that of water using Fieller's theorem, (18.3). The S-PLUS code for doing this can be found in Appendix 18.A.3. The results are displayed in Table 18.5. Using the individual model fits, the D-value of the test substance appears to be higher than that of water for assays 1 and 3 but not for assay 2, indicating the possibility that the D-value of the test substance is greater than that of water.

Table 18.5. For each assay, estimate of and 90% confidence interval for the ratio of the D-value of the test substance to water.

Assay	Ratio	Lower limit	Upper limit
1	1.264	1.068	1.507
2	1.036	0.906	1.186
3	1.216	1.059	1.404

18.5.3   Pooling of Assays

To confirm whether the D-value of the test substance is greater than that of water, the nonlinear model described in Section 18.4.3 is fitted using the nls command, an S-PLUS routine for fitting nonlinear models with the method of least squares. Initial values needed for the modeling are taken from the fit of (18.1) to the complete dataset using the `lm` command. Using the `summary` command, the results of the `nls` fit of (18.4) are extracted and the pooled estimate of the D-value and its standard error are used to calculate a 90% Wald confidence interval as in (18.5). The results are displayed in Table 18.4 (see Appendix 18.A.4 for the code to produce these results). Based on the pooled estimate, we can conclude that the D-value for the test substance is greater than that of water, and also, from a practical point of view, the laboratory may have a problem with sterilization.

Table 18.6. Pooled estimate of the D-value and 90% confidence interval.

$\hat{\theta}$	Lower limit	Upper limit
1.170	1.070	1.269

18.6 Discussion of the Statistical Methodology

An alternative to the log-linear regression model presented in Section 18.4 would be a nonlinear parameterization of the form

$$N = 10^{\alpha + \beta T} + \varepsilon \qquad (18.7)$$

where α, β, and ε have the same interpretation as in (18.1). The models described by (18.1) and (18.7) differ in that the parameters, α and β, enter (18.7) in a nonlinear fashion. The greater flexibility of a curve with respect to a straight line removes the excessive influence of observations at the low and high ends of the time axis, and results in estimates of the D-value that tend to be robust to the inclusion or exclusion of these observations. Alternatively, one could use a weighted nonlinear regression model. Although we like the nonlinear model, it was not pursued further here because the *American National Standard* (1991) and the *International Standard* (1994) guidelines specify the linear model approach.

As a second statistical issue, an alternative method for calculating a confidence interval for the pooled estimate of the ratio of the D-values, θ, would be to use the profile likelihood of θ (see Meeker and Escobar, 1995). For θ near 1, the two methods should provide similar intervals. As θ deviates from 1, the difference between the two methods could become apparent. In this case both methods would most likely indicate that the assay failed even though there might be some numerical difference in the corresponding confidence intervals. Because our interest is in cases where θ is near 1 we chose to keep things simple and use the Wald interval.

Another approach that we have not considered here would be to use Bayesian methods allowing for a hierarchical structure between assays. However, the number of assays is usually too small to allow for an adequate assessment of the between-assay variation.

The statistical methodology presented here is not the most sophisticated available. In those situations where we could choose from among various approaches, some of which we point out above, we elected to keep things as simple as possible. The methods are to be used on a routine basis without a statistician present for interpretation. As usual in applied work, when one deals with nonstatisticians, it is often better to keep the methodology simple.

18.7   References

American National Standard (1991). Guidelines for the use of ethylene oxide and steam biological indicators in industrial sterilization processes. Association for the Advancement of Medical Instrumentation.

Cook, R.D. and Weisberg, S. (1982). *Residuals and Influence in Regression.* Chapman & Hall, New York.

Fieller, E.C. (1954). Some problems in interval estimation. *Journal of the Royal Statistical Society, Series B* **16**, 175–185.

Heintz, M., Urban, S., Gay, M., and Bühlmann, X. (1976). The production of spores of *Bacillus Stearothermophilus* with constant resistance to heat and their use as biological indicators during development of an aqueous solution for injection. *Pharmaceutica Acta Helvetiae* **51**, 1337–1343.

International Standard (1994). Sterilization of health care products—Biological indicators. ISO 11138-1, 2nd ed., 1994-10-01.

Kotz, S., Johnson, N.L. and Read, C.B. (1983). *Encyclopedia of Statistical Sciences*, Vol. 3. Wiley, New York.

Meeker, W.Q. and Escobar, L.A. (1995). Teaching about approximate confidence regions based on maximum likelihood estimation. *The American Statistician* **49**, 48–53.

18.A Appendix

18.A.1   Plots and Fits for Assays 2 and 3 (Section 18.5.1)

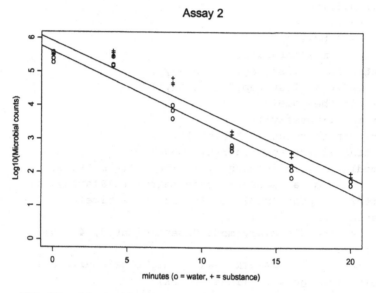

Figure 18.2. Plot of data and least squares fits for test substance and water for assay 2.

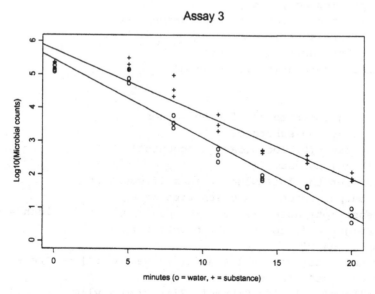

Figure 18.3. Plot of data and least squares fits for test substance and water for assay 3.

18.A.2   S-Plus Code to Test for Linearity (Section 18.5.1)

```
# Fitting a quadratic and a linear regression
# per substance per assay.
# Initializing
alpha <- 0.01
subst <- "abc123"
noassay <- max(data$assay)
SubstLinFit <- list(rep(0, noassay))
WaterLinFit <- list(rep(0, noassay))
# Now fit the models
for(i in (1:noassay)){
  for( trt in c(subst, "water")){
# Setting linear and quadratic models
    mod1 <- lm(log10(counts) ~ time + I(time*time),
            data, subset= (treatment==trt)&(assay==i))
    mod2 <- update(mod1, log10(counts) ~ time)
# Testing the quadratic model
    pv <- round(summary(mod1)$coefficient[3, 4], 5)
    if (pv < alpha)
      message <- "WARNING: assay might be nonlinear!"
    else message <- "(linearity OK)"
    cat("For ", trt, "and assay ", i,
      ",\n the p-value for testing a quadratic term is ",
      pv, message, "\n")
# Keeping the linear fits for further use
    if (trt==subst)
      SubstLinFit[[i]] <- mod2
    else WaterLinFit[[i]] <- mod2
  }
}
# Plotting observed and fitted values
# per assay and substance
ylim <- range(0, log10(data[, "counts"]))
for (i in 1:noassay) {
# Uncomment the following 2 lines if you want
# the output to go to a postscript file
#  postscript(paste('assay', i,'.ps', sep = ""), width = 5,
#    height = 4, append = F, print.it = F)
  temp <- data[
    data[, "assay"] == i & data[, "treatment"] == "water",
      c("time", "counts")]
  plot(temp[, 1], log10(temp[, 2]), ylim = ylim,
    xlab = "minutes (o = water, + = substance)",
```

```
    ylab = "Log10(Microbial counts)", type = 'p', pch = 'o')
  abline(WaterLinFit[[i]])
  temp <- data[
    data[, "assay"] == i & data[, "treatment"] == subst,
      c("time", "counts")]
  points(temp[,1], log10(temp[, 2]), pch = '+')
  abline(SubstLinFit[[i]])
  title(paste("Assay", i))
# Uncomment the following line if you are
# using the postscript device
#   dev.off()
}
```

18.A.3   S-PLUS Code to Model Individual Assays (Section 18.5.2)

```
# Calculating d-values and 90% confidence intervals
conf.level <- 0.9
a <- matrix(0, nrow = noassay, ncol = 17)
dimnames(a) <- list(paste("assay", 1:noassay),
  c("bsubst", "sesubst",
    "dsubst", "dlowsubst", "dhighsubst",
    "bwater", "sewater",
    "dwater", "dlowwater", "dhighwater",
    "ratio", "lowratio", "highratio",
    "rsqsubst", "mssubst", "rsqwater", "mswater"))
# Student's t quantiles for the confidence intervals
tval.subst <- numeric(noassay)
tval.water <- numeric(noassay)
# Calculating the confidence intervals for d-values
# per assay per substance
for(i in (1:noassay)){
  tval.subst[i] <- qt(p = 1 - (1 - conf.level)/2,
    df = SubstLinFit[[i]]$df.residual)
  tval.water[i] <- qt(p = 1 - (1 - conf.level)/2,
    df = WaterLinFit[[i]]$df.residual)
  a[i, "bsubst"] <-
    summary(SubstLinFit[[i]])$coefficient[2, 1]
  a[i, "sesubst"] <-
    summary(SubstLinFit[[i]])$coefficient[2, 2]
  a[i, "rsqsubst"] <- summary(SubstLinFit[[i]])$r.squared
  a[i, "mssubst"] <- summary(SubstLinFit[[i]])$sigma
  a[i, "bwater"] <-
    summary(WaterLinFit[[i]])$coefficient[2, 1]
```

```
  a[i, "sewater"] <-
    summary(WaterLinFit[[i]])$coefficient[2, 2]
  a[i, "rsqwater"] <- summary(WaterLinFit[[i]])$r.squared
  a[i, "mswater"] <- summary(WaterLinFit[[i]])$sigma
}
a[, "dsubst"] <- -1/a[, "bsubst"]
a[, "dlowsubst"] <- -1/
  (a[, "bsubst"] - tval.subst * a[, "sesubst"])
a[, "dhighsubst"] <- -1/
  (a[, "bsubst"] + tval.subst * a[, "sesubst"])
a[, "dwater"] <- -1/a[, "bwater"]
a[, "dlowwater"] <- -1/
  (a[, "bwater"] - tval.water * a[, "sewater"])
a[, "dhighwater"] <- -1/
  (a[,"bwater"] + tval.water * a[, "sewater"])
# Calculating the confidence interval
# for the ratio of d-values according to Fieller's theorem
zval <- qnorm(1 - (1 - conf.level)/2)
a[, "ratio"] <- a[, "bwater"]/a[, "bsubst"]
a[, "lowratio"] <- sqrt((a[, "bwater"] * a[, "bsubst"])^2 -
  (a[, "bwater"]^2 - zval^2 * a[, "sewater"]^2) *
  (a[, "bsubst"]^2 - zval^2 * a[, "sesubst"]^2))
a[, "highratio"] <- (a[, "bwater"] * a[, "bsubst"] +
  a[, "lowratio"])/(a[, 1]^2 - zval^2 * a[, "sesubst"]^2)
a[, "lowratio"] <- (a[, "bwater"] * a[, "bsubst"] -
  a[, "lowratio"])/(a[, 1]^2 - zval^2 * a[, "sesubst"]^2)
# Print results for Substance
cat("Estimate of slope of the linear regression,",
  "its standard error, estimate of the D-value,",
  "lower end of the confidence interval for D-value,",
  "upper end of confidence interval for D-value,",
  "R-square of the fit, and mean square of the fit",
  "for", subst, " \n")
# Print results for Water
print(a[, c(1:5, 14, 15)])
cat("\n Estimate of slope of the linear regression,",
  "its standard error, estimate of the D-value,",
  "lower end of the confidence interval for D-value,",
  "upper end of confidence interval for D-value,",
  "R-square of the fit, and mean square of the fit",
  "for water  \n")
print(a[, c(6:10, 16, 17)])
# Print estimated ratios and CIs
```

```
cat("\n Estimate of the ratio of the D-value of",
  subst, "to water,",
 "lower and upper ends of its confidence intervals  \n")
print(a[, 11:13])
```

18.A.4   S-PLUS Code to Pool Assays (Section 18.5.3)

```
# Pooling assays
alpha <- matrix(0, nrow=2, ncol = noassay)
beta <- rep(0, noassay)
# Building the expression of the formula to be used in nls()
formula <- "log10(counts) ~ "
water <- "water"
for(i in (1:noassay)){
  formula <- paste(formula, "((alpha[1, ", i, "]",
    "+ theta * beta[", i, "] * time)",
    "* I((treatment ==", water, "))",
    "+ (alpha[2, ", i, "] + beta[", i, "]",
    "* time) * I((treatment == subst)))",
    "* I((assay ==", i, "))")
  if (i < noassay) formula <- paste(formula, "+")
}
attach(data)
# A simple model to provide initial values
initial <- lm(log10(counts) ~ time, data)
# Using the formula built above, parsing it
# and embedding it in nls()
modtheta <- nls(parse(text = formula)[[1]], data = data,
  start = list(alpha = matrix(
        rep(summary(initial)$coefficient[1, 1], noassay),
        nrow = 2, ncol = noassay),
      theta = 1,
      beta = rep(summary(initial)$coefficient[2, 1],
        noassay)),
  trace = F)
mtheta  <- summary(modtheta)$parameters["theta", "Value"]
setheta <- summary(modtheta)$parameters["theta",
  "Std. Error"]
cat("\nEstimated ratio of d-values:  theta=",
  round(mtheta, 2),
  "\n90% Wald confidence limits for theta: (",
  round(mtheta - 1.645 * setheta, 3), ", ",
  round(mtheta + 1.645 * setheta, 3), ") \n")
detach("data")
```

18.A.5   Data

	treatment	assay	time	counts
1	abc123	1	0	135000.0
2	abc123	1	0	875000.0
3	abc123	1	0	865000.0
4	abc123	1	5	885000.0
5	abc123	1	5	485000.0
6	abc123	1	5	595000.0
7	abc123	1	10	13700.0
8	abc123	1	10	13150.0
9	abc123	1	10	10400.0
10	abc123	1	15	975.0
11	abc123	1	15	1235.0
12	abc123	1	15	820.0
13	abc123	1	20	148.0
14	abc123	1	20	133.5
15	abc123	1	20	193.5
16	abc123	2	0	310000.0
17	abc123	2	0	305000.0
18	abc123	2	0	290000.0
19	abc123	2	4	390000.0
20	abc123	2	4	275000.0
21	abc123	2	4	345000.0
22	abc123	2	8	40000.0
23	abc123	2	8	62000.0
24	abc123	2	8	45000.0
25	abc123	2	12	1300.0
26	abc123	2	12	1650.0
27	abc123	2	12	1600.0
28	abc123	2	16	360.0
29	abc123	2	16	375.0
30	abc123	2	16	280.0
31	abc123	2	20	89.5
32	abc123	2	20	71.0
33	abc123	2	20	71.0
34	abc123	3	0	205000.0
35	abc123	3	0	140000.0
36	abc123	3	0	220000.0
37	abc123	3	5	135000.0
38	abc123	3	5	190000.0
39	abc123	3	5	295000.0
40	abc123	3	8	33500.0
41	abc123	3	8	21500.0

42	abc123	3	8	90500.0
43	abc123	3	11	6200.0
44	abc123	3	11	3000.0
45	abc123	3	11	1950.0
46	abc123	3	14	545.0
47	abc123	3	14	445.0
48	abc123	3	14	510.0
49	abc123	3	17	225.0
50	abc123	3	17	385.0
51	abc123	3	17	285.0
52	abc123	3	20	65.5
53	abc123	3	20	119.0
54	abc123	3	20	77.5
55	water	1	0	660000.0
56	water	1	0	585000.0
57	water	1	0	645000.0
58	water	1	5	325000.0
59	water	1	5	400000.0
60	water	1	5	300000.0
61	water	1	10	1550.0
62	water	1	10	1300.0
63	water	1	10	2450.0
64	water	1	15	155.0
65	water	1	15	120.0
66	water	1	15	185.0
67	water	1	20	22.5
68	water	1	20	23.0
69	water	1	20	16.5
70	water	2	0	375000.0
71	water	2	0	245000.0
72	water	2	0	185000.0
73	water	2	4	150000.0
74	water	2	4	280000.0
75	water	2	4	160000.0
76	water	2	8	7000.0
77	water	2	8	10000.0
78	water	2	8	4000.0
79	water	2	12	550.0
80	water	2	12	450.0
81	water	2	12	650.0
82	water	2	16	160.0
83	water	2	16	70.0
84	water	2	16	120.0
85	water	2	20	42.0
86	water	2	20	42.0

87	water	2	20	57.0
88	water	3	0	135000.0
89	water	3	0	120000.0
90	water	3	0	180000.0
91	water	3	5	140000.0
92	water	3	5	75000.0
93	water	3	5	55000.0
94	water	3	8	6000.0
95	water	3	8	2500.0
96	water	3	8	3500.0
97	water	3	11	250.0
98	water	3	11	600.0
99	water	3	11	400.0
100	water	3	14	80.0
101	water	3	14	100.0
102	water	3	14	70.0
103	water	3	17	44.5
104	water	3	17	47.5
105	water	3	17	47.0
106	water	3	20	6.5
107	water	3	20	4.0
108	water	3	20	10.5

19

Acceptance Sampling Plans by Attributes

Harry Yang and David Carlin
MedImmune, Inc., Gaithersburg, MD, USA

19.1 Introduction

As the field of quality control enters the next millennium, many pharmaceutical and biotechnology companies have striven to achieve quality in the drug and biologic products they manufacture. This is due in part to increasing public demand for products to be free of defects, as well as the 1987 publication by the international organization for standardization (commonly referred to as ISO) of the ISO 9000 series of standards for quality management and quality assurance. The ISO 9000 standards provide guidance for manufacturers to implement effective quality systems. They are also widely used by customers as a yardstick to measure the adequacy of the producer's quality system. Manufacturers are motivated to be certified by registrars accredited by the ISO so as to enhance potential customer confidence in the quality of their products. An excellent review of ISO 9000 was given by Rabbitt and Bergh (1997). In 1994, the United States Congress passed the current Good Manufacturing Practice (cGMP) Act initiated by the Food and Drug Administration (1995). Since then, lawsuits have been filed by the FDA against a few pharmaceutical companies for failing to appropriately implement cGMP guidelines in their product release. The legal actions added a new tempo to the improvement of producers' quality systems. As one of the key elements in the ISO 9000 standards and an integral part of the cGMP, acceptance sampling inspection before the distribution of a product lot is mandated by law.

Acceptance sampling is a mechanism of selecting a random sample of a lot of finished products and evaluating each component of the sample as meeting or not meeting a specified criteria. A decision is then made to accept or reject the lot based on the sample. A formal definition of acceptance sampling is given in ANSI/ASQC Standard A2-1978 as: "Sampling inspection in which decisions are made to accept or not accept a product or service; also, the methodology that deals with the procedures by which decisions to accept or not accept are based on the results of the inspections of samples" (Wadsworth et al., 1986, Chap. 12). Because the quality of the product has been predetermined by the production process by the time of inspection, acceptance sampling itself does little to the quality improvement. It does, however, provide crucial feedback about the production process. Since a rejected lot is usually 100% inspected and rectified if at all possible, this indirectly improves the overall quality of the product. In addition, frequent rejection of bad lots raises a red flag to a manufacturer that the process might need to be improved. Today, acceptance sampling plans are more and more used as tactical elements in overall strategies designed to achieve desired quality systems by manufacturers. Such strategies usually encompass continual stabilization and improvement of processes through effective process monitoring and quality control based on computer automation, application of statistical and probabilistic theory, and industrial psychology.

Acceptance sampling inspection leads to a decision concerning the acceptance or rejection of a lot under inspection. Because the entire lot of material is not being tested, a degree of uncertainty exists with the decision made. Thus, there are risks to the consumer and producer in making a wrong decision about the disposition of the lot. Specifically, there are two types of errors, namely, rejecting a good lot and accepting a bad lot. In the context of hypothesis testing, they are referred to as Type I and Type II errors, respectively. An ideal sampling plan is the one that minimizes both types of errors. However, a sampling plan aimed at decreasing one type of error normally results in an increase in the other type of error. Therefore, the determination of a sampling plan is a tradeoff between the two types of errors.

Traditionally, a sampling plan is selected to keep both Type I and Type II errors below predetermined levels. In practice, the costs or consequences of making the two types of errors might not be the same. For example, if drug A is the only product on the market that is both safe and effective in improving the survival of kidney transplant patients, then the cost of rejecting a good lot of product is far greater than accepting a lot not as potent as it should be. This is because in rejecting a good lot, delayed treatment could result in a patient's death and in accepting a bad lot, drug A could still have some beneficial placebo effect on the patient's condition. Therefore, different levels of losses are associated with Type I and Type II errors. A sampling plan can be designed to reflect the costs of two kinds of errors; this can be accomplished in the framework of risk analysis as described by Berger (1985).

In general, a sampling plan is determined through assessing all practically feasible sampling plan options. It involves trial and error and requires careful probability calculations. Therefore, this process can be statistically, as well as computationally, intensive. For scientists who do not have a strong background in statistics or programming, it can be a formidable task. This presents the need for an interactive tool that can easily be used to construct acceptance sampling plans. Although there are a wide array of software packages commercially available for quality control, none of them provides tools designed for the development of acceptance sampling plans. In this chapter, we present a tool developed specifically for this purpose, using the statistical software S-PLUS developed by Insightful Corporation. The tool enables nonstatistical users to explore various acceptance sampling plans in an interactive fashion.

In Section 19.2 we introduce the concepts of single, double, multiple, and sequential sampling plans for attribute data and basic statistical concepts such as the operating characteristic (OC) curves and binomial distribution. Attribute data are obtained from situations in which inspection results can be classified into two classes of outcomes. This includes defective and nondefective, as well as other classifications such as measurements in and out of specifications. Section 19.3 describes a risk analysis-based plan. This plan is especially useful in cases where the cost associated with each type of error can be quantified and the inspection history of the same product is well documented. We discuss our S-PLUS tool for a systematic solution to designing sample plans by nonstatistical users in Section 19.4. An example dealing with sampling plan issues encountered in the investigation of out-of-specification clinical laboratory results is also presented. Finally, we summarize the results and discuss possible generalization of the sampling plans and the S-PLUS tool in Section 19.5.

19.2   Acceptance Sampling

19.2.1   Concepts

An acceptance sampling plan is an inspection process used to determine whether a lot of a product meets a set of prespecified criterion to be accepted. A lot is a collection of items manufactured through a production process under essentially the same conditions. The size of the lot is the total number of product units in the lot. For example, a product may be produced in lots of 4000 vials. Each vial represents a unit in the lot and the size of the lot is 4000. One possible inspection plan is to inspect each item of the lot and remove or replace defective items from the lot. Although effective, the 100% inspection is both costly and impractical, especially when the inspection is destructive. As a compromise, a fraction of the lot is usually inspected. Assuming all the products in the lot were manufactured under the same set of conditions (except some random variations in the production process), each item in the lot has a fixed chance to be nondefective. To ensure the fraction tested truly represents the entire lot, a random sample from the lot needs to be selected for testing. Because the acceptance/rejection decision is made based on the test results of a sample of products, there are two types of errors likely to occur: rejecting a good lot (also known as producer's risk (PR)) and accepting a bad lot (also known as consumer's risk (CR)). A lot is called a "good" lot if its fraction of defective units is at or below a certain acceptable level, called the producer's quality level (PQL). A lot is called a "bad" lot if its fraction of defective units is at or above a certain unacceptable level, called the consumer's quality level (CQL). The PQL represents the largest fraction of defects that is acceptable to the customer. The CQL denotes the smallest fraction of defects indicative of poor, unacceptable quality. The PR is the probability that a lot whose quality is no worse than PQL is rejected by the sampling procedure, and the CR is the probability that a lot whose quality is no better than CQL is accepted. A producer desires to keep the rejection rate as low as possible without compromising quality in order to

maximize profits. In contrast, the consumer likes to minimize the possibility
that a lot with low quality is accepted.

The tradeoff between the producer's and consumer's interests is usually
made through the so-called operating characteristic (OC) curve. This curve depicts the acceptance probability of lots with varying quality or fraction of defectives that might be submitted for inspection under a specific acceptance sampling plan. When the quality levels PQL and CQL are specified, the two types
of risks PR and CR can be directly identified from the curve. If both of the risks
are within anticipated ranges, the sampling plan used to construct the OC curve
is appropriate. Otherwise, alternative plans have to be explored, using the OC
curve. Figure 19.1 shows what a typical OC curve looks like. When the two
quantities PQL and CQL are specified, the producer's and consumer's risks PR
and CR can be identified from the plot. The effect of sampling plans on the OC
curve will be discussed in the following sections.

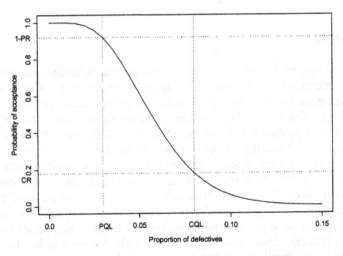

Figure 19.1. Operating characteristic curve.

19.2.2   Single Sampling Plan

Attribute sampling summarizes multiple quality characteristics for each item in a
sample into an overall test outcome (pass versus fail). For example, in releasing
a lot of drug product, there are a series of appropriate laboratory determinations
of satisfactory conformity to final specifications for each unit in a sample, including the purity, particle size, identity, and strength of each active ingredient
and potency. Attribute sampling is widely used in the pharmaceutical and biotechnology industries for product lot inspections. The sampling plans introduced in this chapter deal with product lots with dichotomous test outcomes,
that is, the inspection results are classified into two categories such as in and out
of specifications.

A single sampling plan is one in which a decision concerning the acceptability of a lot is made on the basis of testing one sample of a predetermined size from that lot. A sample of size n is taken and each component is tested as to whether it is defective or not. If the number of defects x is no greater than the acceptance number of failures c, the lot is accepted. Otherwise, the lot is rejected. Figure 19.2 shows a diagram of this sampling plan.

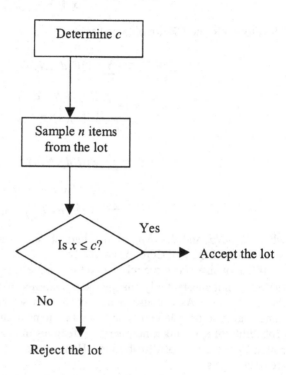

Figure 19.2. Diagram of single sampling plan.

Developing a single sampling plan is equivalent to determining the sample size n and acceptance number c to guarantee that both PR and CR are within prespecified limits. This in essence is a probability problem. Let p_p and p_c denote the PQL and CQL, respectively. The actual PR and CR are calculated as follows:

$$\text{PR} = \Pr\left(\text{Reject the lot, given the true proportion of defects is PQL}\right)$$
$$= 1 - \Pr\left(x \le c \mid p_p\right) \tag{19.1}$$

$$\text{CR} = \Pr\left(\text{Accept the lot, given the true proportion of defects is CQL}\right)$$
$$= \Pr\left(x \le c \mid p_c\right) \tag{19.2}$$

Depending on the lot size N, sample size n, and number of defectives d in the lot, various probability distributions can be used to calculate PR and CR. In theory, the number of defectives x in a random sample of n items follows the hypergeometric distribution. That is,

$$P(x; d, N, n) = \binom{d}{x} \binom{N-d}{n-x} \bigg/ \binom{N}{n}$$ (19.3)

Therefore, PR and CR are obtained as

$$PR = 1 - \sum_{i=0}^{c} \Pr(i; d_p, N, n)$$
$$= 1 - \sum_{i=0}^{c} \binom{d_p}{i} \binom{N-d_p}{n-i} \bigg/ \binom{N}{n}$$ (19.4)

$$CR = \sum_{i=0}^{c} \Pr(i; d_c, N, n)$$
$$= \sum_{i=0}^{c} \binom{d_c}{i} \binom{N-d_c}{n-i} \bigg/ \binom{N}{n}$$ (19.5)

where $d_p = Np_p$ and $d_c = Np_c$ are numbers of defects in the lot when the lot quality levels are PQL and CQL, respectively.

If the lot size N is large relative to the sample size, then the results of y draws $(y \le n)$ do not substantially change the probability of selecting a defective unit in the next draw. As a consequence, each item selected has approximately the same chance to be a defective unit as other items in the sample. This implies the probability of selecting x number of defectives in a sample of n can be approximated by the binomial distribution. Hence, under this assumption, PR and CR are obtained as

$$PR = 1 - \sum_{i=0}^{c} \binom{n}{i} p_p^i (1 - p_p)^{n-i}$$ (19.6)

$$CR = \sum_{i=0}^{c} \binom{n}{i} p_c^i (1 - p_c)^{n-i}$$ (19.7)

19.2.3 Effect of Single Sampling Plan Parameters on the OC Curve

As a function of the sampling parameters, the operating characteristic curve changes with different selections of sample size and acceptance number. A clear understanding about how the change may occur is very useful in designing a sampling plan. Several observations about the potential changes of the OC curve can be readily made. As seen in Figure 19.3, if the acceptance number c is fixed, the OC curve will move to the left as the sample size n increases. This implies a decrease in the consumer's risk but an increase in the producer's risk. A similar effect is observed in Figure 19.4 by keeping the sample size n fixed and increasing the acceptance number c. If n is increased while keeping the ratio between n and c constant, both PR and CR are decreased. This is illustrated in Figure 19.5. Using this knowledge about the OC curve as guidance, designing an acceptance sampling plan is a process of trial and error choosing the sample size and acceptance number. In the experiment, one constructs an OC curve based on one set of n and c. If satisfactory values of PR and CR cannot be identified from the OC curve at the specified values of PQL and CQL, one varies n, c, or both, depending on the current values of PR and CR. Obviously this process can be very time-consuming. However, with the help of an S-PLUS tool introduced in Section 19.4, acceptance sampling design is not a formidable task.

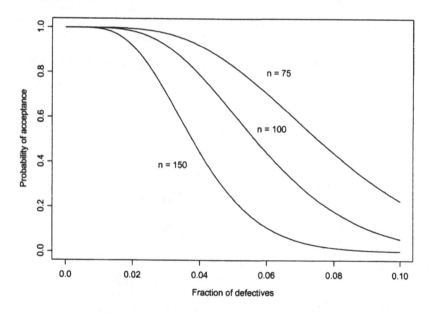

Figure 19.3. Operating characteristic curves of single sampling plans with $c = 5$ and $n = 75$, 100, and 150.

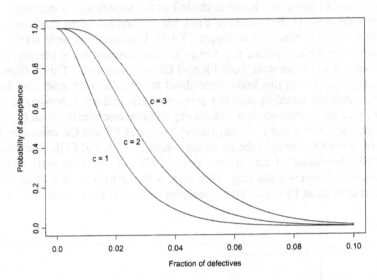

Figure 19.4. Operating characteristic curves of single sampling plans with $n = 100$ and $c = 1, 2$ and 3.

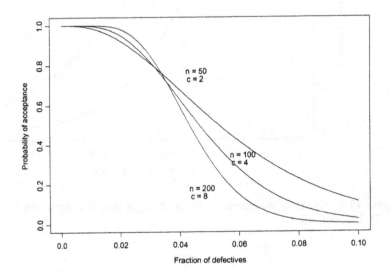

Figure 19.5.   Operating characteristic curves of single sampling plans with $(n, c) = (50, 2), (100, 4)$, and $(200, 8)$.

19.2.4   Double Sampling Plan

As an extension of the single sampling plan, a double sampling plan allows the lot acceptance or rejection decision to be made on the basis of one or two samples. It is characterized by two sample sizes and two sets of acceptance–rejection criteria. The two sample sizes may or may not be equal. Initially a sample of size n_1 is taken. If the number of defectives x_1 is less than or equal to the first acceptance number c_1, the lot is accepted. If x_1 is greater than the rejection number r_1, the lot is rejected. Otherwise, a second sample of size n_2 is taken. If the number of defectives $x_1 + x_2$ from the total $n_1 + n_2$ sample is less than or equal to c_2 (the acceptance number for the combined samples), the lot is accepted. If there are more than c_2 defectives, the lot is rejected. In other words, the second rejection number is $r_2 = c_2$. This kind of sampling plan is depicted in Figure 19.6.

The sample sizes n_1 and n_2 and the acceptance numbers c_1 and c_2 are chosen to obtain desirable PR and CR levels. As with the single sampling plan, the PR and CR of a double sampling plan can be calculated using the hypergeometric or binomial distribution, depending on the relative lot and sample sizes. For example, if N is relatively larger than n, then PR and CR can be estimated as follows:

$$
\begin{aligned}
PR &= \Pr\left(\text{Reject the lot, given the lot has an acceptable } p_p\right) \\
&= 1 - \Pr\left(\text{Accept the lot, given the lot has an acceptable } p_p\right) \\
&= 1 - \left[\Pr\left(x_1 \le c_1\right) + \Pr\left(x_1 = c_1 + 1\right)\Pr\left(x_2 \le c_2 - x_1\right) + \cdots \right. \\
&\qquad \left. + \Pr\left(x_1 = r_1\right)\Pr\left(x_2 \le c_2 - x_1\right)\right] \\
&= 1 - \left\{ \sum_{i=0}^{c_1} \binom{n_1}{i} p_p^{i}(1-p_p)^{n_1-i} + \right. \\
&\qquad \left. \sum_{j=c_1+1}^{r_1} \left[\binom{n_1}{j} p_p^{j}(1-p_p)^{n_1-j} \sum_{k=0}^{c_2-j}\binom{n_2}{k} p_p^{k}(1-p_p)^{n_2-k}\right]\right\}
\end{aligned}
$$
(19.8)

$$
\begin{aligned}
CR &= \Pr\left(\text{Accept the lot, given the lot has an acceptable } p_c\right) \\
&= \Pr\left(x_1 \le c_1\right) + \Pr\left(x_1 = c_1 + 1\right)\Pr\left(x_2 \le c_2 - x_1\right) + \cdots \\
&\qquad + \Pr\left(x_1 = r_1\right)\Pr\left(x_2 \le c_2 - x_1\right) \\
&= \sum_{i=0}^{c_1} \binom{n_1}{i} p_c^{i}(1-p_c)^{n_1-i} + \\
&\qquad \sum_{j=c_1+1}^{r_1} \left[\binom{n_1}{j} p_c^{j}(1-p_c)^{n_1-j} \sum_{k=0}^{c_2-j}\binom{n_2}{k} p_c^{k}(1-p_c)^{n_2-k}\right]
\end{aligned}
$$
(19.9)

484   H. Yang and D. Carlin

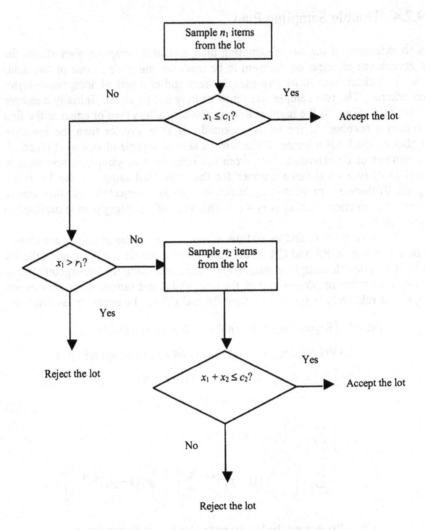

Figure 19.6. Diagram of double sampling plan.

Double sampling inspection reflects the inspector's desire to give a questionable lot a second chance. It is an inspection tool more frequently used than the single sampling plan in the pharmaceutical and biotechnology industries as a cost-effective measure. Double sampling plans can be designed to offer the same protection as a single sampling plan while in general having a smaller average sample size.

19.2.5   Multiple and Sequential Sampling Plans

Doubling sampling can be further extended to allow additional samples to be taken to provide additional flexibility in the evaluation of a lot. Multiple sampling plans inspect up to k successive samples from the same lot under inspection to make a decision. They are also referred to as group sequential plans. Sequential sampling involves the inspection of the lot, item-by-item, as compared to groups in the group sequential plan. Decisions regarding accepting the lot, rejecting the lot, or inspecting one or more items from the lot are made following the inspection of each individual item. Sequential sampling methods may be considered as a special case of multiple sampling plans with sample size 1 and no limit on the number of samples to be evaluated. In this section, we focus the discussion on the design of sequential sampling plans.

The sequential sampling plan is implemented through comparing the cumulative number of defectives x to an acceptance number Y_1 and a rejection number Y_2 determined from two linear equations of the sample size n. If $x \leq Y_1$, then accept the lot. If $x \geq Y_2$, reject the lot. If $Y_1 < x < Y_2$, inspect one more item. This method was initially developed by Wald (1947) during World War II. He demonstrated that for a PR value of a and CR of b, the acceptance and rejection limits take the form

$$Y_1 = an - b_1$$

$$\text{(19.10)}$$

$$Y_2 = an + b_2$$

where the slope a and intercepts b_1 and b_2 are calculated as

$$a = \ln\left[(1-\text{PQL})/(1-\text{CQL})\right]/K$$

$$b_1 = \ln\left[(1-\alpha)/\beta\right]/K$$

$$\text{(19.11)}$$

$$b_2 = \ln\left[(1-\beta)/\alpha\right]/K$$

$$K = \ln\left\{\text{CQL}(1-\text{PQL})/\left[\text{PQL}(1-\text{CQL})\right]\right\}$$

The acceptance and rejection limits are determined so that the PR and CR are equal to prespecified values a and b. Although a general formula for calculating the probability of acceptance exists (Schilling, 1982), the OC curve for a sequential design does not appear as useful as those for single and double sampling plans. The procedure for the sequential sampling plan is illustrated in Figure 19.7.

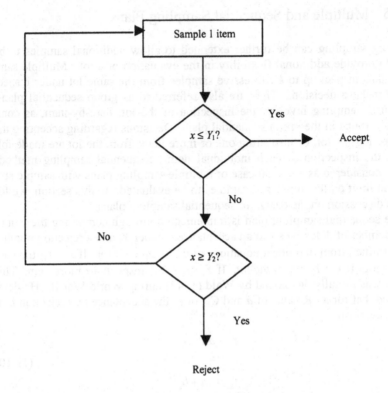

Figure 19.7. Diagram of a sequential sampling plan.

To illustrate the use of the method, suppose we develop a sequential sampling plan to have a 95% chance of accepting a good lot and a 90% chance of rejecting a poor lot, given PQL = 0.02 and CQL = 0.10. We find that $K = 1.695$, $a = 0.05$, $b_1 = 1.329$, and $b_2 = 1.706$. Hence, the acceptance and rejection limits in this example are obtained as

$$Y_1 = 0.05n - 1.329$$

$$(19.12)$$

$$Y_2 = 0.05n + 1.706$$

The operation of the plan can be carried out through the chart in Figure 19.8 in which the cumulative number of defectives x is plotted against the number of items inspected. If x crosses the acceptance or rejection lines, the decision of accepting or rejecting the lot is reached. No decision is made as long as x remains between the two lines and the sampling inspection continues to proceed. As indicated by the acceptance line, the earliest possible stopping time of the sampling inspection is when the first 26 items conform to specifications.

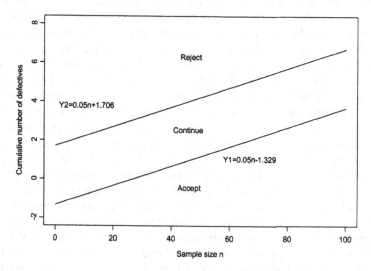

Figure 19.8. Sequential sampling chart.

In comparison with the single and double sampling plans, both multiple and sequential sampling plans have smaller average sample sizes without compromising protections to producer's and consumer's interests, because the decision regarding the disposition of very good or very poor lots may be made earlier with smaller samples. They are often used when the testing is very expensive and a minimum sample is necessary. However, they are operationally more complex and therefore incur more administrative costs. Since sample sizes can be directly translated into costs, they should be accounted for in the determination of single, double, multiple, or sequential sampling plans for a specific acceptance testing.

19.3 Risk Analysis Methods

It is very clear from the above discussions of various sampling plans that selection of a sampling plan is in essence no more than choosing a set of decision rules to provide desirable protection to both consumer's and producer's interests. The acceptance sampling methods developed in the previous sections are aimed at selecting sample size(s) and acceptance–rejection number(s) to keep PR and CR below prespecified levels. Acceptance or rejection decisions are made based on sampling information, namely, number(s) of defectives from the acceptance testing. These decisions, however, do not take into account the costs associated with sampling tests and the consequences of decisions. For example, in the production of a drug, a series of acceptance inspections are performed before the release of the final product. The cost per unit testing might differ at various stages of testing. For instance, it might be much less costly to reject a lot in an earlier stage of drug production than to turn down a finished lot of the product later on that is defective due to poor materials. In the example discussed in the Introduction, rejecting a good lot of a drug for improving kidney transplant patient survival could have more serious consequences than accepting a less potent lot of the drug, especially when the drug is the only effective product on the market. Therefore the risks of making Type I and Type II errors might not be the same and may vary from one situation to another. Another source of heuristic information is the distribution $\pi(\theta)$ of the fraction of defectives, θ, in a finished lot. This distribution is usually obtained based on past inspection experience with the same product manufactured under the same set of conditions. It can be incorporated in the design of an acceptance sampling plan. Risk analysis in the context of statistical decision theory provides a framework for developing decision rules in the presence of statistical and heuristic knowledge. It is very effective in solving a practical problem when the risk for each possible decision can be quantified. In this section, we formulate the sampling plan determination in terms of a problem in decision theory (Berger, 1985). The sample size(s) and acceptance–rejection number(s) are determined to minimize a mean risk index which depends on PR, CR, PQL, CQL, a prior distribution on the fraction of defectives in the lot, and various losses.

19.3.1   Loss Function

We define $f(y, \theta)$ as a loss function of y and θ where y stands for the decision based on the test result and θ is the fraction of defectives of the original lot. The decision y can take the value of either accept or reject the lot while θ follows a distribution of $\pi(\theta)$. We let $l_a(\theta)$ and $l_r(\theta)$ denote the losses associated with accepting and rejecting the lot, respectively, given the lot quality level is θ. Let δ denote the set of decision rules of a specific type of sampling plan, where δ depends on sample sizes, acceptance numbers, and rejection numbers. These decision rules are described by the diagrams in Figures 19.2, 19.6, and 19.7 for single, double, and sequential sampling plans. Let $O(\delta)$ denote the outcome of a decision based on δ. Then the mean risk $R(\delta)$ of using δ as the method of acceptance inspection is:

$$R(\delta) = E\{ f[O(\delta), \theta] \}$$

(19.13)

$$= \int_0^1 l_a(\theta)\pi(\theta)[1 - OC(\theta;\delta)] + l_r(\theta)[1 - \pi(\theta)]OC(\theta;\delta)\, d\theta$$

where $OC(\theta; \delta)$ is the operating characteristic function of the sampling plan, i.e., the probability of accepting the lot based on the decision rule δ given that the fraction of defectives of the lot is θ.

To factor in the cost of acceptance testing, a more general index is given by

$$R(\delta) = E\{f[O(\delta),\theta]\} + an \qquad (19.14)$$

with a being the unit cost of inspecting one item from the lot. An acceptance sampling plan is the one whose decision rule δ minimizes $R(\delta)$.

As a special case of (19.14), we assume θ has two possible values: PQL and CQL. The loss function $f(y, \theta)$ then has four outcomes as illustrated in Table 19.1.

Table 19.1. Four outcomes for a loss function.

	Good lot	Bad lot
Accept	L_0	L_1
Reject	L_2	L_3

The losses L_0 and L_3 are associated with correct decisions. L_1 is the loss of falsely accepting a bad lot and L_2 is the loss of falsely rejecting a good lot. Note that $1 - OC(PQL; \delta)$ and $OC(CQL; \delta)$ are PR and CR, respectively. The mean risk in (19.14) can be simplified as

$$R(\delta) = pL_0(1-PR) + (1-p)L_1 CR + pL_2 PR + (1-p)L_3(1-CR) + an$$

$$(19.15)$$

$$= p(L_2 - L_0)PR + (1-p)(L_1 - L_3)CR + pL_0 + (1-p)L_3 + an$$

where $p = \pi(PQL)$ is the chance that a finished lot is manufactured at the quality level PQL which can usually be estimated using historical inspection data. Assuming the number of defective units from each sample inspection follows a binomial distribution, the PR and CR are determined from (19.6) and (19.7) for single sampling plans, and from (19.8) and (19.9) for double sampling plans.

As an example, suppose that we are interested in designing a single sampling plan using the risk analysis method introduced in this section. Let n, x, and c denote the sample size, number of failures, and acceptance samples as described in Section 19.2.2. Therefore the decision rule δ is as follows: accept the lot if $x \leq c$; otherwise, reject it. It is also assumed that the unit cost a of inspecting one item from the lot is 1. We further assume that the loss of accepting a good lot or rejecting a bad lot is 1, and the loss of rejecting a good lot or accepting a bad lot is 5, i.e., $L_0 = L_3 = 1$ and $L_1 = L_2 = 5$. Lastly, we assume the chance p for a finished lot to have the quality level PQL is 0.95. Note that the producer's and consumer's risks PR and CR are given in (19.6) and (19.7), respectively. Substituting these quantities and expressions for those in (19.15), we obtain:

$$R(\delta) = 3.8 \sum_{i=0}^{c} \binom{n}{i} p_p^i (1-p_p)^{n-i} + 0.2 \sum_{i=0}^{c} \binom{n}{i} p_c^i (1-p_c)^{n-i} + n + 5 \quad (19.16)$$

Our goal is to find a combination of n and c that minimizes the mean risk $R(n, c)$. That is, if n_0 and c_0 satisfy $R(n_0, c_0) = \min[R(n, c)]$ for any combination of n and c, then the best test plan is to test n_0 samples, and count the number of failures x. If $x \leq c_0$ then accept the lot; otherwise, reject it.

19.4   S-PLUS Tools

S-PLUS is a powerful software program for interactive data analysis, creating graphs, and implementing customized routines. Originating as the S language at AT&T Bell Laboratories, its modern language and flexibility make it appealing to data analysts from many scientific fields. S-PLUS manuals (MathSoft, 1998) provide a comprehensive guide to the software. The book by Krause and Olson (1997) explains the basics of S-PLUS in a clear style at a level suitable for people with little computing or statistical knowledge.

19.4.1   Background

As discussed in Sections 19.2 and 19.3, given the restrictions on PR and CR and estimated PQL and CQL, selecting a sampling plan requires choosing the type of the plan, appropriate sample sizes, and acceptance–rejection criteria to either achieve desirable PR and CR, or minimize an average risk function. Because of the large amount of calculation involved in this process, finding the right sampling plan can be very tedious and time-consuming. In order to solve this problem in a systematic fashion and simplify the search process, an interactive analysis tool is needed. Today, with a wide selection of commercially available PC-based statistical software packages and computer programming languages, developing such a tool is an achievable goal. It is often the case in pharmaceutical and biotechnology firms that people who are involved in designing acceptance sampling plan do not have strong statistical and computing backgrounds. This imposes three basic requirements on the interactive tool developed for usage. First the tool should be user-friendly, preferably menu or window-driven. Second, since the OC curve is the essential method used in the design of single, double, multiple, and sequential sampling plans, the tool should have good graphical presentation. Lastly, the tool should be easily expandable for more sophisticated users. These requirements limit the selection of statistical software or programming languages for the development of such an acceptance sampling system.

Our experience with S-PLUS, a statistical software package for exploratory data analysis, statistical modeling, and graphical display, suggests that it is a package suitable for developing a user-friendly computer interface for acceptance sampling. Unlike other data analysis software, S-PLUS provides an object-oriented programming environment in which components of datasets, analysis models, and statistical and graphical output are treated as objects. Therefore they can be extracted separately, examined, and visually explored. In addition, S-PLUS offers an unparalleled flexibility for customizing its statistical tools to fulfill a specific data analysis task. Because of these advantages, we used S-PLUS to develop a function, AccS, to assist in the creation of acceptance sampling designs.

This function allows users to calculate PR and CR for various combinations of sampling plan type, sample sizes, acceptance-rejection numbers, PQL and CQL. The PR and CR are calculated based on a binomial distribution assumption for the number of defectives in each sample. It provides risk analysis-based sampling plans developed in Section 19.3. Options for displaying the operating characteristic (OC) curve with varying quality level of lots, and risk function plots are also provided by the program. The binomial distribution used in the module can be easily replaced by other distributions such as hypergeometric, Poisson, and normal distributions.

19.4.2   The Function AccS: An S-PLUS Tool for Acceptance Sampling Design

Our S-PLUS tool for acceptance sampling design is a command-line function called AccS that produces a menu in the command window for the user. When executed, it provides choices for single, double, and risk analysis plans, as well as OC curves and risk function plots. (The code for this function is given in the Appendix, Section 19.A.2, and is also available on the companion web page.) When a selection is made (e.g., single sampling plan), users are prompted to provide the sample size n and acceptance cutoff value c. After entering those numbers, a table containing PR and CR for various PQL and CQL values is generated. Table 19.2 shows an example of part of a table that is generated for a single sampling plan with the selections of $n = 100$ and $c = 10$. (The steps to reproduce this example and the others in this section are given in the Appendix, Section 19.A.1.)

The user can check the table to see if, for prespecified values of PQL and CQL, desirable values of PR and CR can be achieved through this sampling plan. If acceptable values of PR and CR cannot be found, other combinations of n and c can be tried. The module also provides an option for plotting the acceptance probability or OC curve against the defect rate p. The OC curve can be used to visually aid the determination as to whether or not the current plan generates satisfactory values of PR and CR. If either of the two quantities exceeds its prespecified value, a larger sample size n and/or a different acceptance number c ought to be tried.

Table 19.2. Single sampling plan with $n = 100$ and $c = 10$.

Defect rate (%)	PR	CR
0.00	0	1
0.01	0.000	1.000
0.05	0.000	1.000
0.10	0.000	1.000
0.20	0.000	1.000
0.50	0.000	1.000
1	0.000	1.000
2	0.000	1.000
5	0.011	0.989
6	0.038	0.962
7	0.091	0.909
8	0.176	0.824
9	0.288	0.712
10	0.417	0.583
15	0.901	0.099
20	0.994	0.006

Figure 19.9 shows the result of selecting the option of plotting the OC curve. Suppose PQL = 5% and CQL = 15%. According to Table 19.2 or the plot in Figure 19.9, it can be determined that PR = 0.011 and CR = 0.099. If the pre-specified limits for both PR and CR are 10%, then the single sampling of testing a random sample of 100 items and accepting the lot on the basis of no more than 10 defective samples will warrant desirable protection for both producer's and consumer's interests.

As a second example, suppose we design a sampling plan using the risk analysis method and we further assume that the quantities L_0, L_1, L_2, L_3, p, PQL, and CQL are determined to be 1, 5, 5, 1, 0.95, 0.10, and 0.15, respectively, based on heuristic knowledge of the lot of finished product. Choosing the sampling plan based on risk analysis in the module will prompt the user to provide the program with the values of the above quantities. We also need to provide the program with an upper limit, N, of sample size to restrict the region in which the risk function is minimized. Setting $N = 25$, a sampling plan that achieves the minimum of the risk function is $n = 25$ and $c = 9$, that is, to test 25 items from the lot and accept the lot if the number of failures does not exceed 9. Otherwise, reject the lot. The function AccS also allows one to plot the risk function for all possible sampling plans (note that each plan corresponds to a set of (n, c)).

Figure 19.10 displays the risk function for the selected values of various costs, prevalence of good lot in the product stream, and producer and consumer quality levels. It is evident that the sampling plan with which the risk function obtains its minimum is not unique. The function AccS will select the most economical one or the one with the smallest sample size.

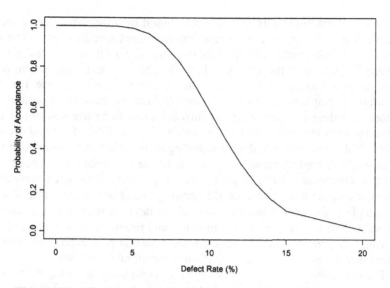

Figure 19.9. OC curve for single sampling plan with $n = 100$ and $c = 10$.

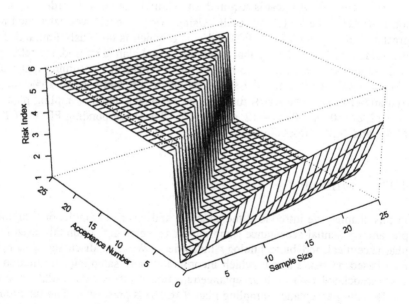

Figure 19.10. Risk function for sampling plan with $L_0 = L_3 = 1$, $L_2 = L_3 = 5$, $p=0.95$, PQL = 0.10, and CQL = 0.15.

19.4.3   Using AccS for Out-of-Specification Investigations

Out-of-specification results are generally caused by either laboratory error or a sample that does not conform to the predetermined specifications. The Food and Drug Administration (FDA) regulations require a full-scale investigation to identify the source of the unexpected results when the initial assessment cannot be documented as laboratory error. This investigation may include a retest of the original sample or a test of a new sample from the same lot of product. Decisions regarding whether or not the current lot should be released will be made based on both the initial and retest results (Vogel, 1993; Barr and Dolecek, 1996). This in essence is a double sampling plan test scheme. Since AccS provides double sampling plans, it can be easily used to assist out-of-specification result investigations. It can be used in two ways. First, if a single sampling plan already exists for testing batches of a certain product or material, using the double sampling plan option in AccS can help to devise a retest plan and calculate PR and CR with results from the initial test and retest. In the event that there is no sampling plan, one can design a double sampling plan initially to serve as a mechanism for potential out-of-specification result investigations.

For the problem we were working on, an acceptance sampling plan was designed by scientists based on regulatory guidelines and standard industry practice. The double sampling plan initially tests three items from a lot of the product. The lot is accepted if there is no failing result. In other words, $n_1 = 3$, $c_1 = 0$, and $r_1 = 3$. A retest is required if the initial test fails. In order to pass the lot, there has to be no failing result. Using AccS, the CR was calculated to be greater than 52% with $n_2 = 1000$ and $c_2 = 0$ which is obviously impractical for the producer. This indicates that unless the initial plan is revised, the retest plan will not add any reassurance to the consumer's interest. Assuming that PQL = 1% and CQL = 15%, and the specification for PR and CR are 15%, through trial-and-error with the AccS function, we reached a double sampling plan with $n_1 = 12$, $c_1 = 0$, $r_1 = 8$, $n_2 = 12$, and $c_2 = 0$. The corresponding PR and CR are 11.4% and 14.2%, respectively.

19.5   Summary

In this chapter, we introduced the concepts and theory of single, double, multiple, and sequential acceptance sampling plans which are commonly used in the pharmaceutical and biotechnology industries. We also discussed a sampling plan based on risk analysis which incorporates both sampling information and costs associated with each acceptance/rejection decision. An S-PLUS function for designing acceptance sampling plans (AccS) is presented. The function encompasses single-, double-, and risk analysis-based sampling plans. It provides users with a tool to interactively explore feasible sampling plans under practical constraints. Because the usage of the function does not require users to have much knowledge in statistics and programming, it is helpful for scientists and engineers in the quality control (QC) departments of pharmaceutical and bio-technology companies to establish acceptance testing schemes. The function is user-friendly and was validated through an Excel spreadsheet utility. Note, however, that sampling plans generated by the module usually represent the theoretical optimum which might not be practical for a specific problem. For example, if a plan requires to test half of the finished lot of product to reassure both consumer's and producer's interest, the company's profit margin would be adversely impacted under the current producer's quality level and consumer's expectations characterized by consumer's quality level CQL and tolerance for risk, CR. In a circumstance like this, either the manufacturer has to streamline his production line and improve the quality of his products, or the consumer has to be more tolerant of possible bad lots of products. In addition, the compromise has to be acceptable to the regulatory authorities. Ultimately, a sampling plan is a tradeoff among the theoretical optimum, practical feasibility, and governmental regulations. The assessment of quantities, such as specifications for PR, CR, PQL, CQL, and costs associated with each type of risks necessary for the application of the module should be supported by data collected from a literature review, relevant experiments, and regulatory agency guidelines.

19.6 References

Barr, D.B. and Dolecek, G.R. (1996). Out-of-specification laboratory results in the production of pharmaceuticals. *Pharmaceutical Technology* **April**, 54–61.

Berger, J. (1985). *Statistical Decision Theory and Bayesian Analysis,* Second edition. Springer-Verlag, New York.

Food and Drug Administration (1995). Current good manufacturing practice in manufacturing, processing, packing, or holding of drug: Amendment of certain requirements for finished pharmaceuticals. *Federal Registry* **60**(13), 4087–4091.

Krause, A. and Olson, M. (1997). *The Basics of S and S-PLUS.* Springer-Verlag, New York.

MathSoft Inc. (1998). *S-PLUS: Programmer's Guide, Version 4.0.* MathSoft, Inc., Data Analysis Products Division, Seattle, WA.

Rabbitt, J.T. and Bergh, P.A. (1997). *The ISO 9000 Book—A Global Competitor's Guide to Compliance & Certification.* Quality Resources, White Planes, New York.

Schilling, E.G. (1982). *Acceptance Sampling in Quality Control.* Marcel Dekker, New York.

Vogel, P.F. (1993). Evaluating out-of-specification laboratory results. *Journal of Parenteral Science and Technology* **47**, 277–280.

Wadsworth, H.M., Stephens, K.S., and Godfrey, A.B. (1986). *Modern Methods for Quality Control and Improvement.* Wiley, New York.

Wald, A. (1947). *Sequential Analysis.* Wiley, New York.

19.A. Appendix

19.A.1   Using AccS for the Examples in Section 19.4.2

In this section, what the user types in is displayed in **bold courier font**.

```
> AccS()
=======================================
CHOOSE AN ACTIVITY
1: Single Sampling Plan
2: Double Sampling Plan
3: Sampling Plan Based on Risk Analysis (SPBORA)
4: OC Curves of Single/Double Sampling Plan
5: Risk Plot of SPBORA
6: Done
Selection: 1


=======================================
ENTER SAMPLE SIZE N: 100
ENTER ACCEPTANCE NUMBER C: 10


Single Sampling Plan
```

	Defect.Rate	PR	CR
1	0.001%	0.000	1.000
2	0.005%	0.000	1.000
3	0.01%	0.000	1.000
4	0.05%	0.000	1.000
5	0.1%	0.000	1.000
6	0.2%	0.000	1.000
7	0.5%	0.000	1.000
8	0.7%	0.000	1.000
9	1%	0.000	1.000
10	2%	0.000	1.000
11	3%	0.000	1.000
12	4%	0.002	0.998
13	5%	0.011	0.989
14	6%	0.038	0.962
15	7%	0.091	0.909

```
16            8%  0.176  0.824
17            9%  0.288  0.712
18           10%  0.417  0.583
19           11%  0.547  0.453
20           12%  0.666  0.334
21           13%  0.766  0.234
22           14%  0.844  0.156
23           15%  0.901  0.099
24           20%  0.994  0.006
```

```
========================================
CHOOSE AN ACTIVITY
1: Single Sampling Plan
2: Double Sampling Plan
3: Sampling Plan Based on Risk Analysis (SPBORA)
4: OC Curves of Single/Double Sampling Plan
5: Risk Plot of SPBORA
6: Done
Selection: 4
ENTER LOWER PLOTTING LIMIT FOR X-AXIS:   0
ENTER UPPER PLOTTING LIMIT FOR X-AXIS:   20
ENTER LOWER PLOTTING LIMIT FOR Y-AXIS:   0
ENTER UPPER PLOTTING LIMIT FOR Y-AXIS:   1
ENTER PLOT TITLE: Figure 3. OC Curve for Single Sampling
Plan with n = 100, c =10
```

```
========================================
CHOOSE AN ACTIVITY
1: Single Sampling Plan
2: Double Sampling Plan
3: Sampling Plan Based on Risk Analysis (SPBORA)
4: OC Curves of Single/Double Sampling Plan
5: Risk Plot of SPBORA
6: Done
Selection: 3
```

```
========================================
ENTER THE UPPER LIMIT FOR SAMPLE SIZE N: 25
ENTER COST OF ACCEPTING A GOOD LOT: 1
ENTER COST OF ACCEPTING A BAD LOT: 5
ENTER COST OF REJECTING A GOOD LOT: 5
ENTER COST OF REJECTING A BAD LOT: 1
ENTER PREVELANCE OF GOOD LOT: 0.95
ENTER PRODUCER'S QUALITY LEVEL: 0.10
ENTER CONSUMER'S QUALITY LEVEL: 0.15
```

```
Sampling Plan Based on Risk Analysis:

  Sample.Size Acceptance.Number Minimum.Risk
1    25                9               1.199872


========================================
CHOOSE AN ACTIVITY
1: Single Sampling Plan
2: Double Sampling Plan
3: Sampling Plan Based on Risk Analysis (SPBORA)
4: OC Curves of Single/Double Sampling Plan
5: Risk Plot of SPBORA
6: Done
Selection: 5
ENTER PLOT TITLE: Figure 4. Risk Function for Sampling Plan
with L0 = L3 =1, L1 = L2 =5, p =0.95, PQL =0.10 and
CQL =0.15


========================================
CHOOSE AN ACTIVITY
1: Single Sampling Plan
2: Double Sampling Plan
3: Sampling Plan Based on Risk Analysis (SPBORA)
4: OC Curves of Single/Double Sampling Plan
5: Risk Plot of SPBORA
6: Done
Selection: 6
```

19.A.2   S-PLUS Code for Developing Acceptance Sampling Plans

```
read.numeric <- function(text) {
### A function to read numeric values
  cat(text)
  x <- as.numeric(readline())
  if(is.na(x))
    x <- read.numeric(text)
  else return(x)
}
```

```
AccS <- function() {
### A function for developing acceptance sampling plans
  spacer <- "\n=======================================\n"
  m1 <- 999
  while(m1 != 6 & m1 != 0) {
## Generate a menu
    cat(spacer)
    choices <- c("Single Sampling Plan",
      "Double Sampling Plan",
      "Sampling Plan Based on Risk Analysis (SPBORA)",
      "OC Curves of Single/Double Sampling Plan",
      "Risk Plot of SPBORA", "Done")  #
## Select an item from the menu
    m1 <- menu(choices, title = "CHOOSE AN ACTIVITY")
    if(m1 == 1) {
## Create a single sampling plan
      cat(spacer)
      n <- read.numeric("ENTER SAMPLE SIZE N: ")
      c2 <- read.numeric("ENTER ACCEPTANCE NUMBER C: ")  #
## p - possible quality levels
      p <- c(0.001, 0.005, 0.01, 0.05, 0.1, 0.2, 0.5, 0.7,
        c(1:15), 20)/100
      CR <- pbinom(c2, n, p)
      PR <- 1 - CR
      DefectRate <- paste(p * 100,
        rep("%", times = length(p)), sep = "")
      sp1 <- data.frame("Defect Rate" = DefectRate,
        PR = round(PR, 3), CR = round(CR, 3))
      cat(" \n")
      cat("Single Sampling Plan\n")
      cat(" \n")
      print(sp1)
    }
    if(m1 == 2) {
## Create a double sampling plan
      cat(spacer)
      n1 <- read.numeric("ENTER INITIAL SAMPLE SIZE N1: ")
      n2 <- read.numeric("ENTER SECOND SAMPLE SIZE N2: ")
      c1 <- read.numeric(
        "ENTER INITIAL ACCEPTANCE NUMBER C1: ")
      r1 <- read.numeric(
        "ENTER INITIAL REJECTION NUMBER R1: ")
      c2 <- read.numeric(
        "ENTER SECOND ACCEPTANCE NUMBER C2: ")
```

```r
  p <- c(0.001, 0.005, 0.01, 0.05, 0.1, 0.2, 0.5, 0.7,
    c(1:15), 20)/100
  DefectRate <- paste(p * 100,
    rep("%", times = length(p)), sep = "")
  CR <- pbinom(c1, n1, p)
  for(i in r1 + 1:n1) {
    CR <- CR + dbinom(i, n1, p) * pbinom(c2 - i, n2, p)
  }
  PR <- 1 - CR
  sp1 <- data.frame("Defect Rate" = DefectRate,
    PR = round(PR, 3), CR = round(CR, 3))
  cat(" \n")
  cat("Double Sampling Plan\n")
  cat(" \n")
  print(sp1)
}
if(m1 == 3) {
  cat(spacer)
  n.up <- read.numeric(
    "ENTER THE UPPER LIMIT FOR SAMPLE SIZE N: ")
  L0 <- read.numeric(
    "ENTER COST OF ACCEPTING A GOOD LOT: ")
  L1 <- read.numeric(
    "ENTER COST OF ACCEPTING A BAD LOT: ")
  L2 <- read.numeric(
    "ENTER COST OF REJECTING A GOOD LOT: ")
  L3 <- read.numeric(
    "ENTER COST OF REJECTING A BAD LOT: ")
  p <- read.numeric("ENTER PREVALENCE OF GOOD LOT: ")
  PQL <- read.numeric(
    "ENTER PRODUCER'S QUALITY LEVEL: ")
  CQL <- read.numeric(
    "ENTER CONSUMER'S QUALITY LEVEL: ")
  R <- matrix(c(rep(100000, times = n.up^2 + n.up)),
    ncol = n.up + 1, byrow = T)
  for(i in 1:n.up) {
    j <- 1
    while(j <= i + 1) {
      CR <- pbinom(j - 1, i, CQL)
      PR <- 1 - pbinom(j - 1, i, PQL)
      R[i, j] <- p * L0 * (1 - PR) +
        (1 - p) * L1 * CR + p * L2 * PR +
        (1 - p) * L3 * (1 - CR)
      j <- j + 1
    }
```

```
      }
      j0 <- c(rep(1, times = n.up))
      for(i in 1:n.up) {
        j <- 2
        while(j <= i + 1) {
          if(R[i, j] < min(R[i, 1:j - 1]))
            j0[i] <- j
          j <- j + 1
        }
      }
      r <- c(rep(0, times = n.up))
      for(i in 1:n.up) {
        r[i] <- R[i, j0[i]]
      }
      i0 <- 1
      i <- 2
      while(i <= n.up) {
        if(r[i] < min(r[1:i - 1]))
          i0 <- i
        i <- i + 1
      }
### j0[i0]-1 represents the true acceptance number
      sp1 <- data.frame("Sample Size" = i0,
        "Acceptance Number" = j0[i0] - 1,
        "Minimum Risk" = R[i0, j0[i0]])
      cat(" \n")
      cat("Sampling Plan Based on Risk Analysis\n")
      cat(" \n")
      print(sp1)
      R[R > 99999] <- max(R[R < 99999]) + 1
      x <- c(1:n.up)
      y <- c(0:n.up)
      R <- round(R, dig = 5)
    }
    if(m1 == 4) {
      xl <- read.numeric(
        "ENTER LOWER PLOTTING LIMIT FOR X-AXIS: ")
      xu <- read.numeric(
        "ENTER UPPER PLOTTING LIMIT FOR X-AXIS: ")
      yl <- read.numeric(
        "ENTER LOWER PLOTTING LIMIT FOR Y-AXIS: ")
      yu <- read.numeric(
        "ENTER UPPER PLOTTING LIMIT FOR Y-AXIS: ")
      cat("ENTER PLOT TITLE: ")
      pt.title <- readline()
```

```
    plot(p * 100, CR, type = "n",
      xlab = "Defect Rate (%)",
      ylab = "Probability of Acceptance",
      xlim = c(xl, xu), ylim = c(yl, yu),
      main = pt.title)
    lines(p * 100, CR)
  }
  if(m1 == 5) {
    cat("ENTER PLOT TITLE: ")
    pt.rtitle <- readline()
    persp(x, y, R, xlab = "Sample Size",
      ylab = "Acceptance Number", zlab = "Risk Index")
    title(pt.rtitle)
  }
 }
}
```

Index